Case Studies in Defence Procurement and Logistics

Volume I: From World War II to the Post Cold-War World

Case Studies in Defence Procurement and Logistics

Volume I: From World War II to the Post Cold-War World

David Moore *et al*.

Contents

Introduction

Foreword

The United Kingdom faces a challenging security environment characterised by:

- the certainty of uncertainty which, when linked to an exponential, breath-taking rate of technological and other change is inherently destabilising;
- increased competitiveness between and within nations where traditional assumptions of who is 'on top' or what it takes to get ahead, and to stay there, will be challenged by new and innovative business, economic and military approaches;
- the phenomenon of decentralisation which challenges the abilities of established, hierarchical societies and organisations to cope with the agility and speed of transition of decentralised, often non-state organisations and movements.

These interwoven trends are compelling and fast moving. They pose particular challenges for those charged with planning a nation's defence and security posture and the capabilities, size, shape and scope of the security and defence forces to ensure that they are relevant, effective, useable and affordable. And there is a concomitant challenge for those charged with equipping and supporting the security and defence forces; their task is to turn policy and broad direction into action, and to exhibit strong acquisition knowledge, planning, decision making and management.

In the UK, defence expenditure is a matter of political, media and public interest. Notwithstanding the fact that more than 95 percent of all equipment projects are delivered to time and cost, every equipment shortfall, delay or cost overrun is seized upon as an example of poor planning, inept management and waste. So it is incumbent on us to do everything we can to improve our performance in what has come to be called 'acquisition'. Structural and process change alone will not take the trick. Ultimately success will come through the knowledge, skills and ability of the people engaged in the business of acquisition. We need to build an effective, responsive, relevant culture of teaching and learning to ensure that we develop a body of professional knowledge spanning the disciplines which underpin an effective acquisition effort – including commercial, finance, project and programme management, identifying and handling risk, logistics, and supply/support chain management.

Case studies are an important tool for encouraging learning and stimulating thinking about the underpinning and enduring issues as well as about the particular circumstances of the time. As a simple infantryman with no background in acquisition I have enjoyed what I have read. What strikes home is the extent to which the acquisition environment and the approaches to acquisition have changed in the last century, and will change again

i

as contractorisation, outsourcing, partnering and third-party delivery on an output basis – all in a globalised context and supporting expeditionary Armed Forces likely to be operating in coalition – become even more integral to the Defence effort for generating effective military capability. Building an effective, qualified, experienced, entrepreneurial cadre of acquisition professionals to lead, drive and manage the acquisition effort is a significant challenge for the UK's Ministry of Defence that we must grasp with both hands. I commend this book to a wide audience for the breadth and depth of its content and analysis in support of that ideal.

A J N Graham
Lieutenant General
Director General, Defence Academy of the United Kingdom,
May 2011

Contributors

Editor

David Moore is currently Director of the Centre for Defence Acquisition within the Department of Engineering Systems and Management, Cranfield University at the Defence Academy's College of Management and Technology in Shrivenham. His varied career has included managerial positions in both the public and private sector. He finished his final tour in the Royal Logistic Corps (RLC) at the rank of Lt. Colonel. David has also developed and implemented Purchasing and Logistics courses for a range of major clients as well as MBA programmes whilst at the University of Glamorgan.

Authors

Peter D. Antill is currently a research assistant working for Cranfield University at the Defence Academy's College of Management and Technology in Shrivenham. Peter has practical experience in the service industry as well as the civil service. A degree holder from Staffordshire University and the University College of Wales, Aberystwyth he also holds a PGCE (Post Compulsory Education) from Oxford Brooks University. A published author, he is currently conducting post-graduate research into British defence policy, historical procurement programmes and expeditionary operations as well as privately collaborating with two colleagues in a military history project, located at http://www.historyofwar.org.

Dr. Jeffrey P. Bradford is a trans-Atlantic defence and national security consultant, formerly with Babcock International Group PLC and prior to that, a specialist with Arthur D. Little Inc. in corporate strategy within the aerospace and defence sector. In addition, Jeffrey holds degrees from Staffordshire, Aberystwyth and Cranfield University at the Royal College of Military Science, Shrivenham.

Jonathan P. Davies spent a year working as a research assistant for Cranfield University in the Centre for Defence Acquisition, working on numerous case studies, conferences, reports and books regarding a variety of acquisition related topics. Prior to this he attended Swansea University, where he attained a first class honours degree reading War and Society. Currently he is spending time in South East Asia educating himself in both the language and culture of China.

Ivar Hellberg is a former officer in the Royal Marines whose thirty-seven year service career saw deployments in a wide variety of roles including logistics and the special forces. During this time he saw active service in the Falklands War (1982) and has commanded the Royal Marines Commando Logistics Regiment as well as serving as defence attaché to Indonesia. Ivar is currently a consultant at Cranfield University, DA-CMT specialising in South East Asia, Logistics, Strategic Mobility and Global Security issues.

Pete Ito has served as a Lecturer and Researcher in Defence Acquisition since September 2008. He is the Academic Leader on the Acquisition Employment Training course, is a lecturer on Through Life Capability Management, and delivers the Introductory Studies and Research Methodology courses in the Defence Acquisition Management MSc. After earning a Bachelor's degree in Political Science from the University of California at Berkeley, and a Juris Doctor (law) degree and a Masters degree in International Affairs from George Washington University in Washington, D.C., Pete worked for 25 years as a Foreign Service Officer for the U.S. State Department. He served in Seoul, South Korea; Copenhagen, Denmark; Bonn and Berlin, Germany; The Hague, The Netherlands and Washington, D.C. His primary focus was political affairs, particularly defence and security policy. He served as the head of the Political-Military Affairs section in Berlin, as Counsellor for Political and Economic Affairs in Copenhagen and as the Deputy Permanent Representative for the U.S. Delegation to the Organization for the Prohibition of Chemical Weapons located in The Hague. He joined Cranfield University in September 2007, working as a researcher in the areas of strategic management and change management, before moving to his current position.

David Jenkins served for twenty-seven years in the British Army on a range of procurement, training and education programmes including being the lead designer and implementation manager on the successful Army Foundation College PFI programme and the Army's principle training advisor to the Defence Procurement Agency (DPA). Dai joined the Centre for Defence Acquisition in July 2008 and has had a crucial role in the design, development and delivery of a range of acquisition training programmes, most notably those focused on Capability Management. He also manages the Welsh Rugby Union 7s Team on the international rugby circuit.

Lt Cdr. Larry Parker, United States Navy, served as a Surface Warfare Officer. He retired in July 2000 and began teaching Navy Junior ROTC. He currently chairs the Naval Science Department at Sun Valley High School in Monroe, NC. In his free time, LCDR Parker pursues a life long passion for military history while his son, Nathan, continues the family tradition of service as a United States Marine. Sea Lion vs. Overlord is the result of extensive research and personal experience as Deck Officer onboard USS Portland (LSD-37).

Dr. Peter Tatham joined the Royal Navy in 1970 and served in a variety of appointments during his career of some 35 years. Highlights include Logistics Officer of the Aircraft

Carrier HMS Invincible in 1994/5 during Operations in Bosnia against the Former Republic of Yugoslavia, and Chief Staff Officer responsible for all high level Personnel and Logistics issues emanating from the 10,000 sailors and 30 surface ships in the Royal Navy (1999-2000). His final three years in the Service were spent in the UK's Defence Logistics Organisation where he was responsible for key elements of the internal programme of Change Management (2000-2004). During this period, he also gained an MSc in Defence Logistic Management. Following his retirement from the RN, he joined the staff of Cranfield University where he lectures in Human Systems and Humanitarian Logistics. He has recently been awarded his PhD for research into the issues surrounding the role of shared values within military supply networks.

Prof. Trevor Taylor was formerly Head of the Department of Defence Management and Security Analysis at Cranfield University's faculty at the Royal Military College of Science in the UK. He was previously Professor of International Relations at Staffordshire University and between 1990 and 1993 was Head of the International Security Programme at the Royal Institute of International Affairs in London. He is also a past Chairman of the British International Studies Association and has been Visiting Professor at the National Defence Academy in Tokyo. He was educated at the London School of Economics (B.Sc(Econ) and PhD) and Lehigh University (MA) in Pennsylvania. He has published extensively on European security and defence industrial issues, and is currently working on the political and defence implications of defence restructuring in Europe.

Christianne Tipping is an independent defence analyst and Research Associate of the Royal United Services Institute for Defence and Security Studies (RUSI). Until July 2008, she was Head of the Defence Leadership and Management Programme at RUSI, researching issues relating to procurement, logistics, personnel, training, ethos, organisational development and leadership. She joined the Institute in 2006 after completing an MBA. Previously, she had spent several years as a Royal Air Force officer, fulfilling training support roles in ground and flying training units and acquiring specialist knowledge in military ethos and leadership development while working as a staff officer in the team which formulated the RAF's strategies in these areas. Christianne joined the RAF after spending some time working in generalist HR in the private sector having completed a BSc in Biological Science and a Postgraduate Diploma in Human Resource Management. She is a Fellow of the Chartered Institute of Personnel and Development.

Anthony G. Williams is an independent consultant specialising in small arms and automatic cannon. He has written many magazine articles and has authored or co-authored the following books: *Rapid Fire* (2000); *Flying Guns* (three volumes, 2003-4); *Assault Rifle* (2004) and *Machine Gun* (2008). He is also co-editor (since 2005) of *Jane's*

Ammunition Handbook and editor (since 2004) of *The Cartridge Researcher*, the bulletin of the European Cartridge Research Association. He maintains a website on military guns and ammunition at www.quarry.nildram.co.uk

Stuart Young joined Cranfield University in 2008 as a Deputy Director in the Centre for Defence Acquisition where he has a particular interest in the relationship between the MoD and Industry across the supply chain and the development of strategies for major acquisition programmes. Stuart joined the Royal Navy in 1977 as a Maritime Engineer Officer, doing a post-graduate level qualification at the Royal Naval Engineering College in Plymouth. He has served in a variety of operational posts at sea but also a number of acquisition-related appointments in the Ministry of Defence. This has included spending three years as the Defence Equipment Marine Engineer with the Defence Staff at the British Embassy in Washington DC and serving as the Electric Ship Programme Manager in the Defence Procurement Agency with direct responsibility for a major UK-France technology development programme. He has also been involved in the selection of innovative technologies for the Type 45 destroyer and CVF programmes and served in the DLO, the Defence Electronic Commerce Service and in the Defence Management and Leadership Centre in Shrivenham.

Acknowledgements

While a number of case studies were specifically written or adapted for this book, the Editor would like to thank the following for their kind permission to include the mentioned material:

Stuart Young, Prof. Trevor Taylor and Dr Peter Tatham for the Rapier case study, a version of which first appeared in the IDEAA 2011 case study book.
Anthony G. Williams for his .256" British article.
Christianne Tipping and RUSI for the Key Issues paper.
Lt Cdr. Larry Parker for his piece on Sea Lion vs. Overlord.
Dr Jeff Bradford for his JSF Alternatives paper.

The Editor would also like to thank Jonathan P. Davies for his contribution in putting this book together. However, this whole project to design, develop and publish the book would not have happened without the total commitment of Peter D. Antill and the Editor wishes to record his heartfelt thanks for all the work that Peter has undertaken.

Preface

This book contains nineteen cases focusing on defence procurement and logistics issues that will provide an excellent learning opportunity to a variety of readers. This wide-ranging audience will include military personnel, those on defence education and training programmes, employees of the defence industry, those in defence agencies and those engaged in defence and security-related research (an example being under- and postgraduate students on international politics and strategic studies courses).

However, such is the nature of these cases that they will also be relevant to and interesting for, not only undergraduate courses, but postgraduate programmes such as MBA, MSc, MA, PgDip, PgCert and continuous professional development in purchasing, logistics, business and supply chain management.

As well as featuring nineteen cases, there is a foreword, notes on contributors, acknowledgements, preface and notes on certain cases to aid teaching and learning. There will also be a chapter on the contemporary nature of defence acquisition, logistics and supply chain management.

The majority of the cases presented will provide a useful source of reference as they are factual, albeit often in a précised form. Others, are based on fact, but are produced in such a manner as to prompt the reader's thinking, conceptualisation and application of concepts into practise.

A key feature is the chronological nature of the cases; although the date and the contextual setting for each case may change, and indeed the scale involved may differ, the overarching theme is that there is much to be learnt from studying experiences from the past and considering them in a contemporary environment. These cases are progressive, and build towards a comprehensive appreciation of major issues and challenges that have presented themselves in the way that the UK has undertaken defence procurement and logistics.

The intention is that the reader should be able to shape his or her thinking in respect of defence procurement and logistics. The reader should be taking into account not only the factual details of each case, but developing conceptual perspectives in order to influence contemporary thinking, doctrine and practise. The chronological nature of the cases, and the broad spectrum of operating environments covered permit a consideration of existing and emergent theories and practise.

Book Overview

The reader will find that there are three sections to this book. Part 1 focuses upon defence logistics and supply chain strategy. The first case *From Helen to Helmand*, provides a historical background to military logistics. It is particularly useful for those seeking a wider understanding of both the evolution, and current situation of, military logistics. Having provided a suitable background, the first three cases cover major World War II operations, initially with a comparative analysis of Operation Sea Lion (the planned German invasion of Britain in September 1940) and Operation Overlord (the Allied invasion of Normandy in June 1944) and provides a view on the relative strengths and weaknesses of German and Allied logistic plans and preparations. This is followed by a case that considers Allied logistics strategy in Southeast Asia with particular emphasis upon those factors that influenced success (or otherwise). Similarly, the need for innovation in the supply chain albeit still in the context of World War II is that involving the Allied invasion of Northwest Europe in Operation Overlord. Although these three cases are set historically, the lessons in respect of concepts and applications, and indeed the wider thinking behind logistics strategy are still pertinent today.

Moving forward in time, Ivar Hellberg, who commanded the Royal Marine Commando Logistics Regiment, writes knowledgably of the logistics effort that enabled the UK to undertake a major amphibious operation to retake the Falkland Islands at a great distance from the home base. This too has resonance with current operations, although current operations are land-based as opposed to sea-based. This is followed by an examination of logistic support to UK forces during Operation Granby (the first Gulf War) in 1990-1. It considers major aspects of logistic support such as the need for fuel and the need for the repair and maintenance of major equipment in an often hostile environment. Logistic lessons learned here are still applicable today, given the recent operations in both Iraq and Afghanistan.

Part 2 has a different focus centring upon defence procurement (although recognising the link with logistic support). The first case concerns *A Rifle Ahead of its Time*, the EM-2 and brings out issues that must be taken into account when undertaking the setting of requirements and the clarification of need, both in a strategic and a practical sense. The second case moves from the 1950s into the 1960s and looks at *The Plane That Barely Flew*, the ill-fated TSR-2. The requirement for this plane was clear, and its technical innovations were to serve aircraft well for many years to come (including Concorde and Tornado). However, many socio-political and economic issues conspired to bring about the cancellation of what was regarded at the time as a world-beating aircraft. This case unlike most of the others contains a number of discussion questions. There are no answers to these, although consideration of the questions should identify issues that are still relevant today.

The procurement of an air defence missile system in the late 1970s, to be used by the Army provides the basis for the next case. It provides and overview of key matters concerning the procurement, and in due course, logistic support for what was to be a vital piece of equipment. Other areas are identified including training and infrastructure and can be seen as the forerunner of what is now considered to be central to contemporary UK defence acquisition – the Defence Lines of Development (DLoDs). Taking a different perspective, there have been numerous attempts at collaborative procurement with varied results. Some have been relatively successful, such as the Panavia Tornado Multi-Role Combat Aircraft (MRCA) and Sepecat Jaguar. In contrast, there are lessons to be learned from those collaborative projects which have not been so successful. The next case examines the European collaborative procurement project (dating from the 1980s and covering most of the 1990s) for a new generation of frigate where conflicting national requirements ultimately led to the UK dropping out and pursuing a unilateral programme that has led to the development of the Type 45 destroyer.

Coming right up-to-date, and recognising the socio-political and economic need for collaborative defence procurement in an age of shrinking defence budgets, the next two cases consider the Joint Strike Fighter project in some depth, although from different perspectives. Firstly, Peter Antill and Pete Ito analyse the major factors involved in the evolution of the project, from a number of disparate single-service programmes and their gestation into a multi-national and multi-service procurement effort. The authors also consider the many issues and challenges that have arisen along the way, including perspectives from the United States, UK and Australia. Maintaining a focus upon the Joint Strike Fighter, Jeffrey Bradford considers the requirements and the alternative solutions that could have been adopted by countries such as the UK and the possible effects of any resultant decision.

Both Part 1 and 2 can be seen as factual cases which are useful for reference documentation. As well as being able to use the text itself, many apposite sources are contained within the references that follow each case. Part 3 takes a different approach. The first case takes a factual approach in considering of the UK's adoption (or otherwise) of small arms ammunition calibres. Whilst it does not pose questions as such, it leaves the reader to ponder whether the UK has made the right decisions at the right time. The reader should consider the final statement posed in the case and its wider implications in a contemporary environment. This is followed by an exercise that considers the outsourcing of an MoD training organisation. It is based upon PFI/PPP initiatives and while it takes a simplified view of what must be taken into account, it indicates that in discussion with others, the reader can gain insight into the complexities and challenges of what, during the last twenty years, has become a major strategic approach in the UK Public Sector.

This is followed by a case that is based upon support chain management in the Royal Air Force. It provides the background to, and challenges of, improving logistic performance through a 'management of change' programme. Indicative issues are provided in a learning and development section. Recognising that procurement and logistics challenges are not confined purely to warfighting scenarios, there follows a case where logistic planning and preparation for a mountaineering expedition is required. Moving forward in time to the late 1990s, the issues of long-range logistics and integration are covered through a paper that notes the challenges of the logistics environment and the identification of different approaches to achieving differing logistic goals. The penultimate case, is actually a paper produced circa 2002 that outlines a comparison between military logistics and the logistics of humanitarian aid organisations.

The final case brings a contemporary perspective to the provision of logistics support to UK Armed Forces in expeditionary operations, developing further, the concepts of integration and optimisation of the supply chain.

Introduction

The Changing and Developing Nature of Procurement, Logistics and Supply Chain Management in the Contemporary Defence Environment

David M. Moore and Peter D. Antill

Centre for Defence Acquisition, Cranfield University, Defence Academy of the UK.

Background

The period covered by this book has seen considerable changes in the way defence procurement and logistic support is undertaken. The two world wars were a major change in the way the UK had fought when compared to the previous hundred-and-fifty years or so. The UK maintained a small but highly professional army, with a large and modern, well-equipped navy to police its Empire, fighting poorly equipped and trained local inhabitants in brush fire wars across the globe.

The world wars saw the UK having to undertake a massive expansion of the British Army, overcome the challenges of training, organising and equipping such a force in order to fight a well-trained, organised and motivated opponent in conventional warfare, primarily on the continent of Europe but with smaller campaigns in subsidiary theatres around the world. The end of World War II brought a return to might have been termed 'normality' i.e. radically reduced armed forces. The following two decades saw a gradual withdrawal from empire, a return to an all-regular force and a further consequent reduction in the overall resources afforded to defence and the armed forces.[1]

In the same period, the reconstruction and recovery of the former Axis countries Japan and Germany occurred, with a move towards enhanced production and manufacturing techniques. Although based upon American ideas, the Japanese in particular seized upon and maximised the potential of these approaches. Hence, at the same time that the UK Armed Forces were getting smaller, Japanese industry was growing apace and in due course, came to dominate global manufacturing. The approaches that were utilised were often adopted by commercial organisations in other countries; those industries that did not adapt were often unable to compete and survive. These diametrically opposed trends epitomised the UK during the 1960s and 1970s.

By the 1980s, the UK had concentrated its armed forces on the defence of West

Germany against the potential threat from the Soviet-controlled Warsaw Pact, a military alliance set up to counter NATO in 1955. This entire decade saw a standoff between East and West, with the UK an important but no longer a central, component of NATO's Central European defence under AFCENT (Allied Forces Central Europe). This relatively static position meant that future conflict would, initially at least, follow a symmetrical pattern (high-intensity conventional warfare). In turn, this permitted a strategy that was relatively stable and consistent for the whole of the 1980s. It was predicated upon a number of 'lines of communication' from the UK into the Low Countries and West Germany that meant there could be relatively large, static stocks that were intended to be expanded using the lines of communication which flowed eastwards (as were the reinforcements to the British Army of the Rhine - BAOR) in a surge manner in order to counter perceived threats. Resupply was conducted by a hierarchical multi-layered system of stock holding, which had inventory management at each level, and provisioning in the previous level.

Material was in essence pushed forward, thus refilling the following level of stockholding to a pre-planned amount. Such a system, involving a 'back-to-front' process invariably, has costs associated with storage, the amount of stock held, transportation onto the next level and double handling. It was therefore a traditional linear or echelon one, consisting of lines of support, which were known as the first, second, third and fourth line. The first, second and third lines were provided by logistic units, which deployed on the operation being carried out, and were either organic to the combat units themselves or attached by higher headquarters. For example, the first line would have been the battlegroups themselves (battalion), second line would have been the parent formation (brigade or division), and third line would have been at the theatre headquarters level. Finally, the fourth line was considered to be the UK Home Base and typically involved the MoD, non-deployable storage locations, defence agencies and the defence industrial base. In effect, this was the holding of stocks, 'just in case' they were needed. This was at the very time, when industry and commercial organisations were adopting strategies that were based upon the opposing concept, i.e. obtaining stocks 'just in time'.[2]

At this time in the UK, many commercial organisations had been unable to compete, and either went out of business or were swallowed up by larger, globally focused organisations. The lesson from commercial organisations from around the world was that there had to be a concentration upon eliminating waste and generating value. Although originally conceived and applied in production and manufacturing organisations, this came to be adopted, or rather adapted to, applications in not-for-profit organisations such as those in the Public Sector. Gradually, moving into the 1990s, organisations, whether in the Public or Private Sectors, were utilising many of the tenets of the 'new' management approaches that had originated in post-war Japan. Initially these were espoused under the heading of 'Just-in-Time' (JIT) or 'Total Quality Management' (TQM), although as time progressed, many variations and adaptations of these initiatives were introduced, all adding to potential management policies that were designed to improve performance.[3]

The UK MoD had not been idle in recognising that the commercial world had made huge strides towards improving performance. These commercial improvements were almost always based upon the supply chain; by supply chain was meant the activities that linked together the identification of need for a product or service, the procurement and contracting activities, and then the inherent need for storage and distribution of the end product or service. Prior to this (1970s – 1990s) commercial organisations had been undertaking a movement towards integration of the management of these activities, often resulting in major enhancements in profitability, service and elimination of 'non-value adding' functions. Whilst recognising these performance-enhancing activities, the MoD had generally acknowledged and discussed their application within the armed forces but had on the whole, not made major changes to the way in which activities to support BAOR were undertaken. It often cited the inherent differences between the military's operating environment, and that of the commercial world as the reason for not taking immediate action.

At the same time, the Public Sector in the UK generally, had been subjected to a range of initiatives that were intended to introduce more commercial awareness into their way of operating. In due course, some of these organisations actually moved from the Public into the Private Sector (for example, utility companies) under the wholesale privatisation agenda of the Thatcher Government. Overall, this did not initially affect the MoD in a major way, except that Sir Peter Levene, under the auspices of the Conservative Government, introduced a strategy of competition between industrial organisations when seeking to place defence equipment contracts.

The situation for the MoD by the beginning of the 1990s was that its logistic support and approach to procurement were under some pressure to change in line with commercial practise. This was especially so, given that the whole basis of what was considered to be the 'threat', i.e. a conventional invasion of West Germany by the forces of the Soviet Union and Warsaw Pact had almost disappeared overnight, with the revolutions in Eastern Europe, the collapse of the Berlin Wall and the disintegration of the USSR itself between 1989 and 1991.

Thus, the scene was set for considerable change in the MoD.

Semantics

In bringing about change to the commercial sector, a fundamental feature of integration was to the fore. What were previously discreet functions had been joined together to optimise a number of inter-related processes. At one time, say in the 1970s, a typical UK organisation in the production or manufacturing environment would have separate purchasing, goods inwards, work-in-progress, finished goods and distribution activities, aided by production lines that were often duplicated with immature reliability and maintenance support. Over the time period indicated above, such manufacturing organisations that were still successful had adopted integration strategies that brought all of these into a coordinated, holistic, total-process approach. Even service organisations have utilised this to enable an enhancement of their performance.

As they have developed, a number of different labels have been given to indicate a gradual but progressive move towards integration. As an example, in the 1970s, 'materials management' was utilised to indicate purchasing and stores coming closer together. Similarly, 'physical distribution management' was the phrase to indicate the movement from finished goods warehouses to customers. Subsequently, these two were incorporated into the concept of 'supply chain management'. It is interesting to note, that as these concepts evolved, the activities claimed within them seemed to have surprising similarities. For example, during the late 1990s and early 2000s, the definition of 'logistics' provided by the Chartered Institute of Logistics and Transport was almost identical to that of 'procurement' put forward by the Chartered Institute of Purchasing and Supply. In the same way, the issues of reliability and maintainability have been seen as a completely divorced activity, yet, similarities between the need to obtain material, store it, utilise it and move it meant that it soon became obvious that the processes involved were so similar that they could no longer be undertaken in parallel to those undertaken for finished goods. The Society for Logistics Engineering (SOLE) produced a definition for the processes that it undertook, that was almost identical to the two previously mentioned. At the present time, the MoD has a supply chain activity and a support chain activity; in order to maximise benefits, it is becoming increasingly obvious that they should be managed in a similar (if not the same) way.

To some, precise definitions and careful application of activities within an area are an essential feature of managing operations; to others, what is important, is not the name applied to something, rather, it is the activity and impact of that work. Whilst not taking either view, for the appreciation of where the MoD is heading in respect of its obtaining and supporting equipment, the only real feature is that it is no longer possible to manage or operate discreet, standalone functions. The enhancement of performance will only come from an integration of all the activities that (in the context of the UK Armed Forces) deliver operational effectiveness. That this has happened can be evidenced in the progressive integration of relevant activities that commenced with the amalgamation of the Principal Accounting Officers for each separate service's logistics branch to form the Defence Logistics Organisation (DLO) and the creation of the Defence Procurement Agency (from the Procurement Executive) that occurred after the Strategic Defence Review of 1998. The need for even greater integration saw the amalgamation of these two agencies to form Defence Equipment and Support (DE&S) in April 2007.[4] Not to be outdone, with most of these activities centred geographically in Bristol, acknowledgement of the necessity to understand 'Through Life Capability Management' (TLCM) has meant that this activity, based in London, must be taken into account when considering any activity to obtain goods and services.

To indicate a more integrated approach, the phrase 'Smart Procurement' was introduced as part of the *Strategic Defence Review* of 1998. However, recognising that this might not be considered totally integrative, this was subsequently amended to 'Smart Acquisition'. Acquisition, being the label that seeks to encompass all of the activities that occur along the supply chain, has been used but is not always recognised in general discussion. For example, the US DoD refers to acquisition *and* logistics.[5] Even within the UK, the definition of acquisition has recently evolved to ensure that the concept of 'capability management' is encompassed. At the present time, the MoD defines acquisition as:[6]

"It covers the setting of requirements; the selection, development and manufacture of a solution to meet those requirements; the introduction into service and support of equipment or other elements of capability through life, and its appropriate disposal. Acquisition is supported by business processes such as:

- *Setting and managing requirements*
- *Negotiating and managing contracts*
- *Programme, project and technology management*
- *Investment approvals*
- *Safety management."*

Therefore, it can be seen that whatever precise phrase is used, and the UK favours acquisition, they all indicate that whatever semantic heading is used, all indicate and integrative, inter-disciplinary approach.

This however, is progressive, over time. This book acknowledges and seeks to reflect that progressive development over a period of many years. Hence the chronological nature of the book and simply because not all can agree on the precise label, it is called 'procurement and logistics' to recognise that it is about obtaining equipment/capability and ensuring support for that that equipment/capability.

Current Situation

It can be seen that over time, things change and this is especially so for the UK MoD. The progressive change that has taken place, and continues to take place, has seen improved integration with resultant smaller forces and support focused upon capability goals. It is a movement that is future focused and must be pursued, for as John F. Kennedy pointed out:

"Change is the law of life. And those who look only to the past or present are certain to miss the future."[7]

A driving imperative for this change has been and continues to be, that of 'value for money', which has been challenging to define and implement precisely at the best of times, is based upon notions of economy, efficiency and effectiveness. The result of this is that the UK Armed Forces has smaller organisational structures and a shrinking asset base. In turn, this has meant and will continue to mean an increased use of contractors to provide logistic service support. A particular challenge arising from this, is that of 'knowledge retention'. As more services are outsourced,[8] the knowledge held previously within the organisation is lost and the commercial organisation that undertakes the outsourced activity gains said knowledge and thus, in subsequent years, is in a position of power. For the future, to counter this, the UK MoD must ensure that it becomes an 'intelligent customer' and is able to make decisions that are knowledge-based and not process-led. There will be considerable challenges, many of which were identified by Bernard Gray and Charles Haddon-Cave in their recent reports.[9] Added to this, the volatility of the international environment and the ill-defined nature of the threat posed to the UK means that there is increasingly, a consideration of cyber warfare and the wider defence and security issues across the acquisition environment. Hence, the changes continue with integration at the heart of the way forward for the UK. For again, as John F. Kennedy said: "Everything changes but change itself."[10] Most recently, the UK conducted a Strategic Defence and Security Review that laid out the approaches that would need to be taken to meet the challenges of the future.

Arguably, to learn about the future, we need to understand where we've come from and how and why decisions were made. The cases in this book provide a progressive chronological insight into real scenarios that impacted upon UK acquisition performance. Within these cases there are relevant issues and challenges which will assist us in facing those that arise in the future.

The Strategic Defence and Security Review

In October 2010, the Prime Minister David Cameron presented the SDSR Report to the House of Commons.[11] It is an HM Government publication, available at [http://www.direct. gov.uk/prod_consum_dg/groups/dg_digitalassets/@dg/@en/documents/digitalasset/ dg_191634.pdf?CID=PDF&PLA=furl&CRE=sdsr.]

This document, whilst providing a clear direction identifies many challenges and prompts a number of pertinent questions, which are:

1. How has SDSR differed from previous reviews?
2. Just how strategic was it?
3. What will be the broader military impact?
4. What will the broader industrial impact?

Consideration of these questions will allow an insight into the future of the UK's procurement, logistics and supply chain management strategies.

1. How has SDSR differed from previous reviews?

In the run up to the May 2010 General Election, both the Conservative and Labour Parties stated they would conduct a defence review as a matter of urgency, given that the last one (not counting the odd update) was published twelve years ago, in 1998.[12] This had followed only seven years on from the previous 'review', entitled *Options for Change*, that had restructured the UK Armed Forces after the end of the Cold War and the dissolution of both the Warsaw Pact and Soviet Union effectively removed the very basis on which they were trained, organised and equipped.[13] *Options for Change* was not a single event however, but a sort of continuous exercise, conducted primarily between 1990 and 1995 with an eye to not only at reducing the defence budget in order to reap a 'peace dividend' and control public expenditure,[14] but also reduced overall personnel numbers by around eighteen percent (for the regular forces – the reserve forces were cut as well) along with amalgamating a number of regiments and cutting the number of submarines, destroyers, frigates, frontline aircraft and tanks.[15] Linked to this was *Frontline First: The Defence Costs Study*[16] and its attempt to rationalise administrative, medical, procurement and logistic support functions.[17] The *Strategic Defence Review (SDR)*, like its predecessor, cut various equipment capabilities and trimmed personnel numbers, although this time the cuts were mainly concentrated in the Reserve Forces.[18] It also looked to increase the MoD's efficiency savings, reduce administrative costs and introduced *Smart Procurement* in order to change the way the MoD procured defence equipment and services to try and avoid the problems of cost overruns and time delays that has plagued UK defence procurement. Since then, *Smart Procurement* has evolved into *Smart Acquisition* and the *Acquisition Operating Framework (AOF)* as part of the overall strategy of Through Life Capability Management (TLCM).[19]

So the question remains, how different has the current review proven to be? The *Strategic Defence and Security Review (SDSR)*[20] has, to all intents and purposes, continued the exact same themes as its two predecessors, that of:

- A reduction in both personnel and equipment numbers – the Army will lose around 7,000 personnel while the Air Force and Navy will lose about 5,000 each. Tanks and heavy artillery will shrink by forty percent, the destroyer and frigate fleet will gradually reduce from twenty-three to nineteen by 2020, the entire Harrier force will be retired – as will older helicopters such as Sea King and Gazelle, the Nimrod programme has been cancelled despite costing £3.6bn to date,[21] HMS *Ark Royal* will be decommissioned and along with HMS *Invincible*,[22] either HMS *Illustrious* or HMS *Ocean* will be sold off.
- Additional 'efficiency savings' – which in this case includes retiring older defence assets, reducing the planned acquisition of new equipment (for example, the number of additional Chinook helicopters has been dropped from twenty-two to twelve – just as the planned number of Type 45 destroyers was reduced from twelve to six), rationalising the defence estate and a reduction in the number of Civil Service personnel by around 25,000.
- Reducing the defence budget – under SDSR, the defence budget will be reduced by eight percent in real terms. Since the end of the Cold War, the defence budget has gradually dropped as a percentage of GDP, from 3.65 percent in 1990, to 2.48 percent in 2000, to 2.15 percent in 2010, representing the shifting priorities in public spending.[23]
- Restructuring the UK Armed Forces – just as the armed forces were restructured after the end of the Cold War for a more general defensive posture and as part of

SDR towards a more expeditionary role, the SDSR will continue the expeditionary theme, moving the armed forces towards the 'Force 2020' structure. This will include withdrawing the British Army from Germany by 2020, reducing the number of deployable brigades to five but retaining the ability to deploy a brigade-sized force anywhere in the world and sustaining it indefinitely or generating a 30,000 personnel force for a one-off operation.

2. Just how strategic was it?

Following on from the argument above, apart from having a major emphasis on cost-cutting, just how much of a strategic reorientation has been laid out by SDSR? Despite the claim that there will be no 'strategic shrinkage',[24] the reduction of the defence budget by eight percent and the cuts in personnel and equipment will inevitably lead to a reduction in the UK's capability to conduct operations abroad and its international influence. Indeed, one of the review's guidelines was a study done by the MoD on the future character of conflict,[25] and so two decades after the end of the Cold War, the character of conflict stubbornly remains as unpredictable as ever, leading to a desire to keep the broadest range of military capabilities as possible.

In the short-term, this led to another round of what some have termed salami-slicing[26] - percentage reductions in a wide-range of military capabilities, but few being eliminated altogether, the most obvious exceptions being the Harriers, three out of the four aircraft carriers and the Nimrod fleet. Added to that, heated inter-service rivalry marked the lead up to the publication of the SDSR[27] with the heads of the three branches outlining their differing views on what priorities should be funded. Despite this, some movement has occurred with an increase in resources for cyber-warfare (£650m), Special Forces and Overseas Aid (the logic being that if we concentrate greater effort towards conflict prevention and stabilisation in the first place, then the Armed Forces may not have to be used).

The overall structure of the armed forces will however be shaped in the medium-term by the Force 2020[28] concept with the Government deciding to keep some of the larger, more expensive programmes (such as the two new aircraft carriers, the Joint Strike Fighter and the Astute-class submarines) and ring-fencing the nuclear deterrent, at least until 2016 when a decision must be taken as to the modernisation of the submarines. The speed at which the review was done and the decision to ring-fence certain capabilities reduced the funds available for alternative acquisition projects which may have provided military capability at far greater value. Some have called it a 'wasted opportunity'.[29]

3. What will the broader military impact be?

The reductions in personnel and equipment will adversely impact on the UK's ability to intervene abroad, especially following the elimination of the Navy's fixed wing aircraft capability, as well as a large proportion of its amphibious capability, with the immediate decommission of *HMS Ark Royal*, the selling off of *HMS Invincible*, the sale of either *HMS Illustrious* or *HMS Ocean*, decommissioning a Bay-class amphibious support ship and the mothballing of the larger 'Landing Platform Dock' (LPD). That, along with the date on which the first of the new Queen Elizabeth-class carriers will enter service has been extended until 2020, will mean a major capability gap for the next decade.

There is also the question of our interoperability with key allies. The UK is not alone in reducing the amount spent on defence in light of adverse economic conditions, especially when there is a lack of a clearly defined and identifiable threat to the UK and other areas of public spending are seeing a cut in their budgets. Looking at where the UK wants to go in terms of defence acquisition collaboration, the approach looks to be somewhat schizophrenic, with on the one hand, two of the programmes that have been ring-fenced having strong links to the USA (JSF and Trident), while on the other, the UK Government helped save the pan-European A-400M (which it still wants to buy)[30] and despite historic political sensitivity over sharing military assets, the UK and France recently signed the Anglo-French Defence Cooperation Treaty.[31]

Until SDSR, the expectation was that the UK would be able to contribute to a US-led multi-national operation, and in sufficient scale and with sufficient capability as to be able to significantly influence its conduct. Just such intent could be found in the defence white paper of 2003:

"Where the UK chooses to be engaged, we will wish to be able to influence political and military decision making throughout the crisis, including during the post-conflict period. The significant military contribution the UK is able to make to such operations means that we secure an effective place in the political and military decision-making processes. To exploit this effectively, our Armed Forces will need to be interoperable with US command and control structures, match the US operational tempo and provide those capabilities which deliver the greatest impact when operating alongside the US."[32]

Such a view was echoed in a follow-on paper published in 2004[33] and another defence white paper in 2005.[34] With US defence spending running at more than twice the rest of NATO combined ($696bn vs. $325bn in 2008)[35] and a R&D budget of $79.5bn[36] there was always a question mark as to how long the UK would be prepared to benchmark its military capabilities against those of the USA and pay for the effort, an effort that was only going to get harder to bear in the current economic environment. For example, membership of the Joint Strike Fighter programme as the only Level 1 partner has so far cost the UK over £2bn and it was something of a shock for Lockheed Martin and the US Marine Corps to learn that the UK would be going for the F-35C (CV) variant as it was considered to be a more capable and cost effective option, instead of the F-35B (STOVL) variant when it was announced in the SDSR.[37] So far the UK is still aiming to acquire in the region of 138 aircraft at an approximate cost of £1.4bn. With the quiet disappearance of this aspiration to benchmark our forces against those of the USA directly, one can only wonder as to what the impact will be on our ability to operate alongside the United States in any future multinational operations and the influence we might be able to exert. There were however continued declarations of the importance of the UK's links with the USA:

"We will reinforce out pre-eminent security and defence relationship with the US. It remains deeply-rooted, broadly-based, strategically important and mutually supportive ... As part of out on-going commitment to working with our US colleagues at all levels, we will strengthen our joint efforts in priority areas".[38]

In the future, such UK contributions to US-led operations are likely to be focused increasingly on specialisation and areas where we hold "areas of comparative national advantage valued by key allies, especially the United States, such as our intelligence capabilities and highly capable elite forces".[39] But this was balanced by recognition that the UK has a need to work more closely with other allies, such as France:

"We will also intensify our security and defence relationship with France. The UK and France are active members of NATO, the EU and the UN Security Council, are Nuclear Weapon States, and have similar national security interests. Our Armed Forces are of comparable size and capability and it is clear that France will remain one of the UK's main strategic partners."[40]

Such recognition was quickly followed by the two countries signing a Defence Cooperation Treaty in November 2010 which emphasises cooperation in the realm of:[41]
- **Strategic Nuclear Deterrence** – this will include stockpile management, by supporting a joint facility in France to design and test nuclear warhead design and in addition, a joint Technology Development Centre in the UK.

- **Naval Operations** – which will include an integrated carrier strike force, seeing French Rafale aircraft flying from HMS Prince of Wales and UK Joint Strike Fighters flying from the Charles de Gaulle, given the UK's new carrier will be fitted with catapult and arrester gear to enable it to fly the JSF carrier variant.

- **Strategic Airlift** – the two countries have put together a common support plan for the A400M, looking to place a single contract with the supplier, Airbus Military, with deliveries to France beginning in 2013 and the UK in 2015. Discussions are also underway over the possibility that France could use some of the spare capacity in the UK's Future Strategic Tanker Aircraft fleet, with fourteen expected to be procured but only nine being needed on a normal day-to-day basis.

- **Joint Expeditionary Force** – is planned to be built around a ground formation, with potential units being of brigade size and joint air and naval exercises being held in 2011. While ground troops from both the UK and France operated together in the Balkans, the French were not involved in Iraq and the two countries are deployed in different sectors in Afghanistan.

- **International Politics** – such an agreement is useful to both parties in the international arena. On the one hand, the UK may see it as a way for it to compensate for the gradual erosion in the Transatlantic relationship, as Washington's focus moves elsewhere (the Pacific and Middle East for example) and for France, as it recognises that Germany will continue to limit its military involvement outside Europe and that the UK is its strongest partner.

- **Industrial Ties** – the agreement sets out plans for collaboration in several types of defence equipment – nuclear weapons aside – including aircraft carriers, submarines, unmanned aerial vehicles (UAVs), transport and tanker aircraft, as well as satellite communication. All of these capabilities will be important to both countries if they wish to project power in the 21st Century, the aim being to foster closer industrial collaboration while maximising capabilities at the best value for money. However, collaboration in defence procurement projects has been far from a resounding success. Success during the 1960s and 1970s (with the Gazelle, Puma and Jaguar for example) were matched with two decades of failure, with the UK dropping out of the Trigat antitank missile programme, the Trimilsatcom satellite programme and more notably, the Horizon Common New Generation Frigate, deciding instead to build the Type 45 destroyer. There has however been one major success – the creation of a Franco-British Joint Venture Company followed by the RAF's decision to buy the Storm Shadow missile led to the formation of MBDA, which is now Europe's leading guided weapons producer. The agreement also looks to capitalise on the bilateral High Level Working Group which has the past few years brought together high-level representatives from both government and industry to help coordinate defence research and planning of defence technology projects.

4. What will be the broader industrial impact?

One immediate impact has been that, due to the cuts made in overall equipment numbers as well as planned acquisitions, the UK defence industry will have to start rationalising itself – the first example being BAE Systems, who announced the potential loss of just under 1,400 jobs due to the SDSR in early December[42] - and moving their focus towards the export market as they reduce their exposure to the domestic UK market.

The Government hopes that export sales will pick up to offset the reduction in opportunities within the UK but given that most countries in North America and Europe are following the same pattern, this hope seems to be misplaced, unless companies look to the Middle East or Asia, or the company in question has a significant presence in the realm of cyber-security, which was promised about £650m worth of additional funding.[43] Future activity is likely to concentrate on the acquisition of such firms in the UK.

In addition to manufacturing, the services sector is bound to be hit as the MoD seeks to cut the number of UK bases currently in operation and sell MoD-owned property, reducing the opportunities for picking up outsourced service delivery contracts, on top of the reductions in the equipment inventory seeing a reduced need for service, maintenance and support, something that has already started to hit BAE Systems and may well hit Babcock, given its provision of services to defence estates and support of the Harrier, Tornado and VC-10 platforms. There has also been the cancellation of the proposed Defence Training College at RAF St Athan, a thirty-year PFI project to bring together all the individual service technical training onto one site, with the Qinetiq-led Metrix Consortium being left to absorb costs of some £37m while the government decides on how to proceed.[44]

SDSR has also raised the possibility of additional privatisation of the UK Government-owned Defence Support Group (DSG) and the Marchwood Sea Mounting Centre as well as the sale of the MoD's stake in the national telecommunications spectrum. Such a sale of UK Government assets could generate around £500m over the next few years and the rationalisation of the defence estate along with changes in how the MoD's infrastructure is managed, is hoped to save up to £350m from its annual costs. DSG was formed from the amalgamation in 2008 of the Army Base Repair Organisation (ABRO) and the Defence

Aviation and Repair Agency (DARA) and subject to takeover speculation very shortly thereafter from the VT Group. Babcock acquired the VT Group for £1.16bn and so whether they would still have an interest in acquiring it, given the current environment and lack of opportunities is questionable – it would retain a very high exposure to the UK domestic market, an exposure Babcock have been looking to dilute in recent years by looking for increased export opportunities abroad and the acquisition of VT Group's American assets.

BAE Systems has also been linked to DSG through its work on the Scout Vehicle programme but it's doubtful the company would want to increase its exposure to UK land systems maintenance especially when its preferred growth strategy is through international services and high technology collaboration. General Dynamics UK (GDUK) is another potential buyer, especially as they both signed a teaming agreement in February 2010 relating to the UK Scout Vehicle programme. DSG is to carry out work related to the production of the GD ASCOD Scout vehicle including assembly, integration and testing (the GD ASCOD forms the basis for the UK Scout Vehicle programme). Beyond those, there's the possibility the MoD could break up the Joint Supply Chain Services Group (until August 2010, the Defence Storage and Distribution Agency – DSDA) which include both the British Forces Post Office (BFPO) and the Disposal Services Authority (DSA) – a move that could attract strong interest from industry.[45]

Such a rationalisation will also impact on the UK's ability to sustain its 'operational sovereignty', as laid down in the Defence Industrial Strategy of 2005.[46] The strategy placed an emphasis on security of supply, an important consideration should the UK choose to act independently (in certain circumstances) to defend either itself or its dependent territories, and recognised that vital support would come from the Defence Industrial Base (DIB). SDSR recognises that the UK needs to retain certain industrial capabilities but fails to specify what these are, indicating that the UK is more likely to buy items off-the-shelf.[47] Indeed, for certain items this will necessarily be the case, for example, when the UK comes to replace its current small arms, including the SA80 assault rifle, the lack of an indigenous small arms manufacturer following the closure of the Royal Small Arms Factory (Enfield) in 1988 and the Royal Ordnance (Nottingham) site in late 2001 will mean that the UK will have to look overseas.[48] For such equipment, this implies there should be a willingness to transfer the appropriate technology on the supplier's part as well as their government to enable the UK to support and upgrade this equipment itself, either by the military or by the UK DIB. For example, the situation regarding the UK's purchase of Joint Strike Fighters has been the focus of a great deal of attention in recent years with doubt still surrounding the USA's willingness to transfer all the requisite technology used in the aircraft.[49] The broader question as to the UK's defence industrial capabilities was passed to those developing a

Green Paper on the subject, which was published in December 2010[50] with a White Paper to follow later in 2011.

A Summary

In the seventy-one years that have passed since the setting of the first case in this book, much has changed in respect of procurement and logistics. Even the semantics used to indicate the activities that are undertaken have changed although this has been in a progressive and integrative manner that has been aimed at enhancing ultimate operational performance. The changes have taken place in a socio-economic and political setting, but have been firmly based upon the adaptation of commercial practices into a Public Sector-based defence environment. It has coincided with the decline in the UK's military power and with that of the country's industrial base, where the move has been away from manufacturing to service-based industries.

This is not however, the whole story, as there has been an increase in global business and the rise of the trans-national organisation. This transformation in the global economic environment, has meant that many organisations have had to focus on what they consider to be their core business, and if the UK MoD has to focus on operations such as warfighting as its core business, then those activities on the periphery must be considered as prime candidates for outsourcing or for delivery by third-party contractors. Whilst this inevitably means a contraction in the number of organisations and personnel within the MoD, it does bring opportunities for the professionalisation of the acquisition workforce. This would be a highly qualified, entrepreneurially-focused cadre that would operate as an intelligent customer able to make knowledge-based judgements. That knowledge will be gained from a balance of conceptual knowledge and experience.

To be effective, it must also utilise process knowledge. Organisations such as the MoD have process knowledge in abundance; what will be required is a wide understanding of the concepts and experiential that is partly personal but much of which will come from examining past experience. A key element of the experiential knowledge will be the study and analysis of situations from the past; the case studies in this book acknowledge how procurement, logistics and supply chain management have changed over this period of time, all contain relevant issues and challenges that will be helpful in meeting the scenarios that will arise in the post-SDSR defence acquisition environment.

Notes

[1] McInnes, Dr Colin. *Hot War, Cold War: The British Army's Way in Warfare 1945-1995,*

London: Brassey's UK, 1996, pp. 3 – 4.

[2.] Antill, P. and Moore, David. 'British Army Logistics and Contractors on the Battlefield' in *The RUSI Journal*, Volume 145, Number 5, pp. 46 – 52.

[3.] Christopher, M. *Logistics and Supply Chain Management*, London: Brassey's UK, 1992.

[4.] House of Commons Defence Committee. *Defence Equipment 2008*, HC295, London: TSO, 27 March 2008, p. 5.

[5.] As an example, the title of the office of the person responsible for defence acquisition in the US DoD is The Office of the Undersecretary for Defence (Acquisition, Technology and Logistics), which is listed on its website, ACQWeb [http://www.acq.osd.mil/index.html].

[6.] Ministry of Defence. *'What is Acquisition?'* on the Acquisition Operation Framework website, located at [http://www.aof.mod.uk/aofcontent/strategic/guide/sg_whatisacq.htm]

[7.] Located at [http://www.brainyquote.com/quotes/authors/j/john_f_kennedy.html]

[8.] For what's happening with regard the US DoD, see Grier, P. (2008) *'Record number of US contractors in Iraq'*, located at [http://www.csmonitor.com/USA/Military/2008/0818/p02s01-usmi.html]

[9.] Gray, Bernard. *Review of Acquisition for the Secretary of State for Defence: An Independent Report by Bernard Gray*, October 2009, located at [http://www.mod.uk/NR/rdonlyres/78821960-14A0-429E-A90A-FA2A8C292C84/0/ReviewAcquisitionGrayreport.pdf]; Haddon-Cave QC, Charles. *The Nimrod Review: An Independent Review into the Broader Issues Surrounding the Loss of RAF Nimrod MR2 Aircraft XV230 in Afghanistan in 2006*, HC1025, London: TSO, 28 October 2009, located at [http://www.official-documents.gov.uk/document/hc0809/hc10/1025/1025.pdf]

[10.] Located at [http://www.famousquotesandauthors.com/authors/john_f__kennedy_quotes.html]

[11.] BBC News. *'Defence review: Cameron unveils armed forces cuts'*, located at [http://www.bbc.co.uk/news/uk-politics-11570593]

[12.] Ministry of Defence. *The Strategic Defence Review*, Cm3999, London: TSO, July 1998.

[13.] House of Commons Defence Committee. *Defence Implications of Recent Events*, HC320, London: HMSO, 1990.

[14.] McInnes, Colin. **'Labour's Strategic Defence Review'** in *International Affairs*, Volume 74, Number 4 (October 1998), pp. 823 – 845.

[15.] See HCDC. [http://www.publications.parliament.uk/pa/cm199798/cmselect/cmdfence/138/13805.htm]

[16.] Ministry of Defence. *Frontline First: The Defence Costs Study*, London: HMSO, 1994.

[17.] See HCDC. [http://www.publications.parliament.uk/pa/cm199798/cmselect/cmdfence/138/13806.htm]

[18.] See HCDC. [http://www.parliament.the-stationery-office.co.uk/pa/cm199798/cmselect/cmdfence/138/13816.htm webpage]. This included cutting the Territorial Army by 17,000, reducing the frigate and destroyer fleet by three, reducing the number of minesweepers by three, cutting the number of attack submarines by two and cutting the number of frontline aircraft by 36. This was however somewhat offset by some additional promised purchases, such as two new aircraft carriers (to replace the three existing ones), four Ro-Ro ships and four C-17 aircraft.

[19.] Ministry of Defence. Acquisition Operating Framework website, located at [http://www.aof.mod.uk/]

[20.] HM Government. *Securing Britain in an Age of Uncertainty: The Strategic Defence and Security Review*, Cm7948, London: TSO, October 2010. See also the Fact Sheets associated with this at [http://www.cabinetoffice.gov.uk/resource-library/strategic-defence-and-security-review-securing-britain-age-uncertainty]

[21.] Kirkup, James. **'£3.6 billion Nimrods dismantled for scrap'** in *The Telegraph*, 10 December 2010 and available at [http://www.telegraph.co.uk/news/newstopics/politics/defence/8191690/3.6-billion-Nimrods-dismantled-for-scrap.html]

[22.] Rayment, Sean. **'Aircraft Carrier HMS Invincible is put up for sale'** in *The Telegraph*, 28 November 2010, also available at [http://www.telegraph.co.uk/news/newstopics/politics/defence/8165331/Aircraft-carrier-HMS-Invincible-is-put-up-for-sale.html]

[23.] See [http://www.ukpublicspending.co.uk]

[24.] William Hague, quoted in Stephens, Philip. **'PM should beware cost of taking risk with military'** in the *Financial Times*, 15 September 2010 and available at [http://www.ft.com/cms/s/0/d6f2c8c2-c0ef-11df-99c4-00144feab49a.html#axzz185zxgpI9]

[25.] Ministry of Defence. *Future Character of Conflict,* Development Concepts and Doctrine Centre, Shrivenham, February 2010.

[26.] Nicoll, Alexander (Ed). **'UK cost-cutting review shrinks military capacity'**, *IISS Strategic Comments*, Volume 16, No. 39 (November 2010), available at [http://www.iiss.org/publications/strategic-comments/past-issues/volume-16-2010/november/uk-cost-cutting-review-shrinks-military-capacity/]

[27.] Cowan, Gerrard. **'UK military chiefs launch SDR debate'** in *Jane's Defence Weekly*, posted 22 January 2010, also available at [http://www.janes.com].

[28.] Ministry of Defence. *Fact Sheet 5 (Future Force 2020 – Summary of Size, Shape and*

Structure), Fact Sheet 6 (Future Force 2020 – Royal Navy), Fact Sheet 7 (Future Force 2020 – British Army) and *Fact Sheet 8 (Future Force 2020 – Royal Air Force)*, available at [http://www.cabinetoffice.gov.uk/resource-library/strategic-defence-and-security-review-securing-britain-age-uncertainty]

[29.] Editorial. **'Ex-defence chiefs slam cuts in House of Lords debate'** in *The News*, 12 November 2010, available at [http://www.portsmouth.co.uk/newshome/Exdefence-chiefs-slam-cuts-in.6624320.jp]

[30.] BBC News. *'Final deal to build Europe's A400M military plane'*, located at [http://www.bbc.co.uk/news/business-11701337]

[31.] Norton-Taylor, Richard. **'Anglo-French defence deal is a triumph of pragmatism over ideology'** in *The Guardian*, 2 November 2010 at [http://www.guardian.co.uk/politics/2010/nov/02/britain-france-defence-pragmatism-analysis?intcmp=239]

[32.] Ministry of Defence. *Defence White Paper: Delivering Security in a Changing World*, Cm6041, London: TSO, December 2003, p. 8.

[33.] Ministry of Defence. *Delivering Security in a Changing World – Future Capabilities*, Cm6269, London: TSO, July 2004, p. 3.

[34.] Ministry of Defence. *Defence White Paper: Defence Industrial Strategy*, Cm6697, London: TSO, December 2005, p. 19.

[35.] International Institute for Strategic Studies. *The Military Balance 2010*, London: Routledge, 2010, p. 463.

[36.] Ibid, p. 22.

[37.] Jennings, Gareth. **'Lockheed Martin, USMC remain committed to F-35 STOVL variant'** in *Jane's Defence Weekly* and available on [http://www.janes.com], posted 2 December 2010.

[38.] Op Cit. *Securing Britain in an Age of Uncertainty: The Strategic Defence and Security Review*, p. 60.

[39.] Ibid, p. 12.

[40.] Ibid, p. 60.

[41.] Nicoll, Alexander (Ed). **'The ambitious UK-France defence accord'**, *IISS Strategic Comments*, Volume 16, No. 41 (November 2010), available at [http://www.iiss.org/publications/strategic-comments/past-issues/volume-16-2010/november/the-ambitious-uk-france-defence-accord/]

[42.] Jones, Alan. **'BAE Systems set to cut 1,350 jobs'**, in *The Independent* and available at [http://www.independent.co.uk/news/business/news/bae-systems-set-to-cut-1350-jobs-2155188.html]

[43.] Op Cit. *Securing Britain in an Age of Uncertainty: The Strategic Defence and Security Review*, p. 47.

[44.] Anderson, Guy. **'SDSR Analysis: UK industry has reasons for pessimism'** in *Jane's Defence Weekly* and available at [http://www.janes.com], posted 20 October 2010.

[45.] Anderson, Guy. **'UK SDSR raises privatisation possibilities'** in *Jane's Defence Weekly* and available on [http://www.janes.com], posted 20 October 2010.

[46.] Op Cit. *Defence White Paper: Defence Industrial Strategy*, p. 33.

[47.] Taylor, Trevor. **'What's New: UK Defence Policy Before and After the SDSR'** in *The RUSI Journal*, Volume 155, No. 69, December 2010), pp. 10-14.

[48.] Antill, Peter. **'Patience is a Virtue: The Case of the British Army's Assault Rifle'** in Moore, David. (Ed). *Case Studies in Defence Procurement and Logistics: Volume Two – From Helen of Troy to Helmand Province*, Cambridge: Cambridge Academic, 2012.

[49.] Antill, Peter and Ito, Pete. **'Multi-National Collaborative Procurement - The Joint Strike Fighter Programme'** in Moore, David. (Ed). *Case Studies in Defence Procurement & Logistics, Volume I: From World War II to the Post Cold War World*, Cambridge: Cambridge Academic Press, 2011.

[50.] Ministry of Defence. *Defence Green Paper – Equipment, Support and Technology for UK Defence and Security*, Cm7989, London: TSO, December 2010.

Part One // Defence Logistics & Supply Chain Strategy

Case 1.1 // From Helen to Helmand

A Short History of Military Logistics

David M. Moore and Peter D. Antill

Centre for Defence Acquisition, Cranfield University, Defence Academy of the UK.

Introduction

Logistics is a relatively new word used to describe a very old practice: the supply, movement and maintenance of an armed force both in peacetime and under operational conditions. Most soldiers have an appreciation of the impact logistics can have on operational readiness. Logistic considerations are generally built in to battle plans at an early stage, for without logistics, the tanks, armoured personnel carriers, artillery pieces, helicopters and aircraft are just numbers on a Table of Organisation and Equipment. Unfortunately, it often seems that the high profile weapon systems have had greater priority in resources than the means to support them in the field, be it ammunition, fuel or spares. For it is logistics that will determine the forces that can be delivered to the theatre of operations, what forces can be supported once there, and what will then be the tempo of operations.

Logistics is not only about the supply of matériel to an army in times of war. It also includes the ability of the national infrastructure and manufacturing base to equip, support and supply the Armed Forces, the national transportation system to move the forces to be deployed and its ability to resupply that force once they are deployed. Thus it has been said, "logisticians are a sad, embittered race of men, very much in demand in war, who sink resentfully into obscurity in peace. They deal only with facts but must work for men who merchant in theories. They emerge during war because war is very much fact. They disappear in peace, because in peace, war is mostly theory."[1]

The Ancient World

The practice of logistics, as understood in its modern form, has been around for as long as there have been organised armed forces with which nations and/or states have tried to exert military force on their neighbours. The earliest known standing army was that of the Assyrians at around 700 BC. They had iron weapons, armour and chariots, were well organised and could fight over different types of terrain (the most common in the

Middle East being desert and mountain) and engage in siege operations. The need to feed and equip a substantial force of that time, along with the means of transportation (i.e. horses, camels, mules and oxen) would mean that it could not linger in one place for too long. The best time to arrive in any one spot was just after the harvest, when the entire stock was available for requisitioning. Obviously, it was not such a good time for the local inhabitants. One of the most intense consumers of grain was the increasing number of animals that were employed by armies of this period. In summer they soon overgrazed the immediate area, and unless provision had been made beforehand to stockpile supplies or have them bought in, the army would have to move. Considerable numbers of followers carrying the matériel necessary to provide sustenance and maintenance to the fighting force would provide essential logistic support.

Both Philip and Alexander improved upon the art of logistics in their time. Philip realised that the vast baggage train that traditionally followed an army restricted the mobility of his forces. So he did away with much of the baggage train and made the soldiers carry much of their equipment and supplies. He also banned dependants. As a result the logistics requirements of his army fell substantially, as the smaller numbers of animals required less fodder, and a smaller number of wagons meant less maintenance and a reduced need for wood to effect repairs. Added to that, the smaller number of cart drivers and lack of dependants, meant less food needed to be taken with them, hence fewer carts and animals and there was a reduced need to forage, which proved useful in desolate regions. Alexander however, was slightly more lenient than his father was, as regards women. He demonstrated the care he had for his men by allowing them to take their women with them.

This was important; given the time they spent away on campaign and also avoided discipline problems if the men tried to vent their desires on the local female population of newly conquered territories. He also made extensive use of shipping, with a reasonable sized merchant ship able to carry around 400 tons, while a horse could carry 200 lbs. (but needed to eat 20 lbs. of fodder a day, thus consuming its own load every ten days). He never spent a winter or more than a few weeks with his army on campaign away from a sea port or navigable river. He even used his enemy's logistics weaknesses against them, as many ships were mainly configured for fighting but not for endurance, and so Alexander would blockade the ports and rivers the Persian ships would use for supplies, thus forcing them back to base.

He planned to use his merchant fleet to support his campaign in India, with the fleet keeping pace with the army, while the army would provide the fleet with fresh water. However, the monsoons were heavier than usual, and prevented the fleet from sailing. Alexander lost two-thirds of his force, but managed to get to Gwadar where he reprovisioned. The importance of logistics was central to Alexander's plans, indeed his mastery of it allowed him to conduct the longest military campaign in history. At the farthest point reached by his army, the river Beas in India, his soldiers had marched 11,250 miles in eight years. Their success depended on his army's ability to move fast by depending on comparatively few animals, by using the sea wherever possible, and on good logistic intelligence.

The Roman legions used techniques broadly similar to the old methods (large supply trains etc.), however, some did use those techniques pioneered by Phillip and Alexander, most notably the Roman consul Marius. The Romans' logistics were helped of course, by the superb infrastructure, including the roads they built as they expanded their empire. However, with the decline in the western Roman Empire in the Fifth Century AD, the art of warfare degenerated, and with it, logistics was reduced to the level of pillage and plunder. It was with the coming of Charlemagne, that provided the basis for feudalism, and his use of large supply trains and fortified supply posts called 'burgs', enabled him to campaign up to 1,000 miles away, for extended periods.

The eastern Roman (Byzantine) Empire did not suffer from the same decay as its western counterpart. It adopted a defensive strategy, which Clausewitz recognised as being easier logistically than an offensive strategy, and that expansion of territory is costly in men and material. Thus in many ways their logistics problems were simplified – they had interior lines of communication, and could shift base far easier in response to an attack, than if they were in conquered territory, an important consideration, due to their fear of a two-front war. They used shipping and considered it vital to keep control of the Dardanelles, Bosphorous and Sea of Marmara; and on campaign made extensive use of permanent warehouses, or magazines, to supply troops. Hence, supply was still an important consideration, and thus logistics were fundamentally tied up with the feudal system – the granting of patronage over an area of land, in exchange for military service. A peacetime army could be maintained at minimal cost by essentially living off the land, useful for Princes with little hard currency, and allowed the man-at-arms to feed himself, his family and retainers from what he grew on his own land and given to him by the peasants.

The fighting ships of antiquity were limited by the lack of endurance while the broad beamed seaworthy merchant ships were unsuited to the tactics of the time that were practised in the Mediterranean. It wasn't until the Europeans put artillery on-board such vessels that they combined the fighting and logistic capability in one vessel and thus became instruments of foreign policy with remarkable endurance and hitting power. They reached the zenith of their potential during the Napoleonic Wars, but with the conversion to coal and steam power, a ship's endurance was once again limited. But they could still carry their ammunition and supplies farther and faster, and were thus more logistically independent than horse-powered armies, despite the need for coaling stations. Fuel oil increased endurance by forty percent, but that was due to its greater efficiency as a fuel source. The coming of the fleet train and underway replenishment techniques during the Second World War enhanced the endurance of modern navies massively, and ships could thus stay at sea for months, if not years, especially with the reduced time between dockyard maintenance services. The coming of nuclear power once again extended the sea-going life of a vessel, with endurance limited to that of the crew and the systems that need a dockyard to be overhauled.

The Middle Ages

The appeal by Emperor Alexius to the Pope for help in clearing Anatloia of the Turks in 1095 paved the way for a series of Western European military expeditions which have become known as the Crusades. As a result of these, the Western Europeans significantly advanced their practice of the military arts.

The First Crusade ran from 1096 until 1099 and ended with the capture of Jerusalem. It didn't start very well however, with the various contingents from Normandy, Sicily, France, Flanders and England having ten leaders, internal friction within the army, which at times was no better than a rabble, and having a strong distrust of the Byzantines, which was reciprocated. The Crusaders had no interest in recapturing lost Byzantine lands, while the Emperor had no interest in Jerusalem. The lack of a supply system almost twice brought it to grief early on, when the Crusaders almost starved while besieging Antioch and after the capture of the city, were besieged themselves. The army advanced south to Jaffa the following year, and appeared to learn the logistic lessons from the previous experience.

There was far more co-operation between national contingents and they had the advantage of the Pisan fleet sailing parallel to their route to provide logistic support. This of course only lasted while they were fairly close to the coast, but the army soon had to turn inland towards Jerusalem. The Crusaders were too small in number to completely surround the city and could not easily starve the city into submission as the governor of Jerusalem had ordered all the livestock to be herded into the city and stockpiled other foodstuffs. The Crusaders also found themselves short of water, and thus time was not on their side. The Crusaders thus attempted an assault as early as possible without siege engines and while they overran the outer defences, could not make any headway against the inner walls. Fortunately, the English and Genoese fleets arrived in Jaffa at this point, but conveying their cargo to Jerusalem was time consuming and expensive in both men and animals.

Additionally, there was a shortage of decent timber with which to make siege weapons with, but some was finally found on some wooded hills near Nablus, fifty miles north of Jerusalem. Again, this was a time consuming and expensive operation. By the time work had started on the siege towers it was mid-summer, the Crusaders were suffering from a shortage of water and word had been received that an Egyptian army was marching to the relief of the city. The Crusaders speeded up their preparations and finally rolled out their siege towers and assaulted the city, on the 13th and 14th of July, which fell that night.

The Second Crusade consisted of a French army under King Louis VII and a German army led by Emperor Conrad III. It was launched to take back Edessa from the Muslims and was a logistic disaster. The German army managed to stir up the local inhabitants once they arrived in Byzantine territory by pillaging, but the French army behaved much better and had little trouble. Unfortunately the German army had taken much of the available food and had so frightened the peasants that they had hid what little they had left. To accentuate this, the Germans refused to sell food to the French when they reached Constantinople. The hostility between the two armies led Conrad to split the two armies and take different routes across Anatolia.

To compound this, Conrad split his own army with both groups being routed at different points along their respective paths. Louis' army faired little better, being defeated at Laodicea as well. The French thus made their way back to Attalia on the coast but found that the inhabitants were short of food as well, and the presence of the Crusaders attracted the Turks who set about besieging the city. Louis was forced to leave, taking his cavalry by sea in two lifts to Antioch, but leaving his infantry to march overland. Needless to say, few survived this example of dreadful leadership. Finally, Louis and Conrad joined by Baldwin of Jerusalem, set about besieging Damascus. Unfortunately, not only set their siege lines against the strongest part of the city's defences but sited their base camp in an area that didn't have any water nearby. Unsurprisingly, the siege failed.

The Third Crusade followed some forty years later and came after the Christian defeat at Hattin and the capture of Jerusalem by Saladin. It involved three kings, Richard I of England, Frederick I of Germany and Philip II of France. Frederick was first on the scene, and after marching through Anatolia and capturing Iconium, was unfortunately drowned and his army badly depleted by enemy action and the twin scourges of hunger and disease. A year later Philip and Richard arrived at Acre, where the Christian armies had been besieging the town for almost two years. Within twenty-four hours the morale of the army had been restored and the tempo of operations increased. A relief effort was beaten off, and the city eventually surrendered. Philip then left Richard in sole command of the army, who started the advance to Jerusalem.

His planning and logistics were far superior to what had gone before. For example, he kept in contact with his fleet off the coast, he kept his marches short to preserve the strength of his soldiers, and even arranged a laundry organisation to keep the clothes clean (helping morale and health). He defeated Saladin at Arsuf, stopped briefly at Jaffa and marched towards Jerusalem in the winter rains. His men suffered quite badly, and recognising his mistake, he returned to Ascalon, on the coast. In the following spring, Richard set out once again for Jerusalem, but Saladin retired before him, destroying crops and poisoning wells. The lack of fodder and water meant that Richard finally halted at Beit-Nuba and concluded that he could not risk his army in besieging Jerusalem. Even if he captured the city, he would have to return to England due to the treasonous actions of his brother, John and it was unlikely that the Christian army would be able to hold until his return. So he retreated to Acre where he learnt that Saladin had taken Jaffa with a surprise attack. On hearing this, Richard set out with a small force by ship to Jaffa, with the rest travelling overland. At the sight of these ships, the Christians in the city took up arms against Saladin's troops and Richard, on the prompting of a local priest who had swum out to the fleet, took his small force and routed the occupation army. He even beat off a second attack by Saladin who tried to catch Richard before his main force arrived.

The Crusades pointed to a number of tactical and military engineering lessons that were vital for the improvement of the Western military art. The most important of these was that of logistics. Western armies had lived off the land when campaigning, and when they had stripped an area, they would have to move on, or starve. The length of campaigns tended to be short, as the length of time that Barons and their retainers could spend away from their fiefs was limited. Most Western armies when faced with the scorched earth policies of the Turks, and with no organised wagon train, limited local knowledge as regards the terrain and climate, thus tended to disintegrate.

With the long campaigns in Western Asia, the generals had to re-learn the lessons learnt by Alexander, plan properly or die. In the first two Crusades, many men and horses died of starvation, but Richard showed that good logistic planning could change the situation around completely. He built a logistic base on the island of Cyprus and used that to his advantage when marching from Acre to Ascalon. His refusal to embark on a protracted siege of Jerusalem shows that he understood the serious logistic situation that would have arisen.

The Modern Era

As the centuries passed, the problems facing an army remained the same: sustaining itself while campaigning, despite the advent of new tactics, of gunpowder and the railway. Any large army would be accompanied a large number of horses, and dry fodder could only really be carried by ship in large amounts. So campaigning would either wait while the grass had grown again, or pause every so often. Napoleon was able to take advantage of the better road system of the early nineteenth century, and the increasing population density, but ultimately still relied upon a combination of magazines and foraging. While many Napoleonic armies abandoned tents to increase speed and lighten the logistic load, the numbers of cavalry and artillery pieces (pulled by horses) grew as well, thus defeating the object. The lack of tents actually increased the instance of illness and disease, putting greater pressure on the medical system, thus putting greater pressure on the logistic system due to larger medical facilities needed and the need to expand the reinforcement system.

Napoleon failed the logistics test when he crossed the Nieman in 1812 to start his Russian campaign. He started with just over 300,000 men and reached Moscow with just over 100,000 excluding stragglers. Napoleon had known the logistics system would not sustain his army on the road to Moscow and keep it there. He gambled that he could force the Russians to the negotiating table and dictate terms. He failed, and so had to retreat. The pursuing Russian army did little better, starting at Kaluga with 120,000 men and finally reaching Vilna with 30,000.

The only major international conflict between that of the Napoleonic War and the First World War was that of the Crimean War, fought between October 1853 and February 1856 and involved Russia, France, Britain and Turkey as its main protagonists. From a British and French point of view the main theater of operations was the Crimean Peninsula, but operations also took place in the Caucasus, around the Danube and the Baltic Sea. The background cause of the war lay in what was known as 'The Eastern Question' which involved the Great Powers in the question of what was to be done with the decaying Ottoman Empire, and in particular, its relationship with Russia. The immediate cause was the territorial ambition of Russia and the question of minority rights (the Greek Orthodox Christian Church) under the Turks.

It was logistics, as well as training and morale that decided the course of the war. All three armies in the Crimea suffered in one way or another in terms of the actual combat capability of the forces, but also the logistics back up received. The Russians were losing ground industrially by both Britain and France, both in terms of Gross National Product (GNP) and GNP per capita. While this did not immediately translate into military weakness, the effects would be felt soon enough, with no railways south of Moscow, Russian troops seriously lacked mobility and could take up to three months to get to the Crimea (as would supplies and ammunition) as opposed to three weeks for the British and French who would come by sea. The majority of Russian troops were still equipped with muskets as opposed to rifles, which were more accurate and had a longer range. With the French revolution still casting a deep shadow over the continent, Governments were worried about the loyalty of their troops, and the lack of a war caused officers to emphasis caution, obedience and hierarchy. Nicholas I encouraged this within Russia and thus military parades and the look of the troops' uniforms became more important than logistics or education.[2]

The British Army had suffered as well, in the forty or so years of peace since the end of the Napoleonic Wars. There were some seven semi-independent authorities that looked after the administration of the Army, and contributed to the "complication, the muddle, the duplication, the mutual jealousies, the labyrinthine processes of supply and control".[3] These 'organisations' were made up of the Commander-in-Chief (located at Horse Guards – a sort of Chief of the Imperial General Staff), the Master General of the Ordnance (equipment, fortifications and barracks), Board of General Officers (uniforms), The Commissariat (supplies and transport, but Wellington's baggage train had been disbanded many years previously and so had no real means of moving said supplies), The Medical Department, the Secretary-at-War (who was responsible for the Army's dealings with civilian contractors), and the Secretary of State for the Colonies.

There was little, if any, coordination between these different 'organisations' and thus the provision of logistic support was rudimentary at best. In 1854, the view of administration and the provision of logistic support to the troops in the field was in the hands of, to a greater or lesser extent, the commanding officers of the regiments, some of whom cared for their men, but most simply looked after their own lot. Logistics is not merely about the supply of men and matériel to the theater of operation, but the application of those resources in a timely manner to affect the outcome of battle, but also the provision of food, clothing, shelter and entertainment to the troops in order to safeguard morale and discipline. There were very few in the Crimea who could visualise this problem, or had the power to do anything about it. The British tended to fight the war first and leave the administration to take care of itself. Unfortunately, it made it difficult for any comprehensive revision of the system. Much of the clothing and equipment was left over from the Peninsula War and thus a lot of it was rusting, decaying or falling apart. It was not the fact that there was no food, equipment, fodder or supplies, there was plenty of it in Balaclava. It was that there was virtually no centrally organised system of getting it to where it was needed.

There was also a very loose and ill-defined chain of command, which had contributed to the Army's difficulties. Many commanding officers looked upon their regiments as their own personal property, and were very reluctant to take them out for exercises with other units, which were held extremely infrequently anyway. Few officers had any conception of military tactics to start with. "This army ... is a shambles."[4] All these faults combined with the terrible winter of 1854 to produce chaos, and the medical system effectively broke down. Into this maelstrom walked Florence Nightingale and thirty-eight nurses. Although there was initially resistance to their presence, the stream of wounded from Balaclava and Inkerman overwhelmed the hospital. With her own budget and working unceasingly to improve the conditions (the washing of linen, issue of clothing and beds, special diets, medicine, hygiene, sanitary conditions etc.) there, the death rate fell from 44 per cent, to 2.2 per cent in six months.

The terrible conditions were reported in the Press from reports of *The Times* correspondent, W H Russell, and also in letters from serving officers. Public opinion became such that the Government of Lord Aberdeen fell, and Lord Palmerston took over, with Lord Panmure as Minister of War and Lord Clarendon as Foreign Secretary. General Simpson was sent out to relieve Lord Raglan of the administrative burden, and gradually, the administrative chaos was overcome. A central system for the supply of provisions to depots on the peninsula was formed, Turkish labour was recruited to undertake construction work, the railway from Balaclava to near Telegraph on the Woronzov Road was completed, transport was borrowed from the French and Spanish mules hired from Barcelona. Mr Filder and Admiral Boxter began to restore order in Balaclava and organise the port. The group of dishonest sutlers and contractors that had been operating unchecked were bought under control, and by February, the army was beginning to heal itself, with games of cricket and football being played in the camps. Britain's main military power of course rested with the Royal Navy, and with the effective withdrawal of the Russian fleets into port, provided the main logistic supply line to the British forces in the Crimean Peninsula.

Of the three main armies to take part in the Crimean War, the best was clearly the French, which retained a measure of its professionalism from the Napoleonic era and many officers and men had served in Algeria. There was still a degree of incompetence and around half the officers were chosen from the lower ranks, so many had trouble reading and writing. The teaching of military skills such as reading maps, strategy and topography was as scorned as it was in the Russian Army. But the French General Staff was much more comprehensive than that of the British, and included officers of the Administrative Service as well as specialised corps. It was the supply and medical arrangements that really stood out, and both were superior to that of the British (initially) and Russian Armies, for at Kamiesh, the French built a logistic support base of some 100 acres in area, and another 50 acres of shops where a large variety of goods could be bought.[5]

The American Civil War foreshadowed future warfare, particularly as regards logistics. Both sides were determined, had reasonably competent generals, with large populations from which to draw recruits from, and the means to equip them. This laid the foundations for a long war, not one that would be determined by one or two battles, but several campaigns, and hinge upon the will to sustain the war-fighting capability (material or morale). This was to be a large conflict between large populations with mass mobilisation armies. This meant that a logistics infrastructure would have to be set up to cater for the training, equipping and movement of these armies from scratch. But it would also have to cater for the supply of food, ammunition, equipment, spare parts, fresh horses and their fodder, and the evacuation of casualties (of which there would be greater numbers than ever) and canned food, introduced in the 1840s. Strategy took into consideration not only the combatants' own logistic requirements, but that of the enemy as well. That principle meant that Grant was able to fix Lee in Virginia, which enabled Sherman to march to Atlanta to destroy the Confederates' major communications and supply centre, and hence onto Savannah. Lastly, it was the first major war in which railways played an important part, speeding up the movement of troops and supplies. They also dictated to a great extent, the axes of advance or retreat, the siting of defensive positions and even the location of battles. But it also warned of the consequences of having a large army tied to the railway system for the majority of its supplies, as McClellan found out in both the Richmond Peninsula campaign and after the Battle of Antietam.

Most European observers had lost interest in the war early on, after the shambles of the First Bull Run, but a few (including a Captain Scheibert of the Prussian Army) were impressed with the support given by the Union Navy to the Union Army, in tactical and logistic terms, and the use of railway repair battalions to keep the rail systems functioning. The two lessons they missed or were forgotten, were the growing importance of fortifications (particularly the trench) to offset the increasing firepower of contemporary weapons and the increasing rate of ammunition expenditure. The Austro-Prussian and the Franco-Prussian Wars confirmed the importance (as well as the limitations) of railways but were similar to the wars of the past in that ammunition expenditure was relatively low. It was thus easier for troops to be supplied with ammunition as compared to food.

The World Wars

The First World War was unlike anything that had gone before it. Not only did the armies initially outstrip their logistic systems (particularly the Germans with their Schlieffen Plan) with the amount of men, equipment and horses moving at a fast pace, but they totally underestimated the ammunition requirements (particularly for artillery). On average, ammunition was consumed at ten times the pre-war estimates, and the shortage of ammunition became serious, forcing governments to vastly increase ammunition production. In Britain this caused the 'shell scandal' of 1915, but rather than the government of the day being to blame, it was faulty pre-war planning, for a campaign on the mainland of Europe, for which the British were logistically unprepared.

Once the war became trench bound, supplies were needed to build fortifications that stretched across the whole of the Western Front. Add to that the scale of the casualties involved, the difficulty in building up for an attack (husbanding supplies) and then sustaining the attack once it had gone in (if any progress was made, supplies had to be carried over the morass of no-man's land). It was no wonder that the war in the west was conducted at a snail's pace, given the logistic problems. It was not until 1918, that the British, learning the lessons of the last four years, finally showed how an offensive should be carried out, with tanks and motorised gun sleds helping to maintain the pace of the advance, and maintain supply well away from the railheads and ports. The First World War was a milestone for military logistics. It was no longer true to say that supply was easier when armies kept on the move due to the fact that when they stopped they consumed the food, fuel and fodder needed by the army. From 1914, the reverse applied, because of the huge expenditure of ammunition, and the consequent expansion of transport to lift it forward to the consumers. It was now far more difficult to resupply an army on the move, while the industrial nations could produce huge amounts of war matériel; the difficulty was in keeping the supplies moving forward to the consumer.

This of course, was a foretaste of the Second World War. The conflict was global in size and scale. Not only did combatants have to supply forces at ever greater distances from the home base, but these forces tended to be fast moving, and voracious in their consumption of fuel, food, water and ammunition. Railways again proved indispensable, but sealift and airlift made ever greater contributions as the war dragged on (especially with the use of amphibious and airborne forces, as well as underway replenishment for naval task forces). The large-scale use of motorised transport for tactical re-supply helped maintain the momentum of offensive operations, and most armies became more motorised as the war progressed. The Germans, although moving to greater use of motorised transport, still relied on horse transport to a large extent – a fact worth noting in the failure of Barbarossa. After the fighting had ceased, the operations staffs could relax somewhat, whereas the logisticians had to supply not only the occupation forces, but also relocate those forces that were demobilising, repatriate Prisoners Of War, and feed civil populations of often decimated countries.

The Second World War was, logistically, as in every other sense, the most testing war in history. The cost of technology had not yet become an inhibiting factor, and only its industrial potential and access to raw materials limited the amount of equipment, spares and consumables a nation could produce. In this regard, the United States outstripped all others. Consumption of war material was never a problem for the USA and its allies. Neither was the fighting power of the Germans diminished by their huge expenditure of war material, nor the strategic bomber offensives of the Allies. They conducted a stubborn, often brilliant defensive strategy for two-and-a-half years, and even at the end, industrial production was still rising. The principal logistic legacy of the Second World War was the expertise in supplying far off operations and a sound lesson in what is, and what is not, administratively possible.

The Cold War

With the end of the Second World War, the tensions that had been held in check by the common goal to defeat fascism finally came to the fore. The Cold War started in around 1948 and was given impetus by the Berlin Blockade, the formation of NATO and the Korean War. The period was characterised by the change in the global order, from one dominated by empires to a roughly bipolar world, split between the Superpowers and their alliance blocs. However, the continued activity by both blocs in the Third World meant that both sides continued to draw on the experience of power projection from the Second World War.

East and West continued to have to prepare for both limited conflicts in the Third World, and an all-out confrontation with the other bloc. These would vary between 'low intensity' counter-insurgency conflicts (Vietnam, Central America, Malaya, Indochina and Afghanistan) and 'medium intensity' conventional operations (Korea, the Falklands) often conducted well away from the home base and an all-out Third World War involving high-intensity conventional and/or nuclear conflict. Both sides had to deal with the spiralling rate of defence inflation, while weapon systems increased in both cost and complexity, having implications for the procurement process, as defence budgets could not increase at the same rate.

The principal concern for the defence planners of the two blocs involved the stand off between NATO and the Warsaw Pact in Europe. The history of the two alliances is closely linked. Within a few years of the end of the Second World War, relations between East and West became increasingly strained to the point of becoming the Cold War and a dividing line being drawn across Europe (the 'Iron Curtain' from Winston Churchill's famous speech at Fulton, Missouri). The Soviet inspired coup in Czechoslovakia, the Greek Civil war and the Berlin Blockade all suggested to the Western nations that the Soviets wished to move the Iron Curtain westwards, which was combined with the Soviet failure to demobilise on a par with the West. Initially, the North Atlantic Treaty was signed in April 1949 building upon the Brussels Treaty of 1948, and was signed by the United Kingdom, France, United States, Canada, Belgium, Netherlands, Denmark, Norway, Portugal, Iceland, Italy and Luxembourg. The outbreak of the Korean War (in June 1950) and the early test of a Soviet nuclear device in August 1949 led to fears of a major expansion in Soviet activity. This prompted the Alliance into converting itself into a standing military organisation, necessitating the stockpiling of large amounts of munitions, equipment and spares; "just in case" it was needed. The original members were joined in 1952 by Greece and Turkey, by West Germany in 1955 and by Spain in 1982.

NATO strategy, by the late 1980s, was based around the concepts of "flexible response", "forward defence" and "follow on force attack". The key element of NATO strategy, that of "flexible response", was adopted in 1967, and took over from "massive retaliation". This strategy demanded a balance of conventional and nuclear forces sufficient to deter aggression, and should deterrence fail, be capable of actual defence. The three stages in response to aggression were "direct defence" (defeating the enemy attack where it occurs and at the level of warfare chosen by the aggressor), "deliberate escalation" (escalating to a level of warfare, including the use of nuclear weapons, to convince the aggressor of NATO's determination and ability to resist and hence persuade them to withdraw) and "general nuclear response" (the use of strategic nuclear weapons to force the aggressor to halt his attack). A key commitment has been to "forward defence" (in deference to German political interests), that is, trying to maintain a main front line as close to the Iron Curtain as possible. To this had been added "FOFA" (follow-on-force attack), derived from the US Army's "Air-Land Battle 2000" strategy where "smart" and "stealth" weapons (as seen in the Gulf War) are used to attack enemy rear areas and approaching forces.

For forty years, the main threat to NATO's territorial integrity was the armed forces of the Soviet Union and Warsaw Treaty Organisation, more commonly known as the Warsaw Pact. This organisation came into being on the 14th May 1955 with the signing of the Treaty of Friendship, Cooperation and Mutual Assistance by Albania, Bulgaria, Czechoslovakia, East Germany, Hungary, Poland, Rumania, and of course the USSR. This was supposedly a response to the rearming of West Germany and its incorporation into NATO. The treaty reinforced a number of bilateral mutual aid treaties between the USSR and its allies, which was also complemented by a series of status force agreements allowing for the positioning of substantial Soviet forces on the allies' soil. The original treaty was valid until May 1975 where it was renewed for ten years and again in May 1985 for twenty years.

The purpose of the Pact was to facilitate the Soviet Army's ability to defend the Soviet Union (not surprising, considering the Soviet post-war concern with security) and to threaten Western Europe, while extracting military assistance from the East European states. Refusals or deviance were not tolerated, as seen in Hungary (1956) and Czechoslovakia (1968) but that is not to say the Soviets had it all their own way. The East Europeans were "reluctant to make all the military efforts demanded of them, and have from time to time, resisted Soviet attempts to extract more resources, and refused to undertake all the exercises demanded or even on occasions, to lend full blooded diplomatic support".[6] As a consequence, the dependability of the Pact forces in a war may have been open to question. Much would have depended upon the nature of the conflict.

Warsaw Pact doctrine called for a broad frontal assault while securing massive superiority at a few pre-selected points. The attacking forces would be echeloned, possibly three or more echelons deep (coming from the expectation that NATO would quickly resort to nuclear weapons to stop any breakthrough) even at Theatre (each Theatre consisted of two or more Fronts) level. To the Pact, only the offensive was decisive. The concept of defence was used as a means to shield reorganising forces getting ready to launch another offensive. Pact formations were modular all the way up to Front level (each Front consisted of two to five Armies, but generally consisted of three). One Pact Army was configured similarly to another Army (each Army was made up of from three to seven Divisions but generally consisted of four or five Divisions). Forces in the front echelon would punch holes in NATO's front line for the Operational Manoeuvre Groups and the second echelon to exploit through and hopefully lead to the collapse of the NATO main line position. The third echelon would then pursue the fleeing enemy forces and complete the assigned objectives.

It must be noted however, that as structured, the Pact was not intended to be used in wartime. The Pact was meant to support the stationing of the various Groups of Soviet Forces, control training and exercises, assist in operational effectiveness and supervise and control military policy. The East European national armies were trained and equipped on the Soviet model because in war they would have been fully integrated into the Soviet Command structure as parts of the various Fronts. An example was the invasion of Czechoslovakia, where the joint invasion was conducted by the military command in Moscow.

The logistic implications of a clash between these two giants would have been enormous. Despite its "economic weakness and commercial and industrial inefficiency, the Soviet Union possessed mighty and highly competent armed forces. Indeed, they were probably one of the few efficient parts of the Soviet Union."[7] Also, despite its high ideals,

NATO had a number of drawbacks, the most serious of which was its lack of sustainability. In a major shooting war, so long as the Soviets performed reasonably well, NATO would probably have lost due to the fact it would have run out of things with which to fight. In a static war, logistics is somewhat simpler in the modern age, as ammunition can be stocked and fuel expenditure is limited (thus allowing one to stock that as well). In a highly mobile war, the main consumable used will be fuel rather than ammunition, but in a highly attritional conflict, the reverse will apply. Ammunition will be used to a larger extent than fuel. For example, Soviet tank armies advancing at a rate of between sixteen and forty-five kilometres a day in 1944 – 5 suffered far lower losses in men and tanks and consumed a third less fuel and one sixth the ammunition of tank armies that advanced at a rate of between four and thirteen kilometres a day.[8] Of course, this requirement will have to be modified to take account of what Clausewitz termed the 'friction of war' – terrain, weather, problems with communications, misunderstood orders etc. not to mention the actions of the enemy.

NATO reinforcement and resupply had been coordinated under SACEUR's (Supreme Allied Commander, Europe) Rapid Reinforcement Plan, and could be expected to work if given adequate time (a big 'if'). However, there were possible clashes in that, for example, if the United Kingdom decided to exercise its national option of reinforcing BAOR (British Army of the Rhine) with the 2nd Infantry Division, its arrival may coincide with the arrival of the III US Corps from CONUS (Continental United States) to draw their equipment from the POMCUS (Pre-positioned Overseas Material Configured in Unit Sets) sites and thus cause major logistic problems given the lack of rolling stock to go around. So, paradoxically, the greater the success the United States had in reinforcing Europe, the greater probability there would have been clashes in priority. The plan depended upon NATO forces limiting the expected interference from the enemy (something the Warsaw Pact definitely planned on doing) and kind weather – only then would the plan have had a good chance of succeeding.

Even if the forces had got there, would the logistics system have worked? Given the extended supply lines from the Channel ports across the Low Countries and the lack of operational coordination, either in defensive tactics or logistics one is left to wonder. For example, if one corps' national logistic capability became critical, the Army Group headquarters may have recommended a transfer of stocks between National Logistic Support Commands. If the national authorities refused to transfer stocks then the Army Group Commander would have to refer the decision to the Commander-in-Chief Central Region (CINCENT) who would then negotiate with the Ministries of Defence concerned. Tactical and logistic responsibility was thus separated and command was divided. CINCENT or

17

the Army Group Commanders had no power to reallocate nationally provided operational support capabilities or resources, and did not have access to logistic information that would have helped them make decisions on redeployment or reinforcement. As logistics was a national responsibility, each national corps has a set of 'tramlines' that ran westwards.

Cross corps-boundary logistics was difficult, if not impossible. While routes for such operations had been thought out, there were three different tank gun ammunition types, different fuzing and charge arrangements for artillery ammunition, different fuel resupply methods and no interoperable logistic support system for airmobile operations. All this would mitigate against a cohesive Army Group battle, particularly in the Northern Army Group. Thus sustainability would have been the NATO Achilles' heel. While the agreed stock level was to thirty days, many nations did not stock even to this. All had different ways of arriving at Daily Ammunition Expenditure Rates. Most members had either non-existent or not published plans to gear up their industrial base to replace the stocks once used. As experience in the Falklands War points out, actual ammunition expenditure rates would have been far above those planned.[9] It is also worth remembering that one British armoured division would have needed around 4,000 tons of ammunition of all types per day.

The Soviets (and hence Warsaw Pact) view was that while a short war was preferable, it was possible that the conflict might last some time and stay conventional. There is no such word as 'sustainability' in Russian, the closest being 'viability'. This has a much broader context, and includes such matters as training, the quality and quantity of weapons and equipment, and the organisation of fighting units, as well as supply, maintenance, repair and reinforcements. The Soviets also relied on a scientific method of battle planning; one that took into account military history, to reduce uncertainty to a minimum and to produce detailed quantitative assessments of battlefield needs. They also had a common military doctrine throughout the Warsaw Pact and standard operational procedures.

Soviet forces still relied on a relatively streamlined logistic tail as compared to their Western counterparts. The bulk of logistic resources were held at Army and Front level, which could supply two levels down if required. This gave a false indication to the West of the logistic viability of the Soviet division. Thus senior commanders had a great deal of flexibility in deciding who to support and who to abandon and which axis to concentrate on. Soviet priorities for resupply, in order, were ammunition, POL, spares and technical support, food and medical supplies and clothing. They regarded fuel as the greatest challenge, but their rear services could still make maximum use of local resources, be it clothing, food or fuel. It is probable though that the Soviets would not have had things all their own way.

Keeping a high tempo of operations would consume large amounts of fuel and ammunition. Thus almost every town and wood in East Germany and Czechoslovakia would have become a depot and every road or track would have been needed to transport it and every possible means to carry it utilised, including captured vehicles. NATO would of course be trying to interdict these supply routes and the density of forces would have made traffic control problematic, not to mention the fact that any significant advance would place the leading forces well away from their supply bases and railheads behind the initial start line. However, the Soviets would endeavour to maintain strict control over supply priorities and a ruthless determination to achieve the objective. To this end, surprise would have been vital, and thus the objectives should have been achievable with forces in being, with the minimum amount of reinforcement.

Also, the first strategic echelon would have been required to maintain operations over a longer period of time. There would thus be no secure rear areas, no forward edge of the battle area or front line. The repair and medical services would thus be positioned well forward, giving priority to men and equipment that can be tended to quickly and sent back into action. The Soviets did not have a 'use and throwaway' attitude to men and equipment, but intended to keep the fighting strength of the unit as high as possible for as long as possible. Once the formation had become badly mauled however, it would be replaced by a fresh one – they did not believe in the Western method of replacing unit casualties with reinforcements thus keeping the unit in action over a prolonged period.

The Post-Cold War World

The ending of the Cold War has had profound effects upon the philosophy of, and approach to military logistics. The long held approach of stock-piling of weapons, ammunition and vehicles, at various strategic sites around the expected theatre of operations and in close proximity to the lines of communications was possible when the threat and its axes of attack were known. It is no longer the optimum method in the new era of force projection and manoeuvre warfare. 'High tech' weapons are also difficult to replace, as the US Air Force demonstrated during the 1999 attacks on Yugoslavia, when they started to run short of cruise missiles.

With pressure on defence budgets and the need to be able to undertake a (possibly larger) number of (smaller) operational roles than had previously been considered there has been a closer examination of the approach of commercial organisations to logistics. For the UK, this pressure has been particularly intense and as part of the Strategic Defence Review (1998) the Smart Procurement Initiative was announced. This was designed not only to improve the acquisition process but also to bring about more effective support in terms of supply and engineering. However, it is pertinent at this point, to briefly examine what commercial practices are being considered.

Just after the Second World War, the United States provided considerable assistance to Japan. Out of this, the Japanese have become world leaders in management philosophies that bring about the greatest efficiency in production and service. From organisations such as Toyota came the then revolutionary philosophies of Just-in-Time (JIT) and Total Quality Management (TQM). From these philosophies have arisen and developed the competitive strategies that world class organisations now practice. Aspects of these that are now considered normal approaches to management include kaizen (or continuous improvement), improved customer-supplier relationships, supplier management, vendor managed inventory, customer focus on both the specifier and user, and above all recognition that there is a supply chain along which all efforts can be optimised to enable effective delivery of the required goods and services. This means a move away from emphasising functional performance and a consideration of the whole chain of supply as a total process. It means a move away from the 'silo' mentality to thinking and managing 'outside the (functional) box'. In both commercial and academic senses the recognition of supply chain management, as an enabler of competitive advantage is increasingly to the fore. This has resulted in key elements in being seen as best practice in their own right, and includes value for money, partnering, strategic procurement policies, integrated supply chain/network management, total cost of ownership, business process re-engineering, and outsourcing.

The total process view of the supply chain necessary to support commercial business is now being adopted by, and adapted within, the military environment. Hence initiatives such as 'Lean Logistics' and 'Focussed Logistics" as developed the US Department of Defense and acknowledged by the UK Ministry of Defence in the so-called Smart Procurement Initiative, recognise the importance of logistics within a 'cradle to grave' perspective. This means relying less on the total integral stockholding and transportation systems, and increasing the extent to which contractorised logistic support to military operations is fanned out to civilian contractors – as it was in the eighteenth century.

Force projection and manoeuvre warfare blur the distinction between the long held first, second and third-line support concept of the static Cold War philosophy and link the logistics' supply chain more closely with the home base than ever before. One of the reasons for the defeat of the British in the American colonies in 1776 may have been the length of, and time involved in, replenishing the forces from a home base some 3,000 miles away. The same was true in the Russo-Japanese War with a 4,000-mile supply line along a single-track railway. Whilst the distances involved may still be great in today's operational environment, logistic philosophies and systems are being geared to be more responsive in a way that could not have been previously envisaged.

The five principles of logistics, accepted by NATO, are foresight, economy, flexibility, simplicity and co-operation. They are just as true today as they were in the times of the Assyrians and Romans. The military environment in which they can be applied is considerably different, and, as can be seen in the Balkans in the late 20th Century, adopting and adapting military logistics to the operational scenario is an essential feature for success. Ultimately a "real knowledge of supply and movement factors must be the basis of every leader's plan; only then can he know how and when to take risks with these factors, and battles and wars are won by taking risks."[10]

Notes

[1.] Foxton, 1994, p. 9)
[2.] Kennedy, 1989, pp. 218 – 228)
[3.] Hibbert, 1999, p. 7
[4.] Quoted from the letters of Captain M A B Biddulph, RA in Hibbert, p. 8
[5.] Blake, 1973, p. 108
[6.] Martin, 1985, p. 12
[7.] Thompson, 1998, p. 289
[8.] Thompson, 1998, p. 291
[9.] Thompson, 1998, p. 310
[10.] Wavell, 1946.

Bibliography

Baumgart, Winfried. *The Crimean War 1853 – 1856*, Arnold, London, 1999.

Blake, R. L. V. F. *The Crimean War*, Sphere, London, 1973.

Brown, I. M. *British Logistics on the Western Front*, Praeger, Westport, CT, 1998.

Christopher, M. *Logistics and Supply Chain Management*, Brassey's, London, 1992.

Engels, Donald W. *Alexander the Great and the Logistics of the Macedonian Army*, University of California Press, Berkeley, 1978.

Faringdon, Hugh. *Confrontation: The Strategic Geography of NATO and the Warsaw Pact*, Routledge & Kegan Paul, London, 1986.

Foxton, P. D. *Powering War: Modern Land Force Logistics*, Brassey's, London, 1994.

Hibbert, Christopher. *The Destruction of Lord Raglan*, Wordsworth Edition, Ware, 1999.

Kennedy, Paul. *The Rise and Fall of the Great Powers*, Fontana Press, London, 1989.

Lynn, John A. *Feeding Mars: Logistics in Western Warfare from the Middle Ages to the Present*, Westview, Oxford, 1993.

Martin, Lawrence. *Before the Day After: Can NATO Defend Europe?*, Newnes Books, Feltham, 1985.

Sinclair, Joseph. *Arteries of War: A History of Military Transportation*, Airlife Publishing, Shrewsbury, 1992.

Thompson, Julian. *Lifeblood of War: Logistics in Armed Conflict*, Brassey's, London, 1998.

Thomson, David. *Europe since Napoleon*, Penguin, Harmondsworth, 1971.

Van Creveld, Martin. *Supplying War: Logistics from Wallenstein to Patton*, Cambridge University Press, Cambridge, 1995.

Wavell, Field Marshal. *Speaking Generally*, Macmillan, London, 1946.

Case 1.2 // Amphibious Operations

Sea Lion vs. Overlord: A Comparative Analysis

Larry Parker

Lecturer, Naval Junior ROTC, Sun Valley High School, North Carolina.

Introduction

One of the favourite topics of alternative history (and one of the scenarios endlessly replayed in war games such as *Axis & Allies* and *3rd Reich*) is what if Germany had attempted Operation Sea Lion. Assuming a Luftwaffe victory over the Royal Air Force in the Battle of Britain was Sea Lion feasible in other respects? Could Hitler have knocked the United Kingdom out of the war in the summer of 1940 or would the attempt have led to his first major defeat? This paper will compare and contrast Operation Sea Lion and Operation Overlord utilizing ten criteria essential to success in amphibious assaults - planning, matériel support, deception, intelligence, combined arms support, command structure, technology, innovation, sustainability and enemy defences.

Background

If the Blitzkrieg into Poland shocked the great powers, the complete collapse of France in just six weeks absolutely stunned the world.[1] In June 1940 all that stood between the seemingly invincible Wehrmacht and total domination of Western Europe was the English Channel and 555,000[2] badly shaken British, French, Dutch and Belgian troops evacuated from the continent. Its tanks, trucks, artillery and other heavy equipment abandoned on the beaches of Dunkirk the shattered remnants of the British army regrouped and frantically prepared as best they could to repel an amphibious assault. In centuries past protection against invaders had depended upon the wooden walls of the Royal Navy (RN). Survival now rested on resolute sailors serving aboard steel men-of-war and upon the indomitable courage of Royal Air Force (RAF) pilots and crews.

Four years later the glory days of Blitzkrieg were over, nothing more than the wistful memories of grizzled veterans shared with ever-younger recruits over evening cook fires. In the East the Soviet juggernaut pressed inexorably upon the borders of the thousand year Reich. In the South the Anglo-American Allies had driven Rommel out of North Africa, captured Sicily and in savage fighting were pushing relentlessly, albeit slowly, up the Italian peninsula. In the West the largest amphibious force ever assembled weighed anchor to breach Hitler's vaunted Atlantic Wall and liberate Festung Europa (Fortress Europe).

Part One: OVERLORD

Planning

Smashing the Atlantic Wall was a daunting task. As anyone who has ever conducted an amphibious exercise will attest just getting soldiers out of their racks, into landing craft and onto the proper beach on time and in some semblance of order is no mean feat. Coordinating Naval Gunfire (NGFS) and Close Air Support (CAS) adds another degree of difficulty. The multitude of organizational and logistic considerations involved in amphibious operations is staggering. Each element follows the preceding component in a strict timeline. Every function in the overall plan is interdependent, relying upon precise execution of all parts for success.

Each factor offers an opportunity for Murphy's First Law (anything that can go wrong will go wrong) to intrude. Logistic complications alone can and do create dangerous confusion. Opposition, no matter how light, almost guarantees chaos. In that regard, although Festung Europa did not live up to Goebbels' propaganda claims, German defenders in many places were well placed and highly motivated. When the first Allied soldier landed in Normandy however he did enjoy certain advantages – among them a comprehensive operational plan.

Planning and preparation for Overlord had been underway for over two years[3] drawing upon the lessons learned in North Africa, Sicily, Italy and the Pacific landings. The result was a massive document of extraordinary detail addressing every conceivable contingency. Prepared in an age well before computers it required a monumental effort and remains a remarkable achievement.

Overlord evolved from Operation Roundup, the proposed cross channel invasion scheduled for 1943. Roundup grew out of Operation Sledgehammer the highly problematic assault set for 1942. Both plans were tabled when President Roosevelt, eager to have American ground forces engaged anywhere in Europe as soon as possible, overrode the recommendations and objections of his Joint Chiefs of Staff (JCS) and cast his lot with Churchill who favoured a 'peripheral strategy' in the Mediterranean. The strategic gains from Torch, Husky, Avalanche and Shingle are debatable but the lessons learned undeniably improved Overlord's chances of success.

In the first iteration of Overlord the Chief of Staff to the Supreme Allied Commander (COSSAC) proposed a three-division assault supported by two Airborne Brigades. When Eisenhower assumed command of Supreme Headquarters Allied Expeditionary Force (SHAEF) one of his first decisions was to postpone the invasion by one month. Assault ships and landing craft were sliding down the ways in English and American shipyards in fantastic numbers. This brief delay allowed the Allies to gather sufficient amphibious vessels to expand the invasion force to five infantry divisions augmented by three Airborne Divisions. After factoring in the requirements for calm weather, a near full moon, a low tide beginning to flood and wide, flat beaches within fighter range of English fields, early June 1944 was selected as the most opportune time. Phase I called for the assault of "an initial lodgement area"[4] and the "capture of Caen."[5] Phase II directed the "enlargement of the area captured in Phase I, to include the Brittany peninsula, all ports south to the Loire and the area between the Loire and the Seine."[6] Phase III envisioned the follow on landing and subsequent breakout of Patton's Third US Army.

Most importantly, every aspect of the landing plan was reinforced with realistic training conducted under the strictest secrecy. When the Allied forces went into combat on 6 June 1944 they were physically and mentally well prepared.

In start contrast to the meticulous Allied preparations the architects of the 'master race' did not have a viable master plan for defending the western portion of the Reich from invasion. The reason for this glaring oversight lies with the Führer. Hitler well understood the Roman dictum *Divide et impera* (Divide and rule) and applied it to friend as well as foe. In the Byzantine world of Nazi Germany the different branches of the military and the various organizations of the state competed with one another for Hitler's favour. This policy kept any one person or group from gathering enough power to challenge the Führer but also resulted in great duplication of effort, waste and inefficiency. The Wehrmacht, Kriegsmarine and Luftwaffe each developed separate defensive plans. There was little cooperation or coordination between the services until one reached the very apex of the ruling hierarchy – Hitler himself, and he was all too fallible. Even within the Army there was no consensus on how best to defend the Reich. Rundstedt favoured a mobile defence. He recommended placing the Panzer Divisions in a central reserve, containing the invasion and then delivering a concentrated counterattack.

Based on his experience in North Africa Rommel had a greater appreciation for Allied air power. He felt the first forty-eight hours were critical, that the invasion would be stopped on the beach or not at all. Consequently he opted for a forward defence supported by immediate Panzer counterattacks and positioned his forces accordingly. In a compromise guaranteed to satisfy no one OKW allocated three Panzer divisions to Army Group B, three

Panzer divisions to Army Group G and three Panzer divisions and one Panzer Grenadier division to OB West. Although nominally under Rundstedt's command the four divisions of the theatre reserve were actually under Hitler's personal control. This lack of cooperation and coordination in both preparation and in the ensuing battle severely handicapped the German defence.

Matériel Support

In *On War* Clausewitz discusses offensive momentum in terms of a culminating or balance point. In any campaign or battle offensive power decreases as resistance increases. This process continues until the aggressor achieves victory or his forces are exhausted and a stalemate ensues. Clausewitz's concept of a culminating point is especially applicable in an amphibious assault. To ensure the Allies achieved and maintained offensive momentum the United States assembled twenty-one divisions (13 infantry, 6 armoured and 2 airborne) for Overlord. By D + 90 there would be 1,200,000 American troops and 250,000 American vehicles in France. In addition the British and Canadians contributed another seventeen divisions (10 infantry, 5 armoured and 2 airborne) to the effort.

To prevent an equally rapid build up of German forces SHAEF dedicated the entire U.S. Army Air Force to Overlord. Under the TRANSPORTATION PLAN portion of Overlord the 2700 fighters, 1956 heavy bombers, 456 medium bombers and 171 light bombers of the 8th strategic air force and 9th tactical air force were committed to isolating the invasion area. In Phase I Luftwaffe planes, aircraft factories, repair depots and storage parks were targeted. In Phase II American planes attacked railroad tracks, trains and yards. In Phase III bridges were destroyed isolating France below the Seine. So complete was Allied air supremacy and so effective were the attacks under the Transportation Plan that German reinforcements were forced to travel only at night. Movements that previously would have taken a few days stretched into weeks.

Deception

As part of the deception plan Allied planes continued to bomb other areas of France and the Low Countries as well as targets in Germany. Another subset of the Overlord Plan, Operation Fortitude involved the creation of an entire and entirely fictitious First U. S. Army Group (FUSAG) allegedly commanded by General Patton. Dummy tank, truck and artillery parks were constructed. Phony encampments were created. Signal Corps personnel filled the airwaves with false information concerning diversionary attacks in southern France and Norway in preparation for the main thrust at Pas-de-Calais. Since this 'intelligence' matched Hitler's preconceived notions of the inevitable invasion Operation Fortitude was noteworthy success. Essential reinforcements were pinned in place during the critical first weeks of the Normandy campaign. In addition severe weather, which delayed the Allies until 06 June, also lulled the German defenders into a false sense of security.

Intelligence

Although there were glaring oversights in Allied intelligence[7] two major intelligence coups significantly contributed to Overlord's success. The first was ULTRA. By 1944 the Allies had broken the Enigma code and were able to read German message traffic giving them a good picture of enemy dispositions and intentions. The second was DOUBLE CROSS. At the time of the Normandy invasion all Axis spies in England had been turned. Now working as double agents these operatives feed their handlers false information about attacks in Norway and the Pas-de-Calais. With its spy ring turned and the Luftwaffe unable to penetrate English air space for reconnaissance Germany operated in an information vacuum. Invasion was inevitable but the crucial questions of 'when' and 'where' remained conjecture. As a result Hitler was forced to defend 2,800 miles of coastline, guaranteeing Allied superiority at the point of attack. As Frederick the Great noted, "He, who defends everything, defends nothing."

Combined Arms Support (CAS)

Although the Allies dedicated 6,000 ships and landing craft and 13,000 aircraft to the invasion CAS was not as effective as it should have been. British landing policy stressed tactical surprise. American amphibious doctrine emphasized overwhelming firepower. In a flawed compromise neither was achieved. Landings were conducted in daylight negating the advantage of shock. A brief preliminary bombardment (only 35 minutes at Omaha beach) did little more than alert the defenders. Saturation attacks from landing craft modified to carry rockets (LCT(R)s) were thrown off target due to high surf conditions. Finally, because of cloud cover, pilots delayed bomb release several seconds. As a result their payloads fell well inland. On the positive side the Allies did gain complete air superiority preventing any counterattacks by the Luftwaffe and daring destroyer captains ran their vessels nearly aground in order to deliver highly effective counter battery fire at point blank range. Overall however, it was junior officer and senior NCO initiative that carried the day.

Command Structure

Allied

Supreme Commander Allied Expeditionary Force – General Eisenhower

Deputy Supreme Commander – Air Chief Marshal Tedder

Chief of Staff – Lt. General Bedell-Smith

Allied Naval Expeditionary Force – Admiral Ramsay

Allied Expeditionary Air Force – Air Vice Marshal Leigh-Mallory

Commander 21st Army Group – General Montgomery

1st U. S. Army – Lt. General Bradley

2nd British Army – Lt. General Dempsey

Second in importance only to the immense scope of the Allied invasion plan and intense preparation prior to D-Day was the rigidly organized and thoroughly integrated command structure of the invasion forces. The disparate elements were subordinated to the Supreme Commander. Outwardly affable, Eisenhower was an iron willed, highly capable officer and a fortuitous choice. Years of experience as a staff officer under MacArthur and Marshall served him and the Allies well in this post. As should be expected on a joint and Allied command personality conflicts abounded and staff meetings were frequently acrimonious. When final decisions were made and orders given however, they were saluted smartly and dutifully executed.

German

Hitler*

OKW

OKM/OKH/OKL

Admiral Doenitz	OB West	Reichsmarschall Goering
Field Marshal von Rundstedt		

Naval Group West[8]		3rd Air Fleet[9]
Admiral Krancke		Field Marshal Sperrle

Army Group B: Field Marshal Rommel   Army Group G: Col. General Blaskowitz

15th Army – Col. General Salmuth   1st Army

7th Army – Col. General Dollman 19th Army

*Hitler rather than Rundstedt maintained direct control of mobile theatre reserves, which could not be released without his personal authorization.

Lack of a coordinated plan of defence and a convoluted chain of command ensured the 58 divisions[10] available to Rundstedt were slow to react during the critical period after the initial landing and were later fed piecemeal into the battle for Normandy.

Technology

In a remarkable achievement German engineers fielded V-1s, V-2s and jet aircraft. These potential war winners came in limited numbers however and far too late to alter the outcome of the conflict. Allied industry far surpassed German efforts where it counted, in mass production of basic but nonetheless critical technology. Two examples will serve. The first was shallow draft landing craft. The availability of thousands of LCV/Ps, LCIs, LCCs, LCAs, LCTs, LCT(R)s and LSTs patterned after the Higgins Boat allowed the Allies to put 150,000 men plus supporting equipment on the beach on D-Day[11] overwhelming the German defenders. More importantly these ubiquitous craft ensured the Allies won the race to build up combat power.[12] Secondly, Hobart's Funnies (British and American tanks adapted for swimming, bridge laying, mortar launching, flame throwing, track laying and mine destroying) added crucial firepower on the beach and greatly assisted rapid movement inland in the 2nd Army sector.

Innovation

Perhaps the most ambitious and impressive Allied innovation in Operation Overlord was the MULBERRY, the creation of all weather artificial harbours at Vierville and Arromanches on the Normandy coast capable of handling 5000 long tons of supplies and 1000 vehicles per day. Eighty-nine GOOSEBERRIES (derelict ships) and 150 PHOENIXES[13] (concrete caissons) were scuttled to form an outer breakwater. BOMBARDONS (steel floats 200 feet long) were anchored to create an inner breakwater. When completed these barriers sheltered twenty-one miles of formerly open shoreline providing protected anchorage for large cargo ships. A constant stream of barges and small craft shuttled to and from these anchored supply ships and the shore. LOBNITZ PIERHEADS (floating docks) created piers that allowed LSTs and small cargo ships to unload directly into trucks. WHALES (steel pontoons) formed causeways from the piers to the beach expediting the movement of supplies to the front. Unfortunately severe gales during the period 19 – 22 June destroyed the Mulberry at Omaha Beach and damaged the harbour at Gold Beach. Only the British Mulberry was repaired exacerbating the need to capture Cherbourg and other French ports as rapidly as possible. In the interim LSTs were grounded at high tide and hurriedly off-loaded before the tide returned. In spite of Herculean efforts supply never kept up with demand. Operations would be hamstrung by logistic shortfalls (especially POL) until Montgomery finally captured Antwerp.

Sustainability

To ensure the offensive momentum gained on D-Day continued the Allies identified fifteen French ports for early capture and reconstruction. The Overlord planners also realized the Transportation Plan, which had so effectively hindered German reinforcement would also impede an Allied breakout. Therefore 1,548,000 man-days were allocated for road, railroad and bridge reconstruction. The staggering amount of material and equipment laid up for this effort is indicative of the forethought and thoroughness that went into Allied preparations:

11,700 long tons of construction equipment
15,800 long tons of asphalt
112,000 tons of bridging
800 standard fixed Bailey sets
250 standard pontoon Bailey sets
175 heavy increment Bailey sets
165 heavy pontoon increment sets

In addition, plans were made to deliver 5,000 tons per day of Petroleum, Oil and Lubricants (POL) at D+20 and 10,000 tons per day by D+90. Part came by tanker and part by PLUTO (Pipeline Under the Ocean) another aspect of the Mulberry project. These figures proved inadequate and POL became the Achilles Heel for the British and Americans. The Allied advance literally ran out of gas in late autumn allowing the Germans to regroup and launch their last major offensive of the war in December.

Enemy Defences

Führer Directive number 40 ordered the creation of an 'Atlantic Wall' stretching from Spain to Norway. Covering some 2,800 miles this series of fortifications was one of the largest construction projects in human history. Special emphasis was given to the Pas-de-Calais, the shortest route from England to France and most direct line of march from the landing site into the German heartland. Significantly this was the only area completed by June 1944. Recognizing the vital importance of harbour facilities to any invasion Hitler also insisted each port city be heavily fortified and strongly garrisoned. Typical of operations in the Third Reich part of the work was undertaken by Organization Todt, part by the Army, part by the Air Force and part by the Navy with no coordination of effort. Consequently when Rommel made his first inspection tour in the fall of 1943 he found the wall to be, in his words, "a figment of Hitler's Wolkenkucksheim" (cloud-cuckoo-land). Of the 23,000 structures erected approximately half followed some standard design. The remaining bastions were built haphazardly per strictly service needs or local commanders' discretion. Little or no thought was given to integrating defensive systems or coordinating efforts.

Rommel set about to immediately rectify the situation. At his command Army units dedicated three days per week labouring to improve fortifications. Hedgehogs[14], Belgian gates[15] and stout wooden posts angled toward the sea, all topped with mines, were erected in the tidal zone. Soldiers strung hundreds of miles of barbed wire and laid millions of mines designed to channel invasion forces into killing zones. Artillery was calibrated, firing arcs established and machine guns placed to sweep the beaches. While not yet complete, by June 1944 the Atlantic Wall had been vastly improved. As necessary as this work was it did impact combat readiness. Work and guard details left little time for training. This was a reasonable trade-off however considering Allied mastery of the sea and air.

Summary

Although intelligence, CAS and sustainability were less than perfect the overall concept and execution of Overlord was overwhelming. The Normandy invasion was a masterpiece of strategic planning made possible by the astonishing capacity of British and, especially, American industry. The sheer scale of the operation staggers the imagination. Its size should have been Overlord's weak link as well. Storage facilities, vehicle parks, encampments, embarkation ports and the invasion fleet underway in the channel were fat targets. A decisive blow at the POL dumps for example would have set the invasion back for several months. Fortunately for the Allies interdiction by the Kriegsmarine was not feasible and the major portion of the Luftwaffe was tied down on the Eastern Front. Thus Hitler lost his best chance to stop the attack. Given the state of the Wehrmacht on the Western Front, and the command problems between OKW, OKH and OB West, once the Allies were ashore eventual defeat was only a matter of time.

Part Two: SEA LION

Planning

When France fell in June 1940 Britain would never again be more vulnerable. A successful invasion of England at that point would have ended the war on German terms. Yet, in spite of massive rearmament programs during the 30s and a dominant tactical doctrine in the form of Blitzkrieg, Germany did not possess the wherewithall to capitalize on its amazing good fortune. No contingency plans had been prepared for such an eventuality. Even if they existed, the Kriegsmarine was totally inadequate to the task. The reasons for Germany's lack of military readiness at sea were political and philosophical. Politically Hitler never envisioned a long-term war with Britain, much less an invasion of Albion. Faced with German mastery of the continent he expected the "nation of shopkeepers" to be sensible and come to terms. To understand the philosophical reasons for the state of the Kriegsmarine it is necessary to digress a bit.

Unlike Kaiser Wilhelm II who studied Mahan and invested heavily in the High Seas fleet, the work of General Doctor Karl Haushofer[16] swayed Hitler's strategic view. According to Haushofer's theories of Geopolitik, the growth of motorized road and railroad transport negated England's historic control of the sea. In Haushofer's view mastery of the European heartland was central to world domination. These geopolitical ideas dovetailed neatly with Hitler's quest for *Lebensraum*, racist theories of Aryan superiority and pathological anticommunism. Since Hitler's strategic aims lay on the continent the Wehrmacht and Luftwaffe received top priority. As an ancillary service the Kriegsmarine rearmed under the much more modest Plan Z. Adopted in 1938 Plan Z called for a balanced fleet of capital ships and submarines.

War was not anticipated until 1943 or 1944. On 27 January 1939 the Naval rearmament schedule was extended to 1949. When Germany invaded Poland in September 1939 naval planners were aghast.[17] Naval construction immediately shifted to U Boats but the war was ten years too early and the change in priorities ten years too late. The Kriegsmarine entered the war with 2 new battleships, 2 old battleships, 3 pocket battleships, 1 heavy cruiser, 5 light cruisers, 17 destroyers and, fortunately for the Allies, only 56 submarines. In addition 2 battleships, 1 aircraft carrier (never completed), 4 heavy cruisers and 1 light cruiser were under construction. While the Army and Air Force may have been adequate for Sea Lion the Navy's assets were insufficient to engage the British fleet or wage economic (U Boat) warfare. Lack of a Naval air arm (Goering covetously guarded all air assets) also hampered the operational effectiveness of the Kriegsmarine.

In addition to the political and philosophical problems resulting in an under strength Navy and inadequate planning and preparation time,[18] inter-service difficulties also emerged. The Army saw Operation Sea Lion as nothing more than a large river crossing in which the "Luftwaffe will do the work of artillery, while the Kriegsmarine will do the work of engineers."

Through sheer audacity and at great cost the Kriegsmarine had barely succeeded in Norway. Consequently, the Navy possessed a far more realistic assessment of the difficulties involved in Sea Lion. All depended upon the Air Force however, and, as always, Goering pursued his own agenda.

The final proposal (the word plan carries too strong a connotation) called for the Navy to:

- Block the west end of the channel with U Boats
- Block the east end of the channel with mines and E Boats
- Sortie the main surface fleet into the North Atlantic to draw off the British Home Fleet

After crossing the channel in open barges the army would land and immediately capture a port in order to land Panzers. Second only to air supremacy the early introduction of armour was critical to victory against a numerically superior foe.

Only one training exercise was conducted. The results are quite revealing. Off Boulogne, in good weather and good visibility, with no navigation hazards or enemy defences to contend with, of fifty vessels, committed less than half managed to land their troops at H Hour. One tug lost its tow. One barge overturned when too many soldiers crowded on one side. Several barges broached in the surf and landed broad side to, unable to lower their ramps. The results of the fifty-barge exercise did not bode well for a 1277 barge assault on England.

Matériel Support

In 1940/41 the average German infantry division required 100 tons of supplies per day while engaged in combat. The average Panzer division consumed 300 tons of matériel per day when on the offensive. To land five divisions, the Allies gathered 6,000 vessels and vast stockpiles of provisions. To move nine divisions and sustain them for the first eight - ten days when the second wave was scheduled to land, the Navy gathered 170 cargo ships, 1277 barges and 471 tugs[19] in French ports. The barges immediately became targets for the

RAF. They needn't have bothered. Most of the barges were designed for river traffic and would sink in anything greater than sea state two. On D-Day men and equipment loaded in open barges were to steam in column until ten miles from the landing site, then turn sequentially and steer parallel to the coast. Upon signal all vessels were to execute a flank turn and proceed in line abreast to the beach. This intricate manoeuvre by barges under tow would take place at night with minimal lighting, controlled and coordinated by loud hailer! Rough weather, the Royal Navy and the Royal Air Force notwithstanding, Sea Lion was a recipe for disaster.

Deception

German capabilities, or the lack thereof determined the possible landing sites as clearly to the British as to the Germans. Landing at night was the only deceptive measure adopted for Sea Lion.

Intelligence

Through their spy network the Germans were well informed as to British dispositions and capabilities. In view of German limitations however, this knowledge was of little value.

Combined Arms Support (CAS)

Since the Kriegsmarine would act to decoy the Home Fleet NGFS was not a factor in Operation Sea Lion. Delivery of Combined Arms Support fell solely upon the Luftwaffe. Simultaneously the Air Force was required to:

- Keep the Royal Navy out of the channel.
- Win total air superiority.
- Interdict British reinforcements moving by rail.
- Conduct a mass attack on London in order to force the civilian population to flee, choking the road network.

This was clearly a tall order, considering at this point in the war the Luftwaffe mustered only 1,260 bombers, 316 dive-bombers and 1,089 fighters, while the RAF carried 672 fighters on its Order Of Battle. Had the Luftwaffe been successful in its bid for air supremacy Air Chief Marshal Sir Hugh Dowding was prepared to pull the 11th fighter group out of range until the landings began. He then planned to sortie from the Midlands with the combined strength of the 10th, 11th and 12th Fighter Groups in conjunction with the Royal Navy. Due to geography the Luftwaffe had no chance to complete its overall mission.

Command Structure

As noted previously each service vied with the other for Hitler's favour. Raeder fought, justifiably, to expand the Navy. Goering schemed to enlarge the Luftwaffe to match his own immense girth. (Dunkirk, the Battle of Britain, air supply of Stalingrad, Luftwaffe field divisions and the Herman Goering Panzer division were monuments to his ego and testimony of his influence with Hitler.)

Within the Army, Panzer generals argued with more conservative infantry generals over strategy and the Waffen SS competed with standard Wehrmacht units for men and matériel. As a result command relationships within and between the services were often strained and operations suffered accordingly. With regard to Sea Lion at a 31 July 1940 coordination meeting called by Hitler himself Luftwaffe representatives did not attend and, as discussion moved to purely Army matters, Admiral Raeder walked out.

Technology

The Germans applied no new technology to overcome the numerous obstacles involved in Sea Lion.

Innovation

Other than paratroops the Germans introduced no innovations that might have improved the chances for success.

Sustainability

Sustainability is defined as massive long-term matériel support. As Theodore Gatchel writes in *At the Water's Edge*, an invasion is a race between the attacker and the defender to build up combat power. Considering the German scheme for the initial landing and follow on support Sea Lion was a race the Wehrmacht could not win.

Enemy Defences

Although faced with the same time constraints the British came up with a far superior defensive plan. The Royal Navy and Royal Air Force plans were identical – pre-emptive attacks on staging areas, interdiction at sea and all out assaults at the landing points. In the same period the Army refitted the survivors of Dunkirk, organized a Home Guard, created beach defences and set up stop lines, backed by mobile reserves. These preparations should have been more than adequate to turn back the surviving barge loads of seasick *soldaten*.

Summary

The Luftwaffe proved inadequate against surfaces ships at Dunkirk sinking only thirteen destroyers and damaging another nineteen in a confined area as they loaded troops. It proved equally incapable against the RAF during the Battle of Britain losing 1887 aircraft of all types in exchange for 1547 fighters. Clearly the Luftwaffe could not stop the RAF or the RN much less both.

The German invasion of Crete is instructive *vis-à-vis Sea Lion*.[20]

Reinforcement and supply by sea proved impossible even though the Luftwaffe had absolute air superiority. The Royal Navy intercepted and utterly destroyed the first flotilla of small boats hazarding the crossing from Greece. No further attempts were made to reinforce by sea. Instead the 5th Mountain Division troops were flown in. Although they eventually prevailed, German parachute and glider troops and the JU-52 transport arm of the Luftwaffe were decimated in the process.

One can imagine the slaughter had the RN and RAF run through 1,277 barges loaded with men and equipment during the proposed Sea Lion channel crossing. Although it did not appear so at the time Sea Lion was never a viable military option. At best it was a propaganda ploy, a political threat that might have brought a timorous leader like Chamberlain to the negotiating table but never a tenacious warrior such as Churchill.

Part Three: Conclusion

As discussed in Parts I and II, in every category except defences – planning, matériel support, deception, intelligence, combined arms support, command structure, technology, innovation and sustainability – Allied preparation for Overlord was far superior to German efforts in the same areas during Sea Lion. Certain specific concerns within those general categories also stand out as tabulated opposite:

	Sea Lion	Overlord
Overall Mission Commander	No	Yes
Coordinated Planning Staff	No	Yes
Sound Amphibious Doctrine	No	Yes
Practical Experience	No	Yes
Air Superiority	No	Yes
Sea Control	No	Yes
Purpose built landing craft	No	Yes
Heavy lift capability	No	Yes

In World War II the Germans were notorious for operating on a shoestring and early on frequently succeeded due to the shock value of Blitzkrieg. This must have been the expectation since, for Sea Lion to be successful; England would have to collapse as France had in June 1940. Great Britain proved more resilient than France however. Consequently, in an extended operation, German planning and preparation would have been woefully inadequate. Nor did Hitler learn from Sea Lion. One year later he invaded the Soviet Union convinced it would fall in one summer campaign. True to form, as the struggle drug on, German planning and preparation, especially in the area of logistics, proved inadequate for a sustained operation.

The Anglo-American Allies tended to the other extreme. Operations were constrained by logistic concerns and senior leaders were frequently criticized for tentative strategic movements and failing to respond quickly to fleeting tactical opportunities. Montgomery, for example, talked boldly but was a meticulous planner and absolutely would not budge until every biscuit was in place.

There are advantages and disadvantage to both methods and valid reasons why Germany operated as it did. When a soldier or marine hits the beach in an amphibious assault however, especially considering what was at stake in 1940 and 1944, he deserves every advantage detailed planning can afford. In this respect the Allies succeeded whereas the Germans were fortunate indeed Sea Lion never came to pass.

Notes

[1.] At the time France was considered the worlds leading military power.

[2.] 340,000 from Dunkirk (Operation Dynamo), another 215,000 during Operations Aerial and Cycle, the evacuation of Le Havre, Cherbourg, St Nazaire, etc.

[3.] Initial planning for Overlord began after the Casablanca Conference. An Anglo-American staff led by Lieutenant General Sir Frederick Morgan completed the bulk of the work prior to Eisenhower's appointment.

[4.] *Outline of Operation Overlord, Section VIII, Part I, Tab I.* U. S. Army Center of Military History, Historical Manuscripts Collection 8-3.4 AA v. 7

[5.] Ibid.

[6.] Ibid.

[7.] Allied intelligence failed to locate and identify important German units in the defensive Order of Battle (OOB) most significantly the 352nd Infantry Division at Omaha Beach. Allied planners also overlooked the military significance of the hedgerow terrain inland.

[8.] To oppose the 6000 ships in the Allied Armada, Naval Group West mustered 17 submarines and a few destroyers and E (torpedo) boats.

[9.] Arrayed against the 13,000 planes that flew on D-Day, 3rd Air Fleet listed 891 aircraft in its OOB of which only 497 were fully operational.

[10.] 33 static defence units, 13 infantry, 2 airborne, 9 Panzer, and 1 Panzer grenadier many of which were new divisions fitting out or units recovering from service on the Eastern Front.

[11.] These numbers swelled to 330,000 men and 50,000 vehicles by 12 June and 660,000 men by 30 June.

[12.] The importance of landing craft cannot be overstated. No less an authority than General Eisenhower stated, "Andrew Higgins is the man who won the war for us. If Higgins had not designed and built those LCVPs, we never could have landed over an open beach. The whole strategy of the war would have been different."

[13.] In preparation for Overlord shipyards in England employed 20,000 men around the clock to build 150 Phoenixes, hollow concrete blocks 200 feet long, 60 feet wide and 60 feet high. These were towed to Normandy, carefully positioned and scuttled to form a breakwater.

[14.] Hedgehogs – star shaped, six-foot obstacles, constructed of steel girders and topped with mines.

[15.] Belgian gates – large pieces of steel ten feet high set perpendicular to the beach and topped with mines.

[16.] Born 27 August 1869 Karl Haushofer was a professional soldier whose superior intelligence earned him an appointment to the General Staff. Retiring in 1919 as a Major General, Haushofer dedicated himself to the regeneration of Germany. Turning to education he combined the theories of Ratzel, Kjellen and MacKinder developing a doctrine he called Geopolitik. In 1922 he founded the Institute of Geopolitics in Munich and through his student, Rudolph Hess, profoundly influenced Hitler during the formative period of the Nazi party.

[17.] "Today the war against France and England broke out, the war which, according to the Führer's previous assertions, we had no need to expect before about 1944. As far as the Navy is concerned, obviously it is in no way very adequately equipped for the great struggle with Great Britain by autumn 1939. The submarine arm is still much too weak to have any decisive effect on the war. The surface forces are so inferior in number to those of the British fleet that they can do no more than show that they know how to die gallantly." Admiral Erich Raeder.

[18.] The order to begin planning was not given until 02 July 1940, allowing only 84 days prior to the proposed invasion date.

[19.] Diverting so many cargo vessels and river ferries greatly disrupted commercial traffic in the Baltic and on the Rhine.

[20.] The parallels between Sea Lion and Operation Merkur are striking. The invasion of Crete also suffered from a very short planning period and relied on commandeered caciques (small two masted fishing boats with an auxiliary engine) assembled at Piraeus for reinforcement and supply. British destroyers annihilated the first flotilla. No second attempt at sea borne landings or supply was made. Although they prevailed the resulting shortage of heavy equipment, transportation and supplies cost the air and glider borne troops dearly.

Bibliography and Suggested Reading

Bibliography

Atkinson, Rick. An Army at Dawn, *The War in North Africa, 1942 – 1943*. New York: Henry Holt and Company, 2002.

Belote, James H. *Typhoon of Steel*. New York: Harper and Row, 1970.

Bradley, Omar. *A Soldier's Story*. New York: Random House, 1999.

Carell, Paul. *Invasion! They're Coming!* New York: Schiffer Publishers, 1995.

Clausewitz, von, Carl. *On War*. NJ: Princeton University Press, 1989.

Creveld, Martin Van. *Technology and War*. New York: The Free Press, 1989.

Delaforce, Patrick. *Smashing the Atlantic Wall*. London, UK: Cassell & Co., 2001.

D'Este, Carlo. *Decision in Normandy*. New York: Dutton, 1983.

Eisenhower, Dwight D. *Crusade in Europe*. New York: John Hopkins University Press, 1997.

Gailey, Harry A. *Peleliu 1944*. Annapolis, MD: Naval Institute Press, 1983.

Gailery, Harry A. *The War in the Pacific*. Novato, CA: Presidio Press, 1995.

Gatchel, Theodore L. *At the Water's Edge*. Annapolis, MD: Naval Institute Press, 1996.

Gray, Colin S. *Seapower and Strategy*. Annapolis, MD: Naval Institute Press, 1989.

Griffith, Samuel B. *Sun Tzu – The Art of War*. New York: Oxford University Press, 1971.

Hanson, Victor Davis. *Carnage and Culture*. New York: Doubleday, 2002.

Hanson, Victor Davis. *Ripples of Battle*. New York: Doubleday, 2003.

Hastings, Max. *Overlord*. New York: Touchstone, 1985.

Jomini, Baron Antoine Henri de. *The Art of War*. Mechanicsburg, PA: Stackpole Books, 1992.

Keegan, John. *Six Armies in Normandy*. New York: Penguin, 1994.

Keegan, John. *The Price of Admiralty*. New York: Penguin, 1990.

Lewis, Adrian R. *Omaha Beach – A Flawed Victory*. Chapel Hill, NC: University of North Carolina Press, 2001.

Morison, Samuel Eliot. *History of United States Naval Operations in World War II, Vol. IX, Sicily – Salerno – Anzio, January 1943 – June 1944*. Boston, MA: Little, Brown and Company, 1968.

Potter, E. B. *Nimitz*. Annapolis, MD: Naval Institute Press, 1976.

Potter, E. B. *Sea Power*. Englewood Cliffs, NJ: Prentice-Hall, Inc., 1960.

Potter, E. B. *Sea Power 2nd Edition*. Annapolis, MD: Naval Institute Press, 1981.

Saunders, Anthony. *Hitler's Atlantic Wall*. London, UK: Sutton Publishers, Ltd., 2001.

Sledge, E. B. *With the Old Breed at Peleliu and Okinawa*. Annapolis, MD: Naval Institute Press, 1996.

Weigley, Russel F. *The American Way of War*. New York: Macmillan Publishing Company, Inc., 1973.

Suggested Reading

<httpp://www.army.mil/cmh>
<httpp://www.cgsc.army.mil>
<httpp://www.gebirgsjaeger>
<httpp://www.naval-history.net>
<httpp://www.usmm.org>

Case 1.3 // Burma

Logistics Strategy in South-East Asia

David M. Moore and Peter D. Antill
Centre for Defence Acquisition, Cranfield University, Defence Academy of the UK.

Dr Jeffrey P. Bradford
Defence and National Security Consultant

Introduction

This case study seeks to consider the challenges of building a strategy for managing the logistical aspects of a military campaign in a harsh environment with limited resources.

It could be suggested that any consideration of military operations in the mid-to-late Twentieth Century should be tempered by logistical realities. For those nations who enjoy immediate territorial security, their national interest in the use of military force are often of a diffuse and geographical distant nature. Historical antecedents often influence political commitments. There therefore exists a corpus of knowledge that can assist the logistician (and indeed the operational commander) in understanding the difficulties in conducting operations far from national territory.

This case-study seeks to examine the Allied campaign in Burma during the Second World War.[1] It will examine the role of logistics by applying what have been regarded as the five principles of successful logistics: foresight, economy, flexibility, simplicity and co-operation.[2]

Foresight

"Burma was not the first, nor was it the last campaign to have been launched on no very clear realisation of its political or military objectives" - Slim

By the late 1930s, British strategic priorities in the Far East focused upon the protection of India, and the naval base at Singapore. Singapore's value derived from a study by the Committee for Imperial Defence which saw it best positioned for the control of sea lines of communication in the Indian and southern Pacific Oceans.[3] The problems for this strategy revolved around the unwillingness of commanders to challenge assumptions regarding the inaccessible nature of the terrain in the Singapore region. This also extended to the creative capabilities of the Japanese. However, it must also be noted that in many ways, the Far East was regarded as an auxiliary theatre to Europe. In order to defend Singapore, a fleet was needed. And as one author noted, "There could be no guarantee that a fleet could be made available, or that if it were, it could arrive in time and be of adequate strength."[4]

The Japanese wish to create a Greater East Asia Co-Prosperity Sphere would enable them to acquire access to resources sufficient to ensure their sustained economic domination of the region. In 1939, Japan imported 30,700,000 barrels of oil and domestically produced 2,300,000 barrels. In 1941, oil imports had dropped to 8,300,000 barrels due to its success on the battlefield.[5]

For the British and their allies, the Far East provided them with resources that would have been essential in the struggle to come. For example, rubber production in Malaya and the Dutch East Indies produced 79% of the total world production in 1940, while accounting for 65% of tin and 8.87% of oil.[6] Other essential resources lost early in the war included the largest supplier of quinine, that of Java. The inability to protect these economic treasures increased pressure on the European Allies and their reliance on the United States. Furthermore, in the campaign ahead, the logisticians would have to rely on basic supplies being transported from other theatres at an obvious opportunity cost in transports, fuel and personnel.

The Japanese war plan was both simple in design and at a fundamental level, driven by logistic considerations. The plan had several constraints. Firstly, because of Japan's reliance on imported raw materials, the war had to be short in duration. As one author noted, "thus economics rather than strategic considerations dictated to large extent the areas to be occupied and points of attack".[7] Additionally, in order to maintain control of these territories, Allied lines of communication had to be neutralised. Therefore, the initial phase of the war plan involved an attack on the American fleet at its Pearl Harbor anchorage, which took place on 7th December 1941, and rapidly overrunning the oil producing centres in the region.

It is reasonable to suggest in hindsight that the Japanese war plan was driven by logistical constraints operationally, and control over access to supplies in the long term. The offensive component of the plan was the denial of these assets to the opposing forces.

For the Allies, control over India remained essential for two reasons. Firstly, for the British, it was of vital strategic importance in terms of resources and political commitment. Secondly, the Indian economy was being used to support the war effort throughout the Middle East and Asia. Its loss would dramatically reduce the options available to defence planners and seriously damage morale. However, India was not self-sufficient in the production of rice. It was dependent on Burma, which was also under the control of the British. Clearly, at a strategic level, Burma had to be defended in order to prevent civil unrest in India and ensure its continued contribution of personnel and material to the broader war effort.[8]

The Invasion of Burma

The Japanese invasion of Burma comprised of two stages. The first was an assault to capture airfields in the south, which would enable Japan to maintain air superiority throughout its subsequent operations in the north. The final perimeter of the new Japanese Empire was to be located at the border of Burma on the threshold of India.

The Allied High Command was focused on the successful defence of Singapore. Subsequently, this led to precious reinforcements being deployed there, rather than Burma.[9] The Japanese assault began in January 1942 with two divisions.[10] Facing them were two Allied divisions. The 1st Burma Division (composed mainly of Burmese troops) and an ad hoc British formation, the 17th Division. Given its geography, the Allies were bound to have difficulty in determining where the invasion would come from. Burma had a long eastern border with its neighbour Siam. Burma's essential infrastructure was a railroad between its two major cities, Rangoon and Mandalay.

Japanese forces sought to break the British forces and capture Rangoon before driving to Mandalay. Control of the key population centres would aid their efforts to capture Burma. A further difficulty for the British was in the nature of the Japanese tactics. The use of the technique known as the 'hook' enabled Japanese troops to place blocking forces in the rear of their opponents, cutting their lines of support, retreat and communications. Slim noted his Chinese ally, General Changsha's observation, "the Japanese, confidant in their own prowess, frequently attacked on a very small administrative margin of risk".[11]

Upon the invasion, many of the Burmese troops dispersed into hiding, leaving the 1st Burma Division below strength. Japanese tactics and operational tempo forced the British into a retreat, but to their credit, they managed to conduct a fighting withdrawal back behind the Chindwin River, a journey of nearly 1,000 kilometres, with only their organic logistic support. In assessing the failure to mount a sufficient defence of Burma in the initial stages of the campaign, Slim cited several factors, including a lack of preparation, logistics infrastructure, host nation support, jungle readiness, as well as the 'hook'.[12]

From the point of view of the logistician, why did the Japanese overrun such a large part of Burma so fast? As well as Japanese tactics, one of the most important factors was the poor delegation of responsibility.

Burma, for bureaucratic reasons, fell under the command of Singapore, rather than India, from where any serious efforts would come from, if a major reverse were suffered.[13] This meant that there had been little time in which to initiate few, if any, preparations. Further, infrastructure problems derived from poor roads, which disappeared in the monsoons, and a natural barrier between Burma and India in the north which was mountainous and covered in thick forest. Meteorologically, there was little relief either. The monsoon season lasted for around five to six months per year, and conventional wisdom regarded the period as impossible for sustaining continuous operations through.

There were no major links between India and Burma to exploit for sometime to come. Because of the rapidity with which the Allied formations were raised and deployed, training for jungle operations had not been as thorough as it should have. Clearly, this would affect their operational readiness when facing a superior opponent in unfamiliar terrain, using unconventional tactics.

Economy

"The answer, as almost all answers in war … . depended on supply and transport" - Slim

The principle of economy refers to the necessity to extract every possible advantage from the finite resources at the logistician's disposal. The escape of British forces across the Chindwin River, enabled the preservation of the most essential military resource, that of personnel. However, a lot of equipment had been lost in the 1,000 kilometre withdrawal, due to enemy action, breakdown and loss.

Clearly, the task facing Slim included the reconstitution of his forces, improvements in their morale, and preparations for the counter-offensive. The challenge for him would be enhanced because of being a low priority theatre for additional resources. The Allied conference in Washington in 1943 set the strategic priority as defeating Germany before Japan.[14]

A further complication for the logistician was the surprising difficulty in getting equipment from Indian ports to the front line in Burma. Port capacity for the whole country was equivalent to that of Southampton, and the railroad stock a fraction of that possessed by Britain.[15] Further inland, only one railroad existed between Bengal and Assam, most of which, was only single track and narrow gauge. This enabled the transport of some 610 metric tons of supplies per day. Allied efforts eventually increased this by four-fold.[16]

The British commander needed to find ways of keeping his opponents occupied while reconstituting the bulk of his command. One ingenious method involved that of using air transportation to infiltrate and re-supply a special force that would harass the occupation forces of the Japanese. The Chindits were thought by some, to be an extravagant waste of resources. Normal formations tended to lose their fittest personnel to these elite formations, as well as valuable equipment.[17]

Regardless of the level of tactical success, it could be argued that the Chindits achieved it at a strategic level. Firstly, they fixed Japanese forces; secondly, they proved that the British could operate successfully in the jungle; and thirdly, at the political level, demonstrated to the Allies that the British were making an active contribution to that theatre.

In the meanwhile, efforts were devoted to improving the logistics infrastructure leading to the frontline. During this period, air transport techniques were pioneered. Methods of translating requirements at the frontline to give timely and appropriate re-supply, required good organisation of supplies for shipment, and good communications for articulating those demands. Given also, the scarcity of silk parachutes in the theatre, the logistician had to improvise with alternatives. [18]

Finally, in providing supplies in this most difficult of theatres, one had to maintain the morale of the forces. Given the ethnic diversity and eating habits, the logistician's artistry was stretched to the limit in supplying diverse foodstuffs to national forces ranging from the Americans to Ghurkas, Africans and Sikhs.

There were however, problems that worked against achieving the best possible economy of logistics distribution. Firstly, as tactical experience of fighting the Japanese increased, it became clear that air transportation of supplies was a viable method of countering the 'hook'. Once the Allied forces realised the Japanese had encircled them, they created a perimeter and awaited supply. Given the slim logistic margins applied by the Japanese, they were forced to maximise their efforts to break through and risk high casualties, or withdraw.

This however, required large numbers of transport aircraft, which the Allies did not possess. As mentioned earlier, the Burma theatre had last call on reserves, but there was also a political imperative, which led to a large segment of the air transport fleet being dedicated to the Chinese. The solution, was the abandonment of peacetime flight and maintenance requirements, with transport planes typically flying twice their normal flying hours.[19] A final consideration was the lack of commonality between equipment used by the Allied forces. Standardisation was a key lesson learned form this campaign.[20]

Additionally, Allied scientists had developed improved procedures to help prevent some of the various tropical diseases from wrecking havoc. This was a serious issue. In 1943, the ratio of soldiers admitted to hospital with combat wounds as opposed to tropical disease was 1:120. By 1944, this figure had fallen to 1:20 by a combination of treatments and ruthless implementation of field hygiene standards[21] which led Mountbatten to utilise these factors as a competitive advantage and seek to engage the Japanese who had not improved their medical capabilities.[22]

In summary, achieving economy in Burma was a severe challenge for the artistic skills of the logistician. Weather hampered the most flexible instrument of re-supply. The terrain was treacherous for vehicles and required new roads cut from the mountains in the north. The infrastructure in India was poor and resources scarce. There were few railway lines, and those that existed required rigorous control to ship the maximum loads.

Flexibility

"Determination by itself may achieve results, while flexibility, without determination in reserve, cannot, but it is only the blending of the two that brings final success." - Slim

Flexibility, as a principle of logistics, relates to the ability to adapt plans to reflect changes in the operational environment. Burma was a complex theatre of operations throughout the Second World War. Compared to other theatres, there were fewer tanks deployed, but still there was a necessity to move forward large and diverse amounts of combat material.

Flexibility impinges on the logistician in two distinct ways, on the physical level and the mental level. In the physical context, it constrains the method and means by which the commander's forces can be sustained during prolonged military activity. Mentally, it manifests itself in the agility necessary to adapt plans to suit the needs of the campaign.

Prior to the outbreak of the war in Burma, the British command utilised the most expeditious route of sustaining the Allied forces present. This grew into a dependence upon the road system, which the Japanese were able to exploit using their 'hook' tactics and cause severe disruption.[23]

The railway was the least flexible system available. Its carriage could be optimised, yet it was still a static target for enemy attack. The roads were of a similar nature, however, their poor quality was rarely perceived to have been affected by enemy action. The Allies built roads in the Manipur province of India through mountains in order to improve logistic distribution. However, as records show, the roads were difficult to navigate due to their location and conditions in the monsoon season.

Air transportation, as suggested, was the most flexible asset available to the Allied forces. The ability to drop supplies to beleaguered troops on campaign, surrounded by the Japanese, provided a boost to morale, apart from their obvious utility. During the Japanese offensive of 1944, Allied re-supply efforts into Imphal and Khohima must have

46

been a demoralising sight for the Japanese Army at the absolute extreme of its lines of communication.

When the Japanese forces were finally broken, the Allied forces were able to pursue them very rapidly because of their proximity to their supplies, and the level of expertise accumulated in air transportation and re-supply. Slim likened the Army to that of being like a fully wound clock, ready to spring forward at the enemy.

During the final stages of the Burma campaign, the attainment of air superiority by the Allies was essential. From this, flowed the ability to re-supply Allied forces without fear of interception by the Japanese. This in turn, enabled rapid marches southwards into Burma, and protection from encirclement. The flexibility of air transport enabled the lines of communication with India and the forward forces to be stretched, and for brief periods, to be ignored.

In mental terms, flexibility of thought enables the logistician to improve the logistic chain as a whole to fit the environment. For example, the construction of barges to shift supplies in preparation for the Allied offensives helped free up resources and move cargoes at minimum fuel cost, thereby enhancing the commander's freedom of action.[24] Table 1.3.1 below details the water traffic in logistics along the Chindwin River in 1945 during the final campaign to recapture Burma.

Table 1.3.1: Transport capacity along the Chindwin River 1945 (in metric tons).

Period	Kalewa - Alon	Kalewa - Myingyan	Alon - Myingyan	Total in Period	Daily Average
01/02/45 - 28/02/45	2174	0	0	2174	77.5
01/03/45 - 28/03/45	5540	0	0	5540	198
29/03/45 - 25/04/45	393	8931	3721	13045	466
26/04/45 - 16/05/45	310	12124	1614	14048	669
17/05/45 - 30/05/45	332	8447	641	9420	673
Totals	**8749**	**29502**	**5976**	**44227**	**416.7**

To some, the logistician's role is to systematically count every item in the theatre and merely use mathematics to optimise the distribution of food, blankets and ammunition.

Clearly, there is an element of artistry to the task. Slim himself regarded one of the key skills to acquire as being "a controlled imagination, and a flexibility of mind".[25]

The logisticians sought to commandeer boats and convert them to burn wood, as well as many other innovations in order to assist their commander in his prosecution of the campaign. For example, the British built several roads using locally manufactured bricks. Every thirty or so kilometres, there was a kiln producing bricks.[26] Clearly, flexibility in its broadest sense has been one of the essential characteristics of logistics.

Simplicity

"The more modern war becomes, the more essential appear the basic qualities that from the beginning of history have distinguished armies from mobs." - Slim

For the logistician, one would assume that the greater the element of simplicity and consultation in the plans to which he is to make a contribution, the greater the chance that such operations will meet with success in their execution.

Slim planned all of the Burma campaign with four criteria in mind. Firstly, all operations should be offensive in nature and take the battle to the Japanese. Secondly, the central idea of the plan must be simple. Third, all subsequent planning for the operation must support the central idea that has been expressed. Lastly, the plan must have within it the ability to surprise the enemy.[27]

The logistician's role was perceived as crucial in turning these criteria into a workable operational plan. Slim assigned his most senior staff officer position to a logistician, which he saw as logical as "administrative possibilities and impossibilities would loom large, larger than strategical and tactical alternatives".[28]

The commander's manner of conceiving plans was initially to consider the possibilities himself, and then to share them and then discuss them with his most senior staff officers, the Major General responsible for administration and the Chief of the Royal Air Force in the Burma theatre. It was clear, that the logisticians had the ability to influence the development of operational plans at the embryonic stage and advise as to their logistic feasibility.

Having designed the principles of the plan, Slim would then be content to write the short statement of intent for the operation and then delegate responsibility for the execution to the subordinate commanders, whom he would go out to the field to brief personally.

Following the retreat from Burma, the British set about reconstituting their forces, and building up the logistics infra-structure necessary to support an eventual counter-offensive into Burma. Their aim was to improve the supply situation as a prerequisite for sheer survival, before building up for the counter-offensive.

To that end, air supply techniques were developed to an extent that during the final campaigning in 1945, air delivered material formed some 75% of daily supplies received by the Allies.[29] Furthermore, oil pipelines were developed to enhance the flow of petrol, oil and lubricants (POL) to forward forces. These pipelines were developed into four systems which combined had a maximum capacity of around 300,000 metric tons per month.[30]

Eventually, the Allied strategy throughout Asia put sufficient pressure upon the Japanese commanders that they knew they had to achieve destruction of the British forces in order to stop India's contribution to the Allied war effort. This led to the Japanese offensive of 1944, which in turn led to the sieges of Khohima and Imphal. Slim's command had already considered this eventuality. By fortifying these vulnerable rear-area locations which were established in an *ad hoc* manner during the headlong flight north at the start of the war, they could be used later to fix Japanese forces.

The Japanese were constrained by their tight logistics margin and the need to break-through before they ran out of supplies. This enabled the revitalised Allied forces to launch its counter-offensive southward against a shattered opponent. In the first six months of 1945, the British pursued a manoeuvre strategy, liberating Rangoon before the start of the monsoon season.

To summarise, the commander's development of simple operational planning concepts, coupled with enabling the logistician access to the commander at an early stage assisted in shaping the operation's effectiveness.

Co-operation

"There is only one thing worse than having allies – that is not having allies." - Slim

The last of the five principles is that of co-operation. This section seeks to contemplate co-operation between the armed services and within the armed services in the context of the Burma experience.

The Burma campaign fell within the theatre of war that saw the British, Americans and Chinese fighting together. Politics would play a large part in the determination of strategy. The Chinese had been fighting the Japanese for several years before the start of the Second World War. They received logistic assistance from the United States, as well as advisors and a limited troop commitment, as the main thrust of American policy was the use of maritime and Allied forces in an 'island hopping' campaign.

It was essential therefore, to maintain the alliance. Following Mountbatten's appointment, one of his first tasks was to create a joint headquarters. This brought together, not only the nationalities, but also the services. The clear benefit to be reaped was in the gradual creation of a joint sense of purpose, and an understanding of each others' problems both between land, sea and air forces, but also between British, American and Chinese commanders. For the logistician, this was a useful development. The ability to commandeer US transport aircraft at short notice could occasionally be justified by the proximity of the decision-makers.[31] Relations between the services were very close. As noted by one participant, "American and British squadrons shared airfields, air traffic control facilities, equipment and messes… this was one of the biggest campaign winning factors".[32]

Importantly, within the theatre, the retreat from Burma and the harsh conditions engendered a sense of camaraderie between the Air Force and the Army. The air transportation drops and Chindit insertions required the closest co-operation and trust between the services. In summery, co-operation was a vital prerequisite, which could not be taken for granted in the Burma campaign. It was something that had to be nurtured and which took up much of Mountbatten's time. For the logistician, co-operation between the allies in a coalition, enabled resources to be shared more equitably, alleviating some of the potential hazards of campaigning on limited resources.

Conclusions

"The Japanese first demonstrated painfully on us that it was not so much numbers and elaborate equipment that count … . but training and morale" - Slim

Having examined the Burma campaign in the context of the five principles of logistics, this final section seeks to analyse the campaign in respect of modern best practise. In military terms, the Burma campaign has, it could be suggested, several lessons for contemporary logistics practice.

The campaign was conducted at long distance, with little external support. The forces were initially demoralised, fighting in terrain for which they were unprepared mentally or materially. Given the near complete lack of infrastructure, the development of a logistics system facilitated greater operational flexibility as witnessed in the final campaigning in 1945 when less than one per cent of Allied supplies was moved by road. Whilst air mobility through the helicopter had not yet come of age, breaking the near exclusive dependency on the road system was indeed a historic achievement. Essential to the success of the campaign was the adaptability of the system's delivery methods, which we would recognise today as effective supply chain management.

The conflict in Burma has implications for business management as well. Foremost among these, is the ability to manage change. The organisation Slim inherited would not have survived, let alone turned defeat into victory, without the ability to motivate, initiate and sustain the change process.

The leadership characteristics exhibited by Slim and his subordinate commanders would find clear parallels among the qualities found in the leading edge organisations today. The realisation was that, given the initial quality of their organisation, competing head-on with their adversary was clearly disadvantageous. However, the adoption of competitive strategies led to an indirect approach, which served to weaken the position of their opponent for the eventual counter-offensive. Slim's employment of Special Forces serve clearly to illustrate the possibilities. This is further emphasised by the innovative pioneering approach taken to sustain the cutting edge of the organisation. In this case, the Chindit Special Forces were infiltrated and subsequently re-supplied exclusively by air.

The framework utilised in this analysis, encompassed key approaches, which, we can now translate into modern business practice. In retrospect, it is possible to consider not merely the effective management planning, but the development of a key strategic vision and culture to realise the change necessary to achieve success.

Economy readily translates into the 'lean' thinking concepts introduced and developed by Womack and Jones.[33] Amongst others, Christopher introduces the notion of the agile supply chain. This concept refers to the flexibility necessary to attain market presence and sustain it in the face of competing demands.[34]

As alluded to earlier in the paper, the development of the logistics infrastructure following the retreat from Burma had at its core, the notion of simplicity. Regrettably, the translation of this experience into business terms has been somewhat misplaced. By considering what can we do? How? And in what order? Writers such as Peters have, in the past few years, developed our conception of management best practice in line with the theme of simplicity.[35]

Global best practice indicates that the development of close relationships between and within, organisations is a means to gain competitive advantage.[36] The definitions within Slim's organisation of who served whom and who the partners were, enabled his staff to rebuild the organisation with a focus to achieving battlefield success. In conclusion, this article has sought to illustrate the need for competitive advantage, which requires quality in strategic leadership, organisational adaptation and innovative strategic decisions. The Allied experiences illustrate, for all of us, whether in defence or commercial logistics, valuable and timely lessons.

Notes

[1] Burma received its independence from Britain in 1948, and today is known as Myanmar.

[2] Foxton, P. D. *Powering War: Modern Land Force Logistics.* London: Brassey's 1994. Ch. 1. p3.

[3] Woodburn-Kirby, S. *The War Against Japan: Volume 1, The Loss of Singapore.* London: her Majesty's Stationery Office, 1957. Ch. 1. p3. The Committee considered Hong Kong as an alternative, but subsequently decided that insufficient protection could be provided to it.

[4] Ibid. p22.

[5] Ibid. Appendix 3. p481.

[6] Ibid. Appendix 1. pp477 - 478. Production emanated from Malaya, the Dutch East Indies and British Borneo.

[7] Ibid. Ch. 5. p90.

[8] Wilson, H. W. *The Second World War 1939-1945: Army Administrative Planning.* London: The War Office, 1952. Ch. 2. p11. As noted "when the recapture of Japanese occupied countries came within the bounds of practicability, rice was a factor which had to be taken into account in the formulation of our strategy".

[9] Op Cit. Woodburn-Kirby. Ch. 15. p260.

[10] Grant, I. L. *Burma: The Turning Point; The Seven Battles on the Tiddim Road, Which Turned the Tide of the Burma War.* Chichester: Zampi Press 1993. Ch. 1. p24. The 33rd Division attacked from Siam Thailand, and the 55th Infantry Division from Malaya Malaysia.

[11] Slim, W. *Defeat into Victory.* London: Cassell & Company Ltd, 1956. Ch. 1. p18.

[12] Ibid. Ch. 6. pp115 - 121.

[13] Op Cit. Woodburn-Kirby. Ch.15. p253.

[14] Op Cit. Wilson. Ch. 5. p29. Mountbatten, L. **"The Strategy of the South-East Asia Campaign"** in *The Journal of The Royal United Services Institute,* XCI 564 London: RUSI, 1946. pp468 - 484. p473.

[15] Ibid. Ch. 7. p42.

[16] Op Cit. Mountbatten. p472.

[17] Op Cit. Slim. Ch. 23. pp546 - 549.

[18] Op Cit. Foxton. Ch. 1. p4.

[19] Op Cit. Mountbatten. p481.

[20] Ibid. pp483 - 484.

[21] Op Cit. Mountbatten. pp472 - 473. See also Op Cit. Slim. p169 & p177.

[22] Ibid. p473.

[23] Op Cit. Slim. Ch. 6. p119.

[24] Woodburn-Kirby, S. *The War Against Japan Volume IV, The Re-conquest of Burma,* London: HMSO, 1965. Appendix 16. pp464 - 469.

[25] Slim, W. **"Higher Command in War"** in *Military Review*, May 1990. pp10 - 21. p15. The article was a reprint of a lecture to the US Army Command and General Staff College in 1952.

[26] Op Cit. Slim. Ch. 9. p172.

[27] Op Cit. Slim. Ch.10. p209.

[28] Ibid.

[29] Op Cit. Woodburn-Kirby 1965. Appendix 24. p517.

[30] Ibid. Appendix 2. p438. The four oil distribution systems were A: The Brahmaputra River system 56,000 t. per month. B: Chittagong to Kawela system 49,000 t. per month. C: Calcutta to Chittagong Tinsukia - Myitkyina system 133,000 t. per month. D: Myitkyina to Kumming system 66,000 t. per month.

[31] Op Cit. Mountbatten. p476.

[32] Ibid. p484. Comment by Air Vice-Marshal Williams in subsequent discussion.

[33] Womack, J. P. & Jones, D T. *Lean Thinking*, New York: Simon & Schuster 1996.

[34] Christopher, Prof. M. **Army Logistics Forum**, Cranfield University, June 1998.

[35] Peters, T. J. & Waterman, R. H. *In Search of Excellence*, London: Harper & Row, 1982, as well as subsequent articles and books by Peters.

[36] For example, see Burnes, B. & Dale, D. *Working in Partnership: Best Practice in Customer-Supplier Relations*, Aldershot: Gower Publishing 1998.

Case 1.4 // Operation Overlord

Supply Chain Innovation during World War II

David M. Moore and Peter D. Antill
Centre for Defence Acquisition, Cranfield University, Defence Academy of the UK.

Jeffrey P. Bradford
Defence and National Security Consultant

Introduction

This case-study seeks to consider the logistical requirements of another Second World War campaign. However this case, as opposed to Burma, was one which had first call on assets and resources available. The Allied operation code named Overlord represented the long awaited launch of a second front against Germany during the Second World War. The Allied effort had to be launched across an expanse of water, the English Channel. This generated a number of challenges for the logistician. Some of the many planning issues that arose, included housing the assault forces prior to the operation, marshalling the necessary equipment and supplies whilst deceiving the opponent as to the true location of the attack, and finally the manner of supporting such an operation at a distance, whilst not being able to rely on host nation support.

This case-study aims to consider the logistics issues of Operation Overlord in four stages. Firstly, the case-study will examine the nature of the planning for such an operation and the mechanisms which evolved for coordinating such efforts on a multi-national basis. Secondly, it will consider logistics issues relevant to the actual crossing on D-Day and following this evaluation of the provision of support for the assault. Lastly, the authors will consider the case study in the light of modern logistics practice and seek to offer an opinion on lessons learned from this examination.

Planning for Overlord

"You will... prepare plans for... a full-scale assault against the Continent in 1944 as early as possible." - Directive to COSSAC planners.

"Eisenhower felt that he had 'missed the boat'." - Sixsmith.

The decision to proceed with planning for Overlord was a consequence of the Allied strategy conferences at Casablanca and Washington in 1943. The objective was clear - to prepare for a re-entry into Continental Europe on the 1st May 1944.[1] Following this decision, it was decided to appoint a British General, Morgan, to the post known as COSSAC – Chief of Staff to the Supreme Allied Commander Designate.[2]

The final word in the title hints at one of the challenges of the post. At this time, no commander had been appointed. COSSAC would have a multi-national staff to plan for an operation without a commander to fight for resources and simultaneously second-guess his concept of conducting military operations. The American General, Dwight D. Eisenhower, was finally appointed on 7th December 1943.

COSSAC set about his task with gusto. His first priorities were to obtain the most senior staff officers possible in order to maximise his ability to obtain resources and minimise time spent making requests through the chain of command. Morgan was keenly aware of the limitations of the British system of organisation and saw the enrollment of high-ranking staff officers as a means of circumventing the more esoteric bureaucratic practices of the War Office.[3]

Further, having witnessed the benefits of an integrated headquarters whilst serving with Eisenhower in the past, Morgan was keen to establish a fully integrated Anglo-American headquarters. In selecting the British component of COSSAC's planning staff, Morgan was able to utilise his rank and former staff officer experience to obtain the cream of the current crop of Army staff officers.

COSSAC's deputy, an American staff officer with experience of the Whitehall machine, skillfully handled the American appointments. The inter-service integration of the staffs took somewhat longer to achieve. In the British case, the Royal Navy was quick to involve themselves in the planning process as they would have the primary task of getting the assault forces to the beaches of France. COSSAC noted the reluctance of the Air Force, in that "the bomber barons remained obstinately aloof." [4]

An important step in welding these disparate groups and cultures together, was made with the establishment of a headquarters in Norfolk House, London. This property housed the planning staff and their work but importantly, had a dedicated floor where staff could relax and vent their frustrations. [5]

Planning for Overlord started from the premise of where in Britain would provide the springboard for launching the assault. Railheads, port facilities and the road infrastructure were key to making the assessment that somewhere in South-east England would provide the concentration point. To assist the planning in greater detail, a supplementary directive gave COSSAC the planning assumption of twenty-nine divisions to use in the assault.[6]

Naturally there were cultural barriers to be overcome. The American officers found dealing with the British committee system of decision-making anathema. As Morgan noted, "it took time and patience to explain .... it was necessary to placate the British by playing."[7] From the British point of view, the American style of planning was very different. One British observer noted their tendency to work from the premise of a conclusion and work out the facts afterwards. The success of this approach, he was convinced, was based on the abundance of resources available to the United States Armed Forces.[8]

At a more technical level, the differences between American and British units of measurement were of an extreme importance in an operation of such magnitude. For example, an American gallon was only eighty per cent that of an Imperial gallon.[9] Further differences in terminology meant that both nationalities had to be aware of the difference in meaning between headquarters functions and responsibilities.

Planning for the subsequent break-out and assault on Germany itself was conducted under three cases using the code names RANKIN A, B and C. A assumed German forces held a defensive line but were stretched thin. Case B assumed that they would evacuate parts of their territories, whilst C assumed a total collapse along the lines of November 1918.[10] The RANKIN exercise proved useful in assisting Overlord planners in considering their plan within the context of the Second Front as a whole.

The Overlord plan was completed on the 15th July 1943. Its highlights were an assault on the beaches in the Normandy region, with a view to a rapid breakout to capture successive ports along the North-west coast with which to strengthen lines of communication and insert the rest of the assault forces for the drive on Germany.

In summary, this section has tried to illustrate the problems experienced by the COSSAC staff in designing a major plan ahead of schedule, without a commander, and on a multi-national basis.

Sustainment

"The hard core of their problem was logistical." - Alanbrooke, Chief of the Imperial General Staff.

At a grand strategic level, the Allies were clearly pre-occupied with logistics. The TRIDENT conference in 1943 set the strategic priorities, as not only planning for an invasion of German occupied Europe, but also directing strategic aerial bombardment against the German defence industrial base and associated supply chain.[11] The central challenge for the planning staffs of Overlord would be to ensure that the assault forces would be able to breakout from the landing areas. This in turn, generated a requirement to ensure the provision of supplies for large-scale mechanised operations and especially the provision of petrol, oil and lubricants (POL). As the Chief of the General Staff Alanbrooke noted, "the hard core of their problem was logistical...it was going to be a race against time to build-up strength fast enough to hold off, and ultimately break the defenders".[12] Amphibious landings are some of the most complex and risky of military operations and that "transporting a huge invasion force to Normandy, and successfully landing it in the face of determined opposition, posed difficult and diverse problems; no less difficult was the task of keeping the troops supplied once they were established ashore." [13]

German planners had considered the Allies likely attack routes and accordingly designed their defences to prevent the Allies capturing one of their ports in an amphibious attack. Further, they had considered the most likely route of attack to be across the shortest part of the Channel, toward the *Pas-de-Calais*. The Allies did their utmost to reinforce German perceptions by assigning the American, General Patton, to a fictitious command in the south-east of England codenamed Operation Fortitude.

Having drawn German interest away from the Allied axis of attack toward Normandy, there remained the problem of the lack of port facilities, "since it was perilous, if not actually impossible, to mount a head-on attack against one of the heavily defended ports such as Cherbourg".[14] In order to be able to ensure the rapid delivery of some 600 tons of supplies, per division, per day, innovative solutions had to be devised. General Morgan attributed the solution to Commodore Hughes-Hallet, who during an intense planning discussion which centered on the requirement for a port, made a comment, "if we can't capture a port we must take one with us."[15] Subsequent thinking led to the design of the artificial harbours known by the code name Mulberry.

Two harbours were planned, one for the American and one for the British-Canadian forces, known as Mulberry A and B respectively. Each had three components, known by their code names as 'Phoenix', 'Bombardon' and 'Whale'. 'Phoenix' was a series of concrete caissons, large hollow concrete structures weighing between 2,000 and 6,000 metric tons. The use for these 60 metre caissons was to form a breakwater when sunk at a depth of about 10 metres offshore. Some 1.1 million tons of concrete were transported to France on D-Day, prompting one Briton to liken Phoenix to "rolling the Athenaeum on its side and towing the damn thing across the Channel."[16]

The next component was 'Bombardon', a steel structure which was designed to be anchored to the seabed to provide a deep breakwater for shipping awaiting access to the

Mulberry port facility. These structures were approximately sixty metres in length and of a cruciform structure. 'Whale' was a series of floating piers constructed of steel and concrete, which could be installed at the beachhead. A final addition, was the use of ships that were sunk to provide added protection from the sea, and was known as 'Gooseberry'.[17]

The Mulberry harbours were planned to require two weeks to become fully operational. The planned through-put rates of supplies and vehicles are illustrated below in Figure 1.4.1:

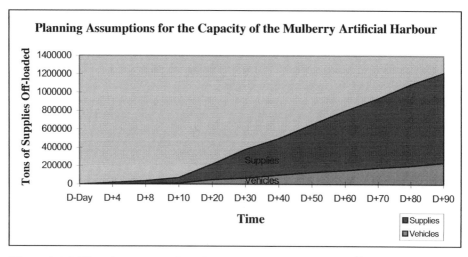

Figure 1.4.1: Planning assumptions for capacity of the Mulberry harbours.

In practice, the ever-present Clausewitzian friction caused some problems in establishing these facilities in Normandy. Bad weather prevented the rapid transport of the three components across the Channel. However, the 'Bombardons' were being laid in an efficient manner. Ten days after the start of the operation, the 'Bombardon' screens were providing a reduction in the effect of the tides by approximately forty per cent.[18]

The two harbours suffered somewhat disproportionate damage. The American Mulberry A was severely damaged, and Eisenhower decided to put further Mulberry resources into strengthening Mulberry B.[19] Among the factors cited for the damage to Mulberry A, was the lack of natural protection in its neighbourhood, and the Gooseberry block ships were not placed with sufficient space in-between. Further, the nature of the seabed caused the Gooseberry ships to sink slightly more, and thus offered less protection above sea level.[20] Finally, the loss of the breakwater caused the assault ships to move about causing extensive damage to the relatively unprotected piers.[21]

Clearly, the planners had provided a significant innovation by designing artificial harbours, which required the involvement of some 50,000 people, yet managed to keep the objective of the project secret. However, it was realised that the provision of petroleum products alone would out-strip the capacity of the Mulberrys once the Allied breakout from the beaches occurred. Further, the requirement for scarce shipping resources to deliver petrol would reduce space for personnel and other vital supplies.

Already, planners had assembled some fifteen million eighteen litre containers for petrol. This was influenced by the planning assumption that no storage facilities in occupied Europe could be counted upon.[22] The plan was for this stockpile to be used in the initial assault as the robust containers would minimise the risk of accidents. However, once the initial invasion had succeeded, much larger deliveries had to be facilitated in some way.

The solution lay in a project which had the acronym PLUTO – Pipeline Under The Ocean. Relatively shortly after the British withdrawal from Europe, some thought had been put into the idea of developing a flexible pipeline which could deliver petrol products to the continent. By 1943, experiments were being undertaken using submarine cable with the core removed to pump petrol across the Severn estuary which had conditions deemed similar, albeit on a smaller scale, to the English Channel.[23]

The success of these experiments led to the decision to use two types of pipes to support the invasion of Europe, 'Hais' and 'Hamel'. The 'Hais' pipes were fifty-four millimetres in diameter while the 'Hamel' pipes were of a larger seventy-six millimetres diameter. The larger pipes were composed of six metre lengths of steel welded together in 1,200 metre lengths. These were coiled onto drums twenty-seven metres long and over fifteen metres in diameter. These coils weighed 1,600 tons which was equivalent in weight to a destroyer. With a true sense of humour, the Royal Navy named the vessels, which carried the drums of this cable, HMS *Conundrums*.[24]

When they were due to be employed, the plan was to deploy the pipelines across two routes known as BAMBI and DUMBO. BAMBI was the longer of the two routes, approximately 120 kilometres from Sandown on the Isle of Wight to Querqueville, west of Cherbourg. The DUMBO route lay from Dungeness to Boulogne, a distance of around 40 kilometres. These two pipelines were to be completed by D+75, approximately a fortnight before the demise of the Mulberrys, which were only due to last for three months.

Following the successful conclusion of a number of trials, the War Office began to plan the potential performance of the PLUTO system. These assumptions are illustrated in Figure 1.4.2 opposite:

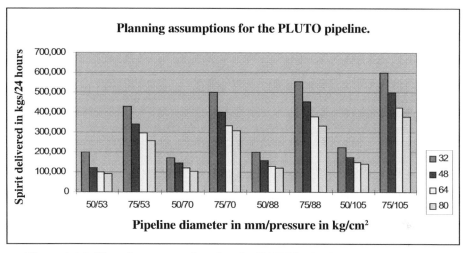

Figure 1.4.2: Planning assumptions for the PLUTO pipeline.

Figure 1.4.2 illustrates the potential throughput of the two pipelines in thousands of kilograms of fuel per twenty-four hour period. The above figures were partly speculative in nature, as the use of cable over sixty-five kilometres in length was not proven in 1943.[25] Planning figures for the development of PLUTO suggested that it would be a month before the system could be made operational on or around D+30. In the meanwhile, a system of ship-to-shore pipelines was devised for the discharge of oil. There were two of these pipelines allocated to each Task Force area, 'Tombola', which had an internal diameter of 150 millimetres, and the larger 'Amethea' which had a larger 250 millimetres diameter. Once installed, this pipeline had the capacity of transferring approximately 600 tons of oil per hour to the shore after D+18. This meant that the transport and reliance on fuel from the canisters would carry on for sometime after the landings.

In practice, the PLUTO operation got underway two days after the start of the invasion. The team initially had to reconnoitre the areas chosen for laying the pipes to ensure their charts were accurate. By D+9, the PLUTO team had deployed operationally to France.[26] The first pipeline was completed on D+19. However, severe weather prevented the early operation of the BAMBI station. It commenced pumping on the 18th September 1944 until the 4th October. In this seventeen day period, the BAMBI station transferred some 4.25 million litres of fuel to Cherbourg. Once the port had been opened by the Allies, it took over the pumping of bulk oil.[27] The DUMBO terminal started pumping on the 27th October 1944 and remained in action until the end of the war in Europe. By the 22nd May 1945, some fifty million litres of fuel had been delivered through DUMBO's sixteen lines without disruption.[28] The PLUTO project filled a gap in the supply line which stretched from the western United States, across the Atlantic by merchant ship to Liverpool, around Britain and then to the continent. Allied engineers built upon this network as the formations advanced to create an uninterrupted fuel supply from the French coast to Germany.

Official historians regard the PLUTO project as "an unqualified success".[29] In its time the project enabled the delivery of nearly 900 million litres. However, the view of the planners was somewhat different. COSSAC personally thought that PLUTO was not worth the effort and material that had been devoted to it.[30] However, it is probably reasonable to suggest that had the Germans pursued a policy of destroying the fuel stocks and distribution nexus of France as it retreated, PLUTO would have been even more significant than it was.

In conclusion, what can be said about the Mulberry and PLUTO projects? Clearly, they had an opportunity cost insofar as scarce resources used for these could have been used elsewhere. With the benefit of hindsight, it can be seen that Cherbourg was captured more rapidly than the Allies had anticipated, and German sabotage of its facilities was minimal. However, in the critical early stage of the operation, had the Allied assault ships been caught in the open without the benefit of any protection, the damage in the American sector especially could have been catastrophic to the lines of supply and communication.

The fuel transfer project PLUTO was a risky innovation that paid dividends. An official history of logistics in the British Army during the Second World War notes that at one point a daily transport requirement of 10,000 tons was saved. Further,

"It is fair to state that, if it had not been possible to pipe petrol from ports of entry to points east of the Rhine, it would have been impossible to maintain the Allied Expeditionary Force in their final offensive."[31]

In summary, the Mulberry and PLUTO projects were a good example of the logisticians ability to 'think the unthinkable' in order to translate the operational commander's plans into reality. Further, they demonstrate ably, the constraints that the issue of logistics places upon the freedom of action of a commander. Without the harbours or pipeline, a strategy of attrition aimed at securing the shortest sea route to the Pas-de-Calais might have presented the only viable alternative.

Execution

"The development of our operations had to be made in spite of unprecedented weather conditions throughout the summer and autumn." - Field Marshal Montgomery

Given the size and scope of an operation which, by any standard, had never been tried on this scale before, it would appear that the Overlord landings were an unequivocal success. The administrative planning performed by the COSSAC headquarters utilised projects in existence to conquer the logistical problems of the exercise. More importantly, the COSSAC staff were also an operational staff for their commander. This was novel, when typically staff officers would provide plans which other groups would execute. This involvement enabled a continuity in the understanding of the problems and probably speeded up their resolution.

As the breakout from Normandy gathered pace, so did the logistical requirements. The Allied formations were advancing at a rate of some sixty kilometres per day. This movement required some four-and-a-half million litres of petrol per day, not to mention ammunition and food.[32]

The American solution was to devise an express route on which only supply vehicles were allowed and was circular in nature so that traffic was one way. The British 21st Army Group resorted, as far as possible, to airlift certain supplies. Combat Service Support continued apace repairing vehicles. Engineers tried to rebuild the railway lines and bridges that were damaged by bombing raids, as well as laying pipeline to keep the oil moving.

Logistics played one further key role in determining strategy in the push toward Germany. It gradually became obvious that the Allied forces were beginning to outrun their lines of supply, which increased the danger posed by any prospective German counter-offensive. The two groups of forces, North British and Allies and Central American could not both be sustained indefinitely. Eisenhower chose to support the Northern group. This led to three US divisions being halted as supplies were diverted.[33]

Clearly, Eisenhower had had to take a difficult decision, which had a political, as well as a military dimension to it. The invasion was an Allied effort, therefore there was considerable domestic pressure to see the Allied forces enter Germany at the same time. Also, there was very likely some creative tension between the two commanders of the formations. The American General George S. Patton was regarded by many observers as a decisive, aggressive officer. Reputedly, his view of logistics was such that he met with his most senior logistician twice, once upon assuming command, and again in the last week of hostilities.[34] Obviously, Patton felt that conservative strategies based upon logistic concerns should not impinge heavily on operations.

However, was his view justified? It could be suggested that his unofficial strategy of obtaining fuel from any source to keep his armoured formations going may have endeared him to his command, but may have hindered the effectiveness of Eisenhower's broad advance strategy. In the contemporary environment, it could be suggested that Patton's cavalier treatment of logistics may have put his formation in danger but also had political repercussions.

One writer considered this strategy, and attempted to determine the nature of the choices available at the time. The thrust into Germany would have secured in 1944, the industrial heartland of Germany, the Ruhr Valley. The risk however, was that Allied forces already on a precariously extended supply line, could be encircled and destroyed. Given the knowledge of German offensive tactics, this is not a totally unrealistic assessment. The blow to this scheme, as opposed to an advance on a broad front apparently came from the realisation that some 1,400 transport vehicles had defective power plants.[35] It was probably fair to say, that Eisenhower, who was keenly aware of the balancing act required when commanding a coalition, knew not only the military risks, but those of the political milieu.

Conclusion

"The outstanding point about the Battle of Normandy is that it was fought exactly as planned before the invasion" - Field Marshal Montgomery

Having considered the planning, execution and sustainment of the Allied invasion of Europe from the standpoint of the military historian, it is intended in conclusion, to view the logistics component of Overlord in the light of modern business and defence logistics practice. Were the missions conducted in a manner acceptable today but with different standards of measurement? Or does modern best practice provide an alternative vision of such an operation if conducted today?

The D-Day scenario in many ways, represents the traditional military scenario beloved of peacetime exercises. In this type of scenario, the enemy, his dispositions and intentions are known, and by channeling sufficient quantities of resources a successful outcome could be achieved. This case study has illustrated clearly the importance of logistics in influencing the course of strategy.

A further consideration in the case of Overlord was its multi-national nature. The role played by American and British planners led to a number of cultural implications. First among these was the concept of the role of logistics in military operations. The US favoured a 'push' approach, channeling resources at the problem. The British, partly influenced by the effects of the long conflict favoured a much more conservative approach to the problem. Secondly, the coalition nature of the operation had political dimensions. Each participating nation wished to be represented throughout the planning stages, and as identified in the case, political considerations influenced the actual implementation of the strategy. The third issue was of a technical nature. The participating nations had their own set of standard measurements which had to be reconciled.

The lengthy planning stage utilised operational research techniques in order to satisfy the requirements. To ensure success, a further factor indicated in the case study, was that there were a multiplicity of channels of distribution from under the sea to the air. Significantly, some of the key innovations used by military planners arose from research outside of the military establishment. The harnessing of the creative potential of other organisations was a major factor in solving what were regarded institutionally as intractable problems.

The case study offers several lessons for the commercial environment. Increasingly, during the last decade, for effective strategic logistics management, training and development now emphasises the ability to manage complexity. As the Overlord case shows, logistics forms an integral part of the strategic conceptualisation of the operation itself. Today, the same is true for commerce. Logistics, rather than being a technical issue following the definition of strategy, is instead becoming a core issue in deciding what can we do? When? And how? The planning for D-Day was of an essentially quantitative nature. However, it could not be denied that there were a number of qualitative aspects as well, and today, managers have available several qualitative frameworks for improving the operations and processes in their organisations. We can only speculate as to their worth during the Second World War.

Following the war, the successful planners and managers of Overlord became the mid-level to senior management in Western industry during the 1960s and 1970s. Arguably, their experience of planning in wartime, had an impact on their practice in peacetime. The implications could be observed in large manufacturing industry's concern with the use of Materials Requirement Planning techniques emphasising the 'push' approach to supply management. The challenge for the subsequent generation is encapsulated in Japanese management philosophies such as 'just in time' and 'total quality management'. These practices utilise a 'pull' approach, which was the foundation for the current interest in 'lean' thinking.

This paper has identified a number of practical issues arising from the examination of past practice. In spite of the scale of the Overlord operation, it was still recognised by senior authorities as a logistical operation, first and foremost. Subsequent cases seek to continue the theme, albeit in a different context.

Notes

[1.] Wilson, H. W. *The Second World War 1939-1945. Army Administrative Planning.* London: The War Office, 1952. Ch. 5. p29.

[2.] Morgan, F. *Overture to Overlord.* London: Hodder & Stoughton, 1950. Ch. 1. p39. Morgan noted with some amusement that official orders confirming his appointment were received on April Fool's Day 1943.

[3.] Op Cit. Morgan. Ch. 2. p42.

[4.] Ibid. Ch. 2. p51.

[5.] Ibid. Ch. 2. p62. One such in-house prank was the development of Operation Overlord classified: American: Stupid, British: Most Stupid.

[6.] Op Cit. Morgan. Ch. 3. p74.

[7.] Ibid. Ch. 3. p84.

[8.] Op Cit. Wilson. Ch. 24. p143.

[9.] Op Cit. Wilson. Ch. 11. p65.

[10.] Op Cit. Morgan. Ch. 5. pp115 - 116.

[11.] Op Cit. Wilson. Ch. 5. p29.

[12.] Bryant, A. *Triumph in the West 1943-1946: Based on the diaries and autobiographical notes of Field Marshal the Viscount Alanbrooke KG OM.* London: Collins, 1959. Ch. 6. p202.

[13.] MacDonald, J. *Great Battles of World War II.* London: Guild Publishing, 1986. p136.

[14.] Ibid. p136.

[15.] Op Cit. Morgan. Ch. 9. p262.

[16.] Ibid. pp265 - 266. The Athenaeum remains one of the most exclusive clubs in London.

[17.] HMSO. *Battle Summery No. 39 Operation "Neptune" Landings in Normandy, June 1944.* London: Her Majesty's Stationary Office, 1994. Section 63. p125.

[18.] Ibid. p127.

[19.] Op Cit. Her Majesty's Stationary Office. Section 67. p139.

[20.] Ibid. p141.

[21.] Ibid. p140.

[22.] Carter, J. A. H. and Kann, D. N. *The Second World War 1939-1945 Army Maintenance in the Field, Volume II: 1943-1945.* London: The War Office, 1961. Appendix 3. p389.

[23.] Op Cit. Morgan. P267. Experiments occurred between Swansea and Ilfracombe after an initial trial across the River Thames.

[24.] Op Cit. Carter and Kann. Ch. 11. p261.

[25.] Ibid. p392.

[26] Op Cit. Her Majesty's Stationary Office. Section 63. p128.

[27] Op Cit. Carter and Kann. p261.

[28] Ibid.

[29] Ibid.

[30] Op Cit. Morgan. p267.

[31] Op Cit. Wilson. Ch. 9. p58.

[32] Op Cit. Wilson. Ch. 15. p89.

[33] Ibid. p90.

[34] van Creveld, M. *Supplying War: Logistics from Wallenstein to Patton.* Cambridge: Cambridge University press, 1977. Ch. 7. pp213 - 214.

[35] Ibid. p216.

Case 1.5 // Operation Corporate

Amphibious Logistics Support in the Falklands

Ivar Hellberg

Centre for Defence Acquisition, Cranfield University, Defence Academy of the UK.

Introduction

The Argentinean invasion of a group of disputed islands known as the Falklands in the South Atlantic brought to the British consciousness a near forgotten piece of history. The subsequent political determination and military effort to project military power to the other side of the globe within appropriate time scales to secure victory has in the subsequent two decades remained in the public psyche.

This case-study seeks to inform the reader as to the Royal Marines contribution to this campaign. More specifically it seeks to inform the student and practitioner of the problems encountered in Operation Corporate on the logistic front and provide useful facts and figures which have not be published so prominently elsewhere.

Mobilisation

The developing situation over the Falkland Islands had continued to worsen. By 1st April 1982 strong elements of the Argentinean fleet and Marine Corps had been reported at sea and it was clear to most people that an invasion of the Falklands was imminent. As Commanding Officer of the Commando Logistic Regiment Royal Marines (part of the 3rd Commando Brigade) I assumed that elements of the Royal Navy were already at sea and in the vicinity of the Falkland Islands and therefore an invasion would be resisted.

Unfortunately this was not the case and despite a reasonable period of warning, no ships (or Nuclear hunter-killer submarines!) had been pre-positioned to prevent an Argentinean landing. Thursday 1st April was a day planning speculation as far as Commando Forces were concerned, but without Government direction contingency planning was unrealistic and we were finally stood down in the evening on the basis that the crisis had been averted through diplomatic agreement. Sadly our respite was short lived as I was woken at 0400 hours on Friday 2nd April to be told that following an emergency meeting of the Cabinet, a complete Naval Task Force, including the majority of 3rd Commando Brigade was to be sent with greatest possible speed to the Falklands. Fortunately, I had not fully stood the Regiment down the previous evening and had arranged an immediate recall system as I somehow felt we had not yet heard the last of the matter. By 0700 hours the whole Regiment was in camp and the first packet of 10 x 4 ton vehicles was heading for the ammunition depot. I felt deeply sorry for the men since the whole Regiment was due to go on Easter leave that very same day – leave that was well earned after some hard exercises and, for many, a long three month deployment in Norway. But I need not have worried as the men accepted the challenge as a real test of their capabilities and set to work with terrific determination and enthusiasm.

The problem of out-loading the entire war stocks of a brigade at short notice should never be underestimated. Problems will always arise to defeat even the best laid plans. Firstly a weekend was approaching and British Rail were unable to reposition their rolling stock in time to meet any of the deadlines, it therefore meant that all the war maintenance reserve (WMR) stocks from the major depots would have to be out-loaded by road rather than the planned rail out-load. In consequence, at very short notice, HQ United Kingdom Land Forces (UKLF) had to provide a massive fleet of Royal Corps of Transport 16 ton vehicles.

Additionally we had to requisition many civilian freight vehicles. Although not planned these additional vehicles (many driven by reservist (TAVR) drivers to augment our own Transport Squadron) provided an excellent service and despite long delays in the depots probably out-loaded more quickly than the rail service could ever have managed! Another serious problem arose from the fact that normally the WMR stocks for at least one complete Commando Group should always remain afloat aboard a Royal Fleet Auxiliary (RFA) supply ship.

Unfortunately about every four years the vessel comes in for maintenance and the stock is turned over and transferred to another RFA ship. Needless to say, as luck would have it, the crisis occurred right in the middle of this transfer period! Another difficulty (which took a very long time to unravel) arose because the unit first line stocks were mixed in with the formation second line stocks within the same depot. But although the odds seemed against us the terrific willingness of all the agencies involved both civilian and military provided the determination to overcome the problems.

In many ways I was pleasantly surprised that in spite of all these snags the Regiment itself was loaded and ready to go by 1600 hours Saturday 3rd April – some thirty-six hours after the alert had been given. The loading of the initial ten days WMR stocks on to the assigned landing ships logistic (LSLs) and RFAs took rather longer but the vast majority of stocks had been loaded by 12 noon on Monday 5th April – some eighty hours after the alert had been given. Most of the amphibious task force shipping departed on the evening of the 5th April or on the morning of the 6th. Throughout the operation I was delighted (and a little surprised) by the terrific cooperation of the 'notorious' dockyard workers – they were magnificent! Once aboard the ships our fate was to a certain extent sealed by the configuration of our loading plan and also by what equipment and men we had brought and more particularly what we had left behind.

The Commando Logistic Regiment Royal Marine is designed to support the Commando Brigade at light scales in almost any environment in the world from the Arctic to the Tropics. The Regiment (which is 80% Royal Marine) had always been under established for the job it has to do, the Squadrons (Headquarters, Medical, Transport, Workshops and Ordnance) rely heavily on reinforcements to bring them to war establishment. In peacetime the strength of the Regiment numbered just over 600 men, however for this particular operation the numbers had been cut right down to 346 officers and men with only 54 prime movers and nine motor cycles. Our only reinforcements were three Surgical Support Teams (SSTs) which were mobilised to provide the vital surgical capability to the medical squadron. Prior to embarkation the force was discouraged from taking any vehicles as there were no roads on the Falkland Islands, the only exception to this rule was the Volvo BV 202 over snow vehicle which was about the only vehicle that could cope with the soft peat to be found on the Falklands. It was only after the strongest representation that I was actually allowed to take ten fuel podded 4 tonners and nine Eager Beavers; at the time I was convinced that civilian gas would become a major problem and that we would pay a serious price for our shortage of pods and jerry cans. I was also concerned that we did not have enough Eager Beavers...

At this stage I should mention that the War Maintenance Reserve (WMR) of 3 Commando Brigade consisted of a total of 30 days stocks of Combat Supplies at Limited War rates with 60 days stocks of technical and general stores. All these stocks (approximately 9,000 tons) were loaded with tremendous speed leaving behind only a small balance of stocks to be loaded onto follow-up ships.

The Journey to the Falklands

The majority of the Regiment including the RHQ, Transport, Workshops and Ordnance squadron elements left Marchwood early on 6th April aboard LSL *Sir Lancelot*. The medical squadron with SSTs 2 and 3 were fortunate enough to travel aboard SS *Canberra* while SST 1 was embarked upon HMS *Hermes*. With hindsight I suppose one could have been forgiven during those early days in believing that we would be seeing a lot of HMS

Hermes and it would be comparatively easy to wrest SST1 back from the Royal Navy once they were needed ashore in support of the land battle – how wrong we were!

Our journey down to the Ascension Islands was spent in accordance with a fairly rigorous training programme covering the following objectives:

- Detailed logistic and strategic planning.
- Physical fitness and circuit training.
- Terrain briefing (to know the Falklands backwards).
- Intelligence briefings (to know the enemy strengths, equipment and capabilities).
- First Aid.
- Weapon Training and test firings.
- Minor tactics/field craft.
- Fire control orders.
- Voice procedure.
- Mess Deck Competitions (all types).

Whilst these basic requirements were being adhered to, I was busy planning how the 3rd Commando Brigade could best be logistically supported in the Falklands. I was firmly convinced that the most effective method would be to support the Landing Force direct from the ships by helicopter (provided that we had control of the air and seas). My conclusion was that we should have two LSLs each loaded with 2 Daily Combat Supply Rates (DCSR) of stocks dedicated to Logistic Support, with SS *Canberra* providing the immediate medical support. This concept could thus easily be adjusted to support two possible options, as well as dividing our assets in case one of the LSLs should be sunk. Options were as follows (extract from Commando Logistic Regiments Concept of Operations):

Option One

Full Brigade option including three Commando Groups with two Parachute Battalions. Option to be supported by two LSLs, the first to be held close inshore to support the land battle and the second to be held in reserve outside the Total Exclusion Zone (TEZ). Each LSL to carry two DCSR with separate Command and Control Teams together with an AWD and AOD (Assault Workshop Detachment and Assault Ordnance Detachment) allocated to each ship. Empty LSLs to replenish from the dedicated Stores Ships MV *Elk* and RFA *Stromness*. BSA (Beach Support Area) ashore to be set up as soon as possible after H Hour to consist of:

1. Command and Control Element (Step up).

2. One Dressing Station Complete (DS).

3. Defence Rifle Company (Composite).

4. Assault Workshop Detachment (AWD).

5. Assault Ordnance Detachment (AOD).

6. Amphibious Beach Unit (ABU).

7. Landing Zone Marshalling Team (Helicopters).

All casualties to be evacuated direct to SS *Canberra* – Dressing Station ashore to be used only in emergency (e.g. non-flying weather).

Option Two

To support two separate operations in different areas with neither force less than one Commando Group. In this case one LSL to be allocated to support each operation carrying two DCSR stocks each (equivalent to 8 x DCSR for one Commando Group). Both LSLs also to carry DS, AWD and AOD, with separate command and control teams. BSA as above.

The above Logistic concept was put to Brigade Headquarters aboard HMS *Fearless* Saturday 10th April and accepted. This led to a complete ship re-stow in Ascension Island in order to achieve the correct split of stocks, men and machinery between the two LSLs (destined to be LSLs *Sir Galahad* and *Sir Percival*); of course the re-stow plan had also to cater for the tactical Concept of Operations. At this time the landing force (3 Commando Brigade (+)) seemed to be expanding by the hour, whilst the Logistic Regiment was still woefully under-strength to support such a force. I therefore clamoured to get the remainder of the Regiment sent out from the UK aboard HMS *Intrepid* (under my Second in Command) as soon as possible – this I finally justified by stressing the need for extra manpower to form a rifle company to defend the BSA.

Before we reached Ascension we began to experience some of the frustrations which bedevil Amphibious Operations. Firstly the Naval Task Force under FOF 1 (Rear Admiral Woodward) accelerated off at great speed taking with him HMS *Hermes* and our SST1.

Also HMS *Fearless* (carrying Brigade HQ) kept vanishing into the blue wastes on long periods of radio silence. Our stay in Ascension lasted 11 days (19 - 30 April) which was spent sweating under the hot sun, shifting cargoes from one ship to another. Ascension is not really a romantic tropical isle, it is a stark heap of volcanic rubble topped only by one mountain (Green Mountain) of any size which supports vegetation. Even the seas are uninviting as they attract millions of horrible, piranha-like fish which voraciously attack and eat anything in sight.

On 30th April we finally left Ascension Islands while diplomatic efforts at finding a settlement were still going on. The Regiment was divided between LSLs *Sir Galahad* and *Sir Percival* with Medical Squadron (less 1 Tp and SST1) still aboard SS *Canberra*. HMS *Fearless* remained behind at Ascension to await the arrival of her sister ship HMS *Intrepid* with the balance of the Regiment aboard (see Annex B).

Thus began what seemed to be a never-ending journey down to the Falklands. During this time the Navy managed to establish complete domination of the seas but seemed to be having considerable problems with the Argentinean Air Force (e.g. the sinking of HMS *Sheffield*). Also the interminable negotiations for a peaceful settlement dragged on, firstly with the USA (Secretary of State) and finally with the United Nations (Secretary General) conducting shuttle diplomacy between London and Buenos Aires. At this time the Argentineans were simply gaining time to reinforce their positions – the sinking of the cruiser *Belgrano* only increased the odds against a settlement.

Eventually the Amphibious Landing Force Fleet rendezvoused with the Naval Task Force outside the TEZ on 16th May. I was given my final orders by our Brigade Commander (Brigadier Julian Thompson) aboard HMS *Fearless* on 17th May which included all the details for the Amphibious Landings including our choice of Ajax Bay for the BSA. He seemed convinced that Operation Sutton (as the Land Battle was to be called) was now inevitable. On 18th May at 1600 hours I gave my final Regimental Orders with all the details of the plan, it only remained to know when 'D' Day and 'H' Hour would be. We did not have long to wait: at 1100 hours on 20th May we crossed the TEZ and 'H' Hour was confirmed to be May 21st at 0630 hours, 1982.

Operation Sutton

The initial landings were brilliantly successful. Complete surprise was achieved and the 'OPGEN MIKE' (detailed plan for landing craft and helicopter timings etc) worked like clockwork. All the main objectives had been achieved by first light on 21st May and all the fighting troops were ashore within 4.5 hours of 'H' Hour. The SBS/SAS diversionary' attacks on Fanning Head and Goose Green together with the Harrier attacks on Port Stanley and Fox Bay seemed to have paid off.

However, the Argentinean Air Force soon made its presence felt. The whole fleet, now tied up in San Carlos water, was subjected to a great number of air attacks. The Harrier Combat Air Patrol simply could not prevent all the attacks from getting through. HMS

Antrim and *Argonaut* were hit and HMS *Ardent* sunk; in return the enemy were seen to have lost 9 Mirage, 5 Sky Hawks and 3 Puccara. On land we took only light casualties with 2 Gazelle helicopters shot down and a total of 4 killed and 28 wounded.

The major conclusion of the day was that all non-essential ships would have to be moved outside the TEZ as soon as possible under cover of darkness to avoid risk of air attack. This included the prestige targets of SS *Canberra* and *Nordland*. The effect of this short notice decision was devastating to Medical Squadron, regrettably there simply wasn't enough time to get everyone off *Canberra* so one complete dressing station (DS2) and SST 3 were left on board and not seen again until 1 June. Also, the LSL support plan was abandoned and LSLs *Sir Percival* and *Sir Galahad* were ordered to run their stocks ashore into the BSA at Red Beach (Ajax Bay) and depart out of danger. MV *Elk*, our other supply ship, was ordered to stay out of the TEZ as the detonation of 2,000 tons of HE might take all the other ships with it!

In consequence all our hard planning had to be changed and our logistic plan was forced to become land based. Hence terrific pressure was applied to get men, equipment and stocks ashore into Ajax Bay (and the old sheep refrigeration plant) as soon as possible. The main danger was that the ground restricted our ability to disperse stocks adequately for which we were to pay a high price later on.

From a logistical point of view our main problem was the lack of dedicated movement sets. Normally Commando Logistic Regiment can depend on its own 4 ton trucks to carry supplies forward to the fighting units and therefore helicopters and boats were just a bonus. However on this operation we were entirely dependent upon the availability of helicopters and landing craft as there were no roads or hard core tracks that we could use. Regrettably the system did not work; we did not control any single movement asset (not even a rubber boat) and every request had to be submitted through Brigade (now ashore) Commander Air Warfare (still aboard HMS *Fearless*) for tasking. Our requests (being of a logistic nature) were usually placed fairly low on the priority table and consequently with the shortage of helicopters almost no re-supply was achieved during the first few days.

On 24th May LSL *Sir Galahad* and *Sir Lancelot* were hit by unexploded bombs and HMS *Antelope* was sunk. This obviously further confirmed that LSLs (or any other ship) were unsuitable to forwarded supply without adequate air superiority. It also confirmed that forward re-supply would have to be by helicopter – sometimes at considerable range. One important factor did not fully register upon me until I got ashore, and that was the size of the Falkland Islands. They are about the size of Wales with long re-supply distances involved (e.g. 93 miles direct from Ajax Bay to Port Stanley). The ground was terrible, very wet tussock and bog with virtually no tracks. Any idea of a quick campaign concluded in about one week was immediately dispelled!

At about this time we managed to lose contact with about 150 of our men who were aboard LSL *Sir Galahad*, including OC Workshops Squadron. Although we knew that *Galahad* had been evacuated because of the unexploded bomb, we simply had no idea which ship they had been taken to and, since we had no means of communicating with the ships, we had no way of finding out! To make matters worse our 'control ship' HMS *Fearless* disappeared again on a mission to pick up Commander Land Forces Falkland Islands' (CLFFI) staff somewhere near South Georgia. On the 25th May HMS *Coventry* was hit and sunk and Medical Squadron's Main Dressing Station (MDS) was very busy dealing with air-raid casualties.

Additionally the Regiment took on the unplanned role of catering for the many displaced Naval personnel ashore – at this stage the crews of *Lancelot* and a number from *Galahad* and *Coventry*. Tents were in very short supply so these people also had to be herded into our one and only building – the old disused sheep refrigeration plant at Ajax Bay (being also used by Medical Squadron).

There can be no doubt that, even in a limited war of this nature, plans have to be changed by the minute and success or failure (even with logistics) often rests with the initiative and determination of some of our youngest and most inexperienced Non Commissioned Officers. One of our greatest difficulties revolved around the re-supply POL (Petrol, Oil and Lubricant) – civilian gas in particular.

The Regiment's Petrol Troop (383 Troop) was Territorial Army and therefore couldn't be mobilised, in consequence, other men from Ordnance and Transport Squadrons had to take their place to become instance 'experts' in fuel of all natures. We found the demand for civilian gas extraordinarily high, in particular the Rapier posts on the top of the hills (our first priority) and unit generators seemed never satisfied. Also the Volvo BVs and raiding craft consumed vast quantities of jerry cans. The problem was not so much a shortage of bulk fuel as enough was available on the ships; the real difficulty was getting the fuel ashore (mostly achieved by PODs floated on the mexe floats) and then transferring the fuel by hand pumps into a very limited number of jerry cans.

Whatever the theoretical exponents of bulk re-fuelling may say, I am now absolutely convinced that in any war situation forwarded units will be screaming for jerry cans. The principle of exchanging a full can for an empty one is good, but in a fast moving battle you cannot expect to guarantee the return of empty jerry cans and you certainly cannot deny them a full can simply because they cannot produce an empty. Our experiments with large flexible tanks were not very satisfactory as they were very vulnerable to aircraft strafing and, for many reasons, the only bulk fuel carried ashore were in the PODs and the APFCs (Air Portable Fuel Containers). The only EFHE (Dracone system) established was AVCAT for the Harriers and helicopters based at Green Beach (Port San Carlos), but even here the problem of setting up such a system without air superiority was both lengthy and dangerous. Just for the record, our civilian gas re-supply for the entire Brigade at Red Beach was run in the early stages by a private solider and 5 assistants working around the clock with minimal supervision.

During 26th and 27th May the Argentinean air attacks continued and despite the horrifying losses their aircraft were taking (it was later confirmed even by the Argentineans that one aircraft out of every two never returned to Argentina). In my view their pilots were very brave to press home their attacks at such low level and with such determination. On 26th May the container ship *Atlantic Conveyor* was hit and sunk by Exocet, tragically she was just about to arrive from the UK carrying our logistic Chinook and other support helicopters.

Then on 27th May at 1930 hours the Regiment received its most severe air attack of the war. The BSA at Ajax Bay was attacked by Sky Hawk and Mirage with virtually no warning, 12 x 400kg retard bombs were dropped, only four of which mercifully exploded. One of the bombs exploded in the area of the Regimental Galley and the echelon of 45 Commando killing 6 men and seriously wounding 26 others – had this happened half an hour earlier the Galley would have been full and there would have been far more casualties. Another bomb exploded amongst the ammunition at the HLS destroying approximately 300 rounds of 105mm and 200 rounds of 81mm, this ammunition together with 45 echelon ammunition went on exploding all night. Three unexploded bombs hit the Medical Squadron Main Dressing Station in the Refrigeration factory one going straight through and two lodging in the building itself – had these exploded then half the Regiment would have been killed. Throughout the raid there were actual operations being performed on the wounded in the Dressing Station: and the daily (AQ) logistic conference was just dispersing from the old factory.

Despite the appalling carnage, shock and sorrow, most of the men got straight on with rescuing the wounded away from the exploding ammunition and attempting to put out the fires. To a few, with undue imagination, the horror of the scene and wanton loss of life created a state of numb shock which lasted for quite some time. One of our men was killed by enemy cannon fire whilst engaging a Sky Hawk with his GPMG at point blank range. One thing is quite clear to all of us at Ajax Bay and that is, even with Rapier, we are woefully short of effective air defence weapons. The Regiment had a very large number of General Purpose Machine Guns which we loaded with 1 in 2 tracer; these guns were very useful in distracting the enemy pilots' aim – even occasionally bringing him down! However, what is clearly needed is a few 20mm AA guns (e.g. Rheinmettal systems) preferably with radar control guidance but firing tracer. Rapier is no substitute for these weapons, as has clearly been related to me by Argentinean Pilots who came to stay in our Prisoner-of-War camp.

I believe logistic units everywhere should have these weapons and learn to fire them themselves – they will definitely prove more cost effective in the saving of men and stocks in the long run. Rapier is very effective but usually only strikes the attacking aircraft after it has attacked its target. Another matter which saved many lives was the infantry training of all the men in the Logistic Regiment. Not only can they fight as part of a rifle section/ troop should the need arise, they also know how to dig in effectively with a good two feet of overheard cover – without this our casualty toll would probably have been three times as high.

On 28th May, 2 Parachute Battalion unleashed their attack on Goose Green and Darwin. It was a vicious and hard fought battle which set the tone of things to come and undoubtedly had a terrifying effect on the Argentineans. The Argentinean forces in the area surrendered on the morning of 29th May, approximately 1,400 prisoners were taken and over 200 Argentinean dead lay on the battle field. 2 Parachute Battalion lost 18 dead with 53 wounded. From a logistic point of view the ammunition rates of fire were incredibly high, for example 105mm HE actually ran out at one point. Generally speaking approximately 4 DAER (Daily Ammunition Expenditure Rate) was expended in 24 hours at limited user rates – 5 DAER in the case of 105mm High Explosive and 81 mm Mortar. Clearly we would have had great re-supply problems had battle been sustained at this intensity for any further length of time.

At this point I feel that a word or two about the press would not go amiss. As is now common knowledge, our designs on Goose Green were made known to the world by the media 24 hours before the attack went in. It is a fact that as a result of this knowledge the Argentineans reinforced Goose Green/Darwin with 12 Infantry Regiment. It is also a fact that the media told the world (and of course the Argentineans) that their aircraft bombs were not detonating; they consequently made great efforts to overcome this failing. When *Galahad* was hit the world was told of this fact long before anyone had a chance of notifying families/next of kin of the casualties, thus causing great alarm amongst all families who thought their men might be on board (in this particular case the whole of the Commando Logistics Regiment, Royal Marine). What saddens me is that some members of the press/ BBC were very responsible and were equally horrified by these blatant breaches of security and lack of consideration or understanding. The media enjoyed privileged information, but this must be bound by responsibility; equally I believe that the Ministry of Defence Press Office must be held responsible for some major indiscretions. The media may consider that the public have an unrestricted "right to know", but this cannot be allowed at the expense of men's lives.

During the next few days the Brigade started to move forward and take the outlying settlements. 45 Commando and 3 Parachute Battalion took out Douglas and Teal settlements without opposition. At the end of all these actions the Regiment found itself with another sometimes forgotten responsibility – that of burials. Most of our own dead were brought back to Ajax Bay for burial and we prepared the graves and very often arranged a service and provided the bearer parties. We held a particularly special service for our own dead on 27th May taken by our own Chaplain. It was a sad, yet moving, occasion, however it is my perception that it caused many people to think more deeply about a personal faith.

As the Brigade moved forward into the area of Mount Kent, we followed up by creating a forward BMA (Brigade Maintenance Area) at Teal settlement (commanded by OC Transport Squadron) with a DP forward near Estancia House. The plan was once again to move large quantities of supplies (particularly 105mm and 81 mm) forward to Teal by LSL supported by helicopters for immediate requirements. Supplies from Teal could then be moved forward to Estancia either by boat (LCVP or Rigid Raider), by Volvo BV202 or helicopter. Unit Echelons could then pick up direct from the DP (Delivery Point) or the forward BMA as they would be co-located at either one of these locations. The system worked well, although once again the helicopter availability caused shortages of some items, particularly packed civilian gas.

Our logistic communications worked extraordinarily well throughout the operation even over these extended distances and the Clansman VRC321/PRC32O sets proved their worth as well as the expert determination of our Signals Troop.

The 1st and 2nd June saw the arrival of HQ LFFI (with Major General Jeremy Moore MC* (CLFFI) and his staff aboard HMS *Fearless*) together with 5 Infantry Brigade (1st Battalion Scots Guards, 1 WG and 7 Gurkha Rifles). Regretfully, 5 Brigade brought very little logistic support with them and Commando Logistic Regiment was directed to become 'Divisional Troops' and take on logistic re-supply for the whole division including 5 Brigade. To help us we were given elements of 81 and 91 Ordnance Company under operational command; unfortunately, although they came with enough stock they brought with them virtually no extra movement assets (e.g. podded 4 ton vehicles, Eager Beavers, Landing Craft or logistic support helicopters).

In order to support 5 Brigade (who were ordered to move forward to Fitzroy) we established a forward BMA at Fitzroy with a DP at Bluff Cove to support their Echelons. The support of this Brigade did not go as well as we had hoped as their logistic communications and procedures were as well tested as 3 Commando Brigade. Nevertheless we simply made the logistics work even if it meant 'skyjacking' helicopters. Just as we were setting up the BAA at Fitzroy (on 8 June) the two LSLs (*Galahad* and *Tristram*) were attacked by Sky Hawks and set on fire, both vessels were abandoned and resulted in tragic loss of life on *Galahad* (43 killed and over 200 injured – mostly from the Welsh Guards). Also most of the surgical equipment of 16 Field Ambulance was destroyed on the *Galahad*. Once again a terrible price was paid for our lack of air defence equipment – in this case even the Rapier Posts had not had enough time to be properly established. Our medical squadron MDS at Ajax Bay was inundated with all the injured (mostly with burns and in great pain) since the MDS of 16 Field Ambulance had effectively been destroyed. Fortunately by this time the hospital ship *Uganda* was permitted to come in daily to Grantham Sound and was soon able to absorb the overcrowding. Virtually all the men in the Regiment were involved with the wounded at this time – often simply by sitting beside the stretchers and comforting those fearful pain. The men were magnificent.

At this point I must make mention of Medical Squadron (MDS) which was based in the A refrigeration plant at Ajax Bay. The teamwork achieved by the 'Red and Green Life machine' was fantastic and a great many friends and foes alike owe their lives to the dedication of the skilled surgeons and lowly MAs. It is also a confident opinion of doctors and surgeons that without the very high standards of first aid practiced by the men at the time of wounding in the front line (buddy-buddy system) then once again many more would never have reached the MDS alive.

Once the men reached the MDS alive then their chances of survival were exceptionally high as virtually all the wounded are both very fit and determined to survive. Out of just over 1,000 casualties received the MDS and the two Forward Dressing Stations (FDSs) (Teal and Fitzroy) only 3 subsequently died. In all 202 major operations were performed at Ajax Bay (MDS) and altogether 108 such operations were performed at Teal, Fitzroy and SS *Uganda*. Of these figures approximately 30% of the casualties treated were Argentinean.

Another group of people that were working for the Regiment (although not always under command!) were the mexe floats and landing craft. There is no doubt that the vast capacity of the mexe float was of enormous value for off-loading supplies. The RCT operators are to be congratulated on their standard of seamanship and their cheerfulness in the face of appalling weather on exposed decks. Also the LCMs of Landing Craft Squadron RM were tremendously useful and could carry a really worthwhile load. These craft, together with LCVPs, Rigid Raiders and Geminis (Raiding Squadron) were absolutely invaluable. Nevertheless at least a proportion of these craft should have been placed under the operational command of Commando Logistics Regiment, Royal Marines.

The final plan for the capture of Port Stanley involved a 3-phase assault. 3 Commando Brigade was to take Mount Longdon, Harriet and 2 Sisters early in the morning of Saturday 11th June, followed by a 5 Infantry Brigade attack on Mount William and Tumbledown. Wireless Ridge was then to be taken by 2 Parachute Battalion as a preliminary to the final assault on Sapper Hill and Port Stanley by 3 Commando Brigade, Royal Marines. From a logistics point of view it meant that at first line every man would have to carry two day's supply of ammunition and food and all vehicle fuel tanks had to be full. Guns and mortars were to have 500 rounds per gun actually on the gun lines prior to each phase, with the ability to re-supply a further 500 rounds straight from the BMAs at Teal and Fitzroy. The plan was completely successful and only two phases were necessary before the Argentineans decided to surrender. Although 500 rounds per gun a day sounds a lot of ammunition it should be noted that some guns and mortars did actually run out (as also happened at Goose Green). Other ammunition items which were used far in excess of their theoretical scales were 7.62 link and tracer, L2 grenades, 66mm High Explosive Anti-Tank, 84mm HEAT and MILAN. Another item constantly asked for but not held was 2" mortar High Explosive (maybe this should be looked at again for future operations). I have listed in the Annex some logistic statistics for further information.

The final surrender of the Argentineans took place during the evening of Monday 14th June (this also included their surrender of West Falkland). The Regiment had to move round from Ajax Bay to Port Stanley as soon as possible to support the force now concentrated in that area. We took up residence in the offices of the Falkland Islands Company.

We left behind a DP at Ajax Bay to look after 40 Commando and other units in the area, also we had to leave behind a large guarding party and administrative facilities to look after the 500 'special category' Prisoners-of-War (mostly officers). Although we attempted to look after these prisoners as best we could (and certainly they were more comfortable than my own men since they were inside out of the cold). I received a most remarkable 'document of redress' made out to me as 'Commanding Officer of the 'Concentration Camp'. The document was drawn up by one of their very senior officers, citing many trivial complaints under the terms of the Geneva Convention. It was clear to me that the Argentinean officers were not used to any discomfort and certainly they avoided mixing with their men.

What particularly interested me from various informal discussions I had with some of their most senior commanders was their deep regret that this invasion of the Falklands had ever taken place! They believed that they had now totally alienated the inhabitants of the Falklands for at least the next generation (certainly true!). They also believed that they would not repeat the adventure again for a long time! To a man they blamed the Americans for not discouraging the invasion; they maintained that the Americans knew about their intentions well in advance and did nothing positive to stop them – this they took as a sign of encouragement.

Once in Port Stanley the Regiment spent a lot of time helping to clear up the most appalling mess left by the Argentinean troops. Their troops (all 11,000 of them) were confined to the airfield area until ships became available to take them back to Argentina.

On the 28th June the Regiment finally departed from Port Stanley aboard LSL *Sir Percival*. The logistic support responsibility for the garrison force had been handed over to 81/89 ordnance company ROAC, 10 Field Workshop REME and one RCT Transport Troop. Sadly we had to leave behind much of our equipment to allow them to fulfil their function properly (e.g. podded fuel vehicles and Eager Beavers).

Our return journey was very well arranged, LSL *Sir Percival* made us most welcome as far as the Ascension Islands (apart from a devastating force 12 storm just after leaving the Falklands) then most of the Regiment took a VC 10 flight from Ascension to Brize Norton via Dakar (Senegal). Medical Squadron once again managed to squeeze a voyage back on the SS *Canberra*. Our reception back into England on 9th July was overwhelming and most exciting – the subsequent receptions and memorial services served only to conclude what had been a remarkable operational experience.

Conclusions

Rather than finish this account by drawing 'conclusions and recommendations' from what has been in effect our first full-scale amphibious operation and limited war for at least 30 years, I have placed the 'logistic problems and recommendations' in the annex. I can now only hope that the will can be found and the money made available to rectify some of the main deficiencies and flaws in our planning and equipment. Nevertheless it is equally clear that without adequate shipping, helicopters and air power the operation could not have been mounted or sustained. It was a close run thing and even a year later may not have been able to contemplate the operation from a logistic point of view. We must now seek to replace the LSLs that have been lost as they were critical to the logistic success of the operation. Finally I must pay tribute to the men who were fantastic. Without their tenacity and enthusiasm nothing would have been possible; I am deeply indebted to their support, as well as all those who supported us back home – in particular I would remember our long suffering families.

Annex

The Royal Marine Commando Logistics Regiment

The peacetime establishment of this formation differed somewhat as to the war establishment formation sent to the South Atlantic as part of Operation Corporate. Table 1.5.1 shows a comparison in raw personnel terms:

Table 1.5.1: Peacetime versus war establishment figures for RM Logistics Regiment.

Squadrons	Peacetime Establishment		Wartime Establishment	
	Enlisted	Officer	Enlisted	Officer
Head Quarters	101	9	135	10
Medical	103	6	258	30
Transportation	113	4	211	6
Workshop			174	5
Ordnance	82	5	114	8
Regimental Total	399	24	892	59

The logistics regiment itself was composed of 5 main squadrons as indicated in the table above. Within the squadrons specialist detachments existed to provide specific services such as casualty evacuation. The personnel strength of the regiment increased in stages. Initially on the support ship Sir Lancelot there were a total of 256 persons on the 4th April 1982. Additional reinforcements on other vessels totalled 261. By the time of the landings at Ajax bay in the Falkland Islands the figure totalled 673.

In terms of the volumes of matériel moved some 17,000 tons of stock had to be moved from the United Kingdom of which ammunition composed some 8,600 tons or some 50%. Within the Falkland Islands themselves movements of supplies can be broken down as follows in table 1.5.2 (overleaf):

Table 1.5.2: Ammunition, Rations and Miscellaneous supply movements

Ammunition	Weight (tons)
Ajax Bay unloading from shipping	3,500
To Teal from Ajax Bay	1,200
To Fitzroy from Ajax Bay	1,000
To DP (Estancia House) from Teal	36
Rations	
Total rations movements	1,200
Hexemine (stove fuel tablets)	300
Biscuits	90
Petrol, Oil, Lubricants (POL)	**1,414**
Defence stores	280
Technical general stores	60
Grand Total	**9,080 tons**

Summary of Logistics Difficulties and Recommendations by the Royal Marines Commando Logistic Regiment

Initial Deployment Considerations

- Firstly, outloading by use of the railway was impractical. British Rail were unable to pre-position rolling stock in sufficient time for rapid deployment.

- Secondly, there was an abundance of vehicle delays at depots and poor administrative processes for drivers.

- Thirdly, the volunteer Royal Marine logistic contingent could not be mobilised. The regular formation departed without any Petrol, Oil & Lubricants expertise.

84

Operational Lessons and Recommendations

- Logistics helicopters allocated to the logistics regiment needed to be placed under the operational command of the regiment from the beginning. This was essential where there was a lack of developed infrastructure otherwise timely delivery of supplies could not be guaranteed.

- The on-loading and off-loading of supply and support shipping needs to be controlled by the logistics regiment in order that the goods carefully packed arrive in the correct order and availability.

- Jerry can and petrol distribution infrastructure holdings proved insufficient. A problem which has dogged modern military operations such as the Overlord landing in World War II. A policy for flexible storage of fuel needs to be reviewed.

- 'Eager Beaver' support vehicles performed magnificently but more could have been used.

- Logistic landing craft were essential in the amphibious landing and losses must be replaced.

- Ammunition requirements need to be reviewed. Especially small arms ammunition (linked and tracer), grenades, anti-tank rounds guided missiles and mortar ammunition.

- Ammunition levels for limited war suggested for 3 Commando Brigade, Royal Marines are unrealistically low.

- Logistic support units require better air defence weapons. Rapier missile batteries need to be complemented by good anti-aircraft guns of 20mm calibre at least.

- Skills training necessary for logistic troops. Especially in the areas of camouflage, field craft, defence, patrolling and use of ground to disperse stocks.

- Dedicated Chaplain required for the regiment.

• Organisation of the regiment needs to be considered in order to handle large numbers of captured prisoners.

Conclusions

The 1982 Falklands War provided many lessons for the armed services, which were becoming optimised to fight a potential conflict against the Warsaw Pact in Europe. For the Royal Marines who reacted magnificently to the challenges of the arduous campaign this case-study has sought to draw out important points regarding the structure and organisation of their formation for providing logistics support to the Commando Brigades in the future.

Bibliography and further reading

Brown, D. *The Royal Navy and the Falklands War*. Barnsley: Pen & Sword Books Ltd, 1987.

Calvert, P. *The Falklands Crises: The Rights and Wrongs*. London: Continuum International Publishing, 1982.

Carrington, P. *Reflect in Things Past: The Memoirs of Lord Carrington*. Glasgow: Collins, Reprint edition, 1988.

Coll, A. & Arend, A. (eds.). *The Falklands War: Lessons for Strategy, Diplomacy, and International law*. London: HarperCollins Publishers Ltd, 1985.

Dillon, G. *The Falklands, Politics and War*. Basingstoke: Palgrave Macmillan, 1989.

Freedman, L. *Britain and the Falklands war*. Oxford: Basil Blackwell, 1988.

Freedman, L. & Gamba-Stonehouse, V. *Signals of War: The Falklands Conflict of 1982*. London: Faber & Faber, 1990.

Gamba, V. *The Falklands/Malvinas War: A Model for North-South Crisis Prevention*. London: HarperCollins Publishers Ltd, 1987.

Hastings, M. & Jenkins, S. *The Battle for the Falklands*. Suffolk: Richard Clay, 1983.

Her Majesty's Stationery Office. Cmnd. 8758 *The Falklands Campaign: The Lessons*. London: HMSO, 1982.

House of Commons Defence Committee. HC345-1 *Implementing the Lessons of the Falklands Campaign*. London: HMSO, 1987.

House of Commons Defence Committee. HC345-2 *Implementing the Lessons of the Falklands Campaign*. London: HMSO, 1987.

Case 1.6 // Operation Granby

Logistics Support during the Gulf War

David M. Moore and Peter D. Antill
Centre for Defence Acquisition, Cranfield University, Defence Academy of the UK.

Jeffrey P. Bradford
Defence and National Security Consultant

Introduction

This case-study discusses the impact of logistics on military operations and seeks to examine the relatively recent conflict in the Persian Gulf between Iraq and the United Nations coalition in 1991. The case will seek to evaluate the importance of logistics in determining the force structure deployed and strategy employed in the liberation of Kuwait during February 1991. The case will examine the role of logistics by applying what have been regarded as the five essential principles of successful logistics; foresight, economy, flexibility, simplicity and co-operation.[1]

The Invasion of Kuwait (August 1990)

Operation Desert Sword, the ground war to liberate Kuwait, was the culmination of a multi-stage military plan set in motion following the invasion of Kuwait by Iraq in August 1990. The invasion itself was the result of a dispute between the two countries the origins of which go back generations but were fuelled by the dispute over the Rumaila oil field. Iraq claimed that Kuwait was pumping more oil from the field than it was entitled to and that it was ignoring the Organization of the Petroleum Exporting Countries (OPEC) production quotas set in 1980 and selling more than it should on the open market thus lowering the price of oil and Iraq's income. Iraq also had historic claims to the land, claiming it was once part of the Basra Province and should have been made part of Iraq in 1932.

Things came to a head when the Iraqis demanded $16.5 billion in compensation (for the illegal pumping and loss of revenue) and cancellation of the $12 billion debt that Iraq had run up in the war with Iran. The Kuwaitis refused. As occupying Kuwait would wipe out much of its debt, provide extra funds and control of major oil reserves, Iraq invaded.

The Iraqi invasion was preceded by a military build-up on its border with Kuwait in a show of force. This was detected early on in the crisis, but the intentions of Iraq were cloaked by diplomatic moves, which led other Arab leaders to believe that one Arab country would not fight another.[2] The attack itself was spearheaded by three divisions of the elite Iraqi Republican Guards, the best equipped, trained and politically reliable of the Armed Forces. These divisions were compared psychologically by Danis (1991) directly with Napoleon's Old Guard.[3] At 02:00 a.m. local time the attackers crossed the border, rapidly overwhelming the Kuwaiti Defence Forces, in an attack which captured the capital, Kuwait City in under twenty-four hours. Whilst a success, the failure to capture the Royal Family which escaped to Saudi Arabia was a fatal flaw, as it left an organised source of legitimate opposition to the aggression, out of reach of the Iraqi leadership.

Following the invasion, regular Army divisions entered Kuwait, and moved to the southern border with Saudi Arabia where they preceded to erect defensive positions, a task at which the Iraqi Army excelled after spending the majority of its war with Iran on the defensive. This reflected Iraq's strategic culture, aiming to defeat attempts at re-taking Kuwait by preparing to fight a defensive war of attrition, along the lines of the eight year war with Iran. The Republican Guards were withdrawn to the Kuwaiti border with Iraq, where they formed the core of the theatre reserve, available to perform operations elsewhere prompting American concerns of a further attack on the oil fields of Saudi Arabia.[4]

The Political Response

The unique nature of the international environment at this time, facilitated the wishes of the United States and its European allies to come to the aid of Kuwait. The end of the Cold War enabled a series of United Nations Security Council resolutions to be passed facilitating "the greatest military deployment and logistical effort since World War II, followed by the most rapid and decisive victory of its scale."[5]

Desert Shield and the Logistics Effort (Aug. 1990 – Jan 1991)

These resolutions enabled the commencement of Operation Desert Shield on the 7th August 1990. Cold War American contingency planning for combating Soviet military operations in the Middle East had recently been updated to take account of the changes in the international environment. The revised plan had been designated Operational Plan (draft) 1002-90 and had been tested in a command post exercise only a few weeks before the crisis. With the Iraqi invasion, the crisis USCENTCOM had organised, trained and planned for had come, and once King Fahd of Saudi Arabia had asked for assistance, it was rapidly implemented. The deployment, code-named Operation Desert Shield began on the 7th August, and as well as the rapid deployment of American troops to the Persian Gulf it sought to build a coalition to enforce the United Nations sanctions placed upon Iraq.[6]

The paratroopers of the 82nd Airborne Division started arriving on the 8th, with elements of the 101st Air Assault and 24th Mechanised Divisions leaving CONUS for Saudi Arabia by the 13th August and the forward elements of the 1st Marine Expeditionary Force (the 1st Marine Division plus its associated air-wing) arriving in the peninsula on the 12th August. These units had completed deployment by September 20th. From the end of September until he end of January, saw a massive build-up of military resources by a number of states. This is underlined by a number of impressive statistics. In the first forty-five days of Desert Shield, more American personnel were deployed than in the first year of the Vietnam commitment.

The American military airlift in the eight months of the campaign delivered nearly six times more supplies than the entire 1948 – 1949 Berlin Airlift.[7] Logistical requirements included some 1.44 million meals, 34 million gallons of water and 1.7 billion barrels of petrol, oil and lubricants per day.[8] The United States Armed Forces utilised not only their military airlift capability, to reinforce their units in the Persian Gulf, but also resorted to commercial carriers. Figure 1.6.1 below shows the percentage cargo carried by aircraft type throughout the crisis.[9]

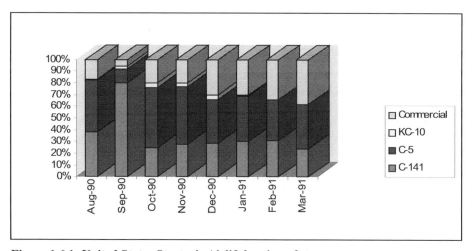

Figure 1.6.1: United States Strategic Airlift by aircraft type.

As can be clearly observed from an analysis from the table that in spite of the threat of conflict, commercial aircraft provided between a fifth and a quarter of the total airlift during the months of preparation and the campaign itself. The British airlift requirement as part of Operation Granby, was supplemented by sorties from other NATO countries as well as obtaining significant support from the United States. The latter support was necessary for the larger helicopters which needed shipment to Saudi Arabia.[10]

It is important to note however, that airlift is expensive and certain cargoes such as main battle tanks (which can weigh up to sixty tons) cannot be carried efficiently by this mode of transport. Sea-lift was, as ever, a vital medium for transporting these cargoes to the Theatre of Operations. Figure 1.6.2 below shows a month-by-month analysis of US deliveries by sea.[11]

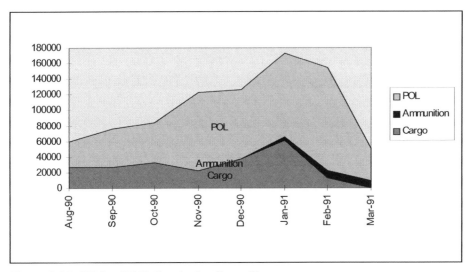

Figure 1.6.2: US Sea-lift Deliveries by Cargo Type

When considering the table above, it is important to note the requirement for Petrol, Oil and Lubricants (POL). This could be seen as a rather surprising requirement, seeing as the conflict was in the most oil-rich region on Earth. But, when one considers the sophisticated nature of the products in question and the need to manufacture them at specialist facilities which are located in the oil consuming nations, the requirement becomes clear.

For the British Armed Forces, sea-lift was not so plentiful. The House of Commons Defence Committee noted that "the sea-lift was heavily dependant in some categories of ships, from a handful of countries".[12] Furthermore, it was noted that there was a distinct reluctance on the part of shipping companies to disrupt their commercial activity to assist in fulfilling the transportation requirements.

However, it is important to realise that this logistical build-up and deployment of forces was allowed to proceed uninhibited by the Iraqi Armed Forces. Such a deployment to a distant country's air and sea ports while being free from interdiction and interception should be viewed as a unique situation, which is unlikely to be repeated in the future.

Operation Desert Storm: The Air Campaign (January – February 1991)

Following the expiry of the United Nation's deadline for Iraq to vacate its self-declared '19th Province', the coalition moved from a defensive stance to an offensive one. On January 17th at 02:38 a.m. (local time), the air campaign, known as Operation Desert Storm commenced. The plan comprised four distinct phases. The first, envisaged a strategic air campaign aimed at twelve different types of target including those facilitating Iraq's ability to support and communicate with it's occupation force in the Kuwait Theatre of Operations, Iraqi infrastructure and its military/industrial complex. The first phase lasted some six days. Phase two sought to gain air supremacy over the Kuwait Theatre, an enterprise which took a mere day. The third and longest phase involved the preparation of the battlefield, by degrading Iraqi formations in the Theatre. The final phase was aimed at supporting the eventual ground offensive.[13]

To impart some appreciation of the scope of this endeavour, some 2,700 aircraft of several nations flew some 130,000 sorties over forty three days, expending 87,679 tons of air-to-ground munitions and consuming 1,953 million litres of aviation fuel (averaging 45.42 million litres a day).[14] This contrasts with the air effort to assist the Allied forces break out of the Normandy beach head in 1944, which involved the use of 1500 bombers dropping 3,400 tons of munitions.[15] Iraq once more declined to respond directly to the air attacks, and offered only token resistance at best, before the Allies rapidly whittled down the Air Force's ability to detect and respond to Allied incursions. Iraq eventually sent the majority of the most modern components of its Air Force to seek sanctuary in Iran, in an attempt to preserve them.

Iraq's response to the Coalition was to resort to using Soviet supplied surface-to-surface missiles, which Iraq had re-engineered, sacrificing explosive payload for increased range. Known as Scuds (or locally as Al' Hussayns), over eighty of these weapons were fired from hard to locate platforms in the twenty-eight thousand square mile area of Iraq, against the capitals of Israel and Saudi Arabia.[16] These weapons, it was hoped, would draw Israel into the war with Iraq, enlarge the dispute into a pan-Arab Jihad, and fracture the Arab component of the Coalition. This event however, did not happen. The United States used diplomatic pressure to ensure Israeli restraint, and many Coalition aircraft were reassigned to carry out what euphemistically became known as 'scud-busting'.

During this period however, more crucial developments took place. The main components of Coalition military strength were moved westwards in great secrecy, to an area opposite which the defensive positions of the Iraqi Army were lighter, and where the Coalition could exploit its decisive advantage in technology and manoeuvrability. Iraq's decision to lightly defend the western area was derived from a historical reputation of the area in swallowing armies, due to featureless nature of the terrain. For the Coalition forces with advanced satellite navigation equipment, known as GPS (Global Positioning System), they could use this error of perception to their best advantage.[17]

By the time the Allies were ready to commence Operation Desert Sword in mid-February, they had cut off the Kuwaiti army of occupation, severing its lines of communication and logistics. The Coalition had also utilised deception techniques to move two corps formations (VII and XVIII) 402 and 204 kilometres west, respectively.[18]

The multi-national forces were arrayed in five organisational areas from west to east; XVIII Corps (comprising American and French mobile forces), VII Corps (comprising the heavy armoured formations of the USA and UK), Joint Force Command - North (JFC - North) which comprised Arab coalition members which were to attack Iraqi units inside Kuwait itself, avoiding the charge of directly attacking another Arab state, Marine Forces Central Command (MARCENT) which by now consisted of the 1st and 2nd Marine Divisions, and Joint Force Command - East (JFC - East) comprising Kuwaiti, Saudi and Qatari forces.

The missions assigned to the five formations were as follows. The XVIII Airborne Corps in the far west was to fly deep into Iraq and seize control of the strategic Highway 8. Capture of this road would cut communications between Baghdad and Kuwait, preventing the dispatch of reinforcements to the beleaguered 19th Province. The VIIth Corps was detailed to march north and then east along the Wadi-al-Batin, and to seek, engage and destroy any elements of the Republican Guard Force Command (RGFC) that were encountered. Given the similarities of language, both culturally through English and professionally through NATO, common logistic requirements and tradition, this most demanding task was assigned to the US and British force.

JFC - North, East and MARCENT were organised to attack the Iraqi forces entrenched in Kuwait by frontal assault. This move would aid in the deception planning (the movement of two corps westwards) and would prevent Iraq in using its forces to cut the coalition lines of supply.[19]

It must be noted however, the role of logistics in determining the operational possibilities to the Commander-in-Chief of Central Command. Initial planning for the liberation was centred upon a full-scale frontal assault across the defensive barriers constructed by the Iraqis on the border of Saudi Arabia and Kuwait. This is a good example of the logistics being insufficient to allow flexibility in planning a modern campaign. Following political disquiet as to the risks involved, the military made their case for a much larger force structure in the Middle East in order to engage in a more mobile activity. They were able to get permission to engage in a logistical build-up sufficient to mount a more mobile offensive into Iraq to destroy key units and restore Kuwaiti sovereignty.[20]

Operation Desert Sword: The Ground Offensive G Day: Sunday 24th February 1991

The analysis of Operation Desert Sword will be structured as follows. For each day of the conflict, the case-study will firstly provide a brief summary of events. Facing the Coalition on the eve of the war, were a total of forty two Iraqi divisions, comprising of around 570,000 personnel. In terms of major equipment holdings, estimates suggest that after the intensive air bombardment, Iraq possessed some 2,865 tanks, 1,955 armoured personnel carriers and 1,772 artillery pieces.[21] The figures used in this data set are higher in these categories due to a decision to treat conservatively any claims regarding the effectiveness of coalition air power.

The Coalition fielded less personnel on the battlefield than Iraq. Although published figures to date are vague, the authors have established a working figure of approximately 315,000 for the thirteen divisions, eleven brigades, three regiments (two artillery) and one commando group.

In the XVIII Airborne Corps area, the war started at 07:00 with the French Daguet Division striking out towards As-Salman airfield, ninety kilometres inside Iraq. This was followed by the 101st Air Assault Division's first step towards capturing Highway 8 — an air assault to a forward operating base 120 kilometres inside Iraqi territory. The base was captured by midday.

In the centre, VII Corps schedule was drastically revised following successes elsewhere, the Corps attack started at 15:00 hours. By the end of the first day, VII Corps units had advanced up to eighty kilometres, whilst the 1st Cavalry Division conducted several feigning attacks to convince the Iraqi commanders that the allies were committing themselves to a frontal war of attrition, rather than a war of manoeuvre. The weather conditions in the theatre, on the 24th February are worthy of mention, as the description by Kindsvatter (1992) dispels stereotypical notions of the desert:

"Cold, misty, cloudy, with rain showers in the morning. Rain stopped in the afternoon, but the winds picked up from the south-east, and blowing sand reduced visibility to 1,000 metres or less. In the evening, winds subsided, but overcast skies made for a dark night."[22]

G + 1: Monday 25th February 1991

In XVIII Corps area, a four hour air assault placed Coalition forces onto Highway 8, 225 kilometres inside Iraq. The 24th Mechanised Division secured its supply routes. The VII Corps meanwhile, continued its march north to meet the Republican Guards. The 1st (UK) Armoured Division attacked a reserve brigade around Objective BRONZE. In the east, the Marines continued their drive towards Kuwait, becoming engaged in a battle at the Magwa Oil Field. Both Arab commands continued their advance towards Kuwait.

The meteorological conditions were still very poor. Kindsvatter (1992) describes conditions in the VII Corps area of operations as "overcast, cold, frequent showers, at times heavy".[23] Eldridge (1991) also describes the atrocious combat conditions of MARCENT where, "rain and fog reduced visibility to 200 – 500 metres".[24] For the XVIII Corps, the weather forecast was to determine the timing of the assault on Highway 8, as that evening "rain, sand storms and very high winds had grounded all helicopters with the exception of the AH-64 Apache."[25]

In summary, the effects of the weather appeared to hamper air transportation and scouting suggesting that either pilots lacked training for desert flying in adverse conditions, or that the night vision equipment was not as effective as had been hoped.

G + 2: Tuesday 26th February 1991

XVIII Corps operations were severely hampered by poor weather conditions, resulting in a number of attacks and operations being cancelled. 24 Mechanised Division attacked an airfield, supply base and numerous other rear echelon targets.

The VII Corps destroyed a major supply base (Al-Busayyah), before finally turning east. That evening a key engagement took place between elements of the 2nd Armoured Cavalry Regiment and the Tawalkana Republican Guard Division, which has become known as 'The Battle of 73 Easting'. 1st (UK) Armoured Division attacked Objective BRASS and the Iraqi 52nd Armoured Division.

Following the Radio Baghdad announcement many units were preparing to withdraw. One of the units which failed to act upon the broadcast become engaged in a major battle around the Kuwait International Airport with the US Marines. The Joint Force Commands continued their advance. The weather once again was universally bad. Kindsvatter (1992) described rain in VII Corps' area, which by the afternoon had become a sand storm.[26] In the west, Dempsey notes that the bad weather continued to hinder operations until the early hours of the 27th February.[27]

G + 3 Wednesday 27th February 1991

With the improvement in meteorological conditions, XVIII Corps continued operations with a new found vigour. A new forward operating base (VIPER) was established to interdict Iraqi forces in and around the city of Basra. The 24th Mechanised Division attacked the Jelibah airfield.

VII Corps continued its drive eastwards, becoming deployed to cover the Iraqi withdrawal, including elements of the Adnan and Medina Republican Guard Divisions. In the east, Kuwait City was liberated by units of the Joint Force Commands, North and East.

Visibility in the VII Corps area was described in Eldridge (1991) as a mixture of fog and smoke from the burning oil wells, which created a "near impenetrable haze".[28]

G + 4: Thursday 28th February 1991

Following the announcement of an 08:00 cease-fire, Coalition operations in VII and XVIII Corps areas altered, according to Kindsvatter (1992) and were "aimed to destroy or capture as much Iraqi equipment as possible."[29] Following the cease-fire, the British and Americans were able to take stock. Having executed a mission which would result in the use of superlatives in terms of time, distance and activity undertaken, what state were those forces in?

The British Army had a severe problem in terms of improving the reliability of its tank fleet in the Gulf. It was noted that:

"At the outset of Granby, what had long been known within the confines of the Army became public knowledge: that many of the Challenger I tanks were not battle-worthy. Challenger availability in 1990 in BAOR was just 23 per cent."[30]

When the tanks reached the Gulf, there was even more cause for concern. Official predictions regarding Challenger engine failures reckoned that a breakdown would occur every 1,235 kilometres. The actual figure in practise was nearer 723 kilometres. As one observer noted, "for the given fleet size, every 5 km one Challenger would require a power pack change".[31] Figure 1.6.3 overleaf shows the mean distance between engine failures during the period November 1990 – February 1991 for Challenger I power pack (P/P), engines (ENG), gear boxes (G/BOX) and Auxiliary Power Units (APU):[32]

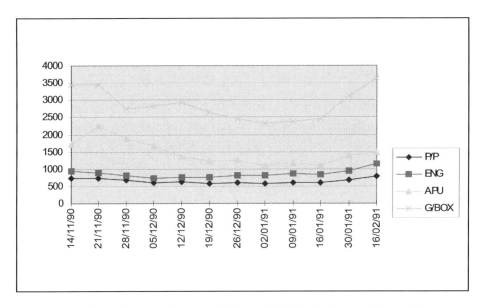

Figure 1.6.3: Mean Distance Between Failure (MDBF) Challenger I Tank Fleet.

The above figure illustrates the classic conundrum in peacetime. Equipment is used infrequently in order to minimise wear and tear, yet this leads to lower availability rates. The Challengers deployed to the Gulf were used constantly by the 4th and 7th Armoured Brigades in training and soon began to improve. Further, the deployment of a support team from Vickers to assist in problem solving, enabled rapid and effective communication of problems along the supply chain.

Conclusions for the Logistician: The Gulf War as a Case of 'Best Practice'?

Having considered the impact of logistics issues in the Gulf campaign of 1991, what lessons can we draw based upon the five principles of effective logistics planning? In terms of foresight, the logistics planning available for supporting large armoured formations was based upon NATO practise for deterring a Warsaw Pact threat.

When it came to the transplantation of this establishment into the desert, a number of problems emerged. These problems revolved around the differences in environment. The desert was clearly larger than Central Europe, unfamiliar to many of the troops and essentially featureless. Without access to satellite navigation technology, the enterprise would have been much more difficult to co-ordinate.

In terms of training, having NATO forces in static defensive positions for several decades, led to logistics becoming a qualitatively planned exercise, which was repeated for some period of time. This, it could be suggested, has led to a certain degree of atrophy, in terms of remembering the notion of logistics as an art, rather than a science. For the movement of formations, one lesson was re-learned.

The development of mobile forces has been constrained for decades on the limitations of air transport. While it is fast, it is restricted in the amount of heavy equipment it can carry quite severely. Also, formations that tend to have very good strategic mobility, tend to be limited when it comes to tactical mobility. Thus, heavy equipment will still have to, by and large, be moved by sea, in the 21st century. Rapid deployment, it could be suggested, is only rapid in terms of making a political commitment in a particular situation.

Regarding the second characteristic of economy, how did the experience in the desert measure up? Economy of logistics can be seen by the use of precision to keep the land forces momentum up during Operation Desert Sword. However, it is clear that the high tempo of operations could not be maintained much longer than the 100 hours of conflict due to the fact that the forces were massively over-extended and force sustainment needed to catch up.

Manoeuvre warfare tends to use fuel rather than ammunition when compared to attritional warfare techniques.[33] Desert operations required many fuel tankers, especially for the fuel hungry tanks deployed by the US Army.[34] Despite these difficulties, it could be suggested that, logistics, as employed during the conflict, were of an economical nature. The supply chain was built up from the continental USA through to the desert for mission critical items and spares.

Materials were on the whole, effectively catalogued into and out of the theatre after the conflict. Increasing the precision of this activity will gradually become of greater importance as stocks of supplies shrink and the forces deploy increasingly sophisticated equipment.

The flexibility of the logistics operation in the desert was amply demonstrated. The senior American logistician noted that the key lesson for him was the creation of a flexible organisation to get the tasks completed to the satisfaction of the theatre commander.[35] The transformation of the logistics establishment from its static European form to a desert sustainment force is a tribute to the skills of the logistics planners and personnel involved.

The Gulf operation was both large and complex. However, it could be argued that at its heart, the logistics operation was simple. The forces in the desert were dependant on logistics support to survive in what was a hostile environment. Host Nation Support was mobilised in order to provide basic supplies with the home country taking responsibility for moving the forces necessary to the Gulf.

The co-operative aspect of the Gulf operation mirrored the success of the political coalition. The coalition forces in the Gulf received assistance through commercial air and sea-lift and as detailed earlier, helped one another move particular items of equipment to the theatre. For the Western forces, co-operation was essential to access the most basic items such as bottled water. In designing the ground campaign, clearly co-operative logistics were a major factor in the British and American forces of VII Corps working together.

The case along with those previously have sought to demonstrate that modern operations are indeed constrained, if not defined, by the logistics infrastructure available, as well as testing what are regarded as the essential practises of good logistics.

The five factors themselves appear to remain a relevant framework for considering military logistics. Future research in this field could be targeted towards a clearer definition of the philosophical underpinning of military operational planning with a view to establishing the link between logistics issues and operations. Such a paradigm would relate the husbanding of logistics with the expenditure of military resources.

Notes

[1] Foxton P. D. *Powering War: Modern Land Force Logistics*. London: Brassey's, 1994. Chapter 1. Page 3.

[2] Heikal M. *Illusions of Triumph*. London: HarperCollins, 1993. Part II. Chapter 2. Page 227.

[3] See Danis A. **"Iraqi Army Operations and Doctrine"** in *Military Intelligence,* April-June 1991, Page 6.

[4] Ibid.

[5] Rottman G. and Volstad R. *Armies of the Gulf War*. London: Osprey Elite Series No. 45, 1993 Page 3.

[6] Department of Defence. *Final Report to Congress on the Conduct of the Persian Gulf War.* Washington: GPO, April 1992, Chapter 3. Pages 32-33, and Craft, D. W. *An Operational Analysis of the Persian Gulf War.* Carlisle Barracks: Strategic Studies Institute, US Army War College, 1992, Pages 16-17.

[7] Commander-in-Chief, U.S Central Command, unclassified Desert Shield/Storm Facts, Figures, Quotes and Anecdotes 7 Aug 1990 - 11 Apr 91 Page 13.

[8] Ibid. Page 19.

[9.] Menarchik D. *Powerlift – Getting to Desert Storm: Strategic Transportation and Strategy in the New World Order.* Westport CT: Praeger Publishers, 1993 Chapter 2. Page 56.

[10.] House of Commons. HC287/I *Defence Committee 10th Report: Preliminary Lessons of Operation.* Granby London: Her Majesty's Stationery Office 17/07/91. *Written Evidence.* Pages 80 - 81. Especially paragraphs 79, 80 and 92 - 95.

[11.] Op Cit. Menarchik. Page 116. Figure 2.30.

[12.] Op Cit. House of Commons HC287/I. Page 21. Paragraph 37.

[13.] Op Cit. Department of Defence. Chapter 6. Pages 95 - 101.

[14.] Op Cit. Commander-in-Chief, US Central Command. Page 18.

[15.] Sheffield G. D. **"Blitzkrieg and Attrition"** in McInnes and Sheffield: *Warfare in the Twentieth Century: Theory and Practice.* London: Unwin Hyman 1988. Page 74.

[16.] Op Cit. Commander-in-Chief, US Central Command. Page 6. Forty-one Scuds were fired at Riyadh Saudi Arabia and forty at Tel Aviv Israel.

[17.] Rip M. R. and Lusch D. P. **"The Precision Revolution: The Navstar Global Positioning System in the Second Gulf War"** in *Intelligence and National Security*, 1994. Page 177.

[18.] Ibid. Page 175.

[19.] Englehardt J. P. SSI Special Report: *Desert Shield and Desert Storm: A Chronology and Troop Lists for the 1990 - 1991 Persian Gulf Crisis.* Carlisle Barracks: Strategic Studies Institute, US Army War College, 1991. Page 5.

[20.] Atkinson R. *Crusade: The Untold Story of the Gulf War.* London: HarperCollins, 1994. Chapter 4. Pages 105 - 139.

[21.] Op Cit. Commander-in-Chief, US Central Command. Page 2.

[22.] Kindsvatter P. S. **"VII Corps in the Gulf War: Ground Offensive"** in *Military Review,* February 1992. Page 24.

[23.] Ibid. Pages 25 – 26.

[24.] Eldridge B. **"Desert Storm: Mother of All Battles"** in *Command,* Nov – Dec 1991. Number 13. Page 27.

[25.] Dempsey, T A. **"On the Wings of the Storm: Heliborne Manoeuvre during the Gulf War"** in *Defence Analysis*, 1994. Volume 10 Number 2. Page 172.

[26.] Op Cit. Kindsvatter. Page 29.

[27.] Op Cit. Dempsey. Page 173.

[28.] Op Cit. Eldridge. Page 36.

[29.] Op Cit. Kindsvatter. Page 37.

[30.] House of Commons. HC43 Defence Committee Fifth Report: *Implementation of Lessons Learned from Operation Granby.* London: Her Majesty's Stationary Office 25/05/94. Page 14. Paragraph 39.

[31.] Campbell A M. **"Equipment Support"** in White M. S. (ed.) *Gulf Logistics: Blackadder's War.* London: Brassey's, 1995. Chapter 7. Page 143.

[32.] Ibid. Page 144. Figure 7.1.

[33.] Thornton S F. **"Supply"** in White M. S. ed. *Gulf Logistics: Blackadder's War.* London: Brassey's, 1995. Chapter 6. Pages 124 - 125.

[34.] The United States M1 Abrams MBT uses a rotary gas turbine engine that ran on petrol rather than diesel, which is what the European MBTs ran on.

[35.] Pagonis W. G. *Moving Mountains: Lessons in Leadership and Logistics from the Gulf War.* Boston MA: Harvard Business School Press, 1992. Page 223.

Further Reading

Armees, **'Soutien d'avant garde'** in *Armees d'Aujourd'hui,* 161 (Jun. - Jul. 1991). pp. 76 - 79.

Army Logistician. Burnishing the Desert Shield, (Jan. - Feb. 1991). pp. 20 - 22.

————. *AMC's Role in Operation Desert Shield.* (Mar. - Apr. 1991). pp. 2 - 5.

Augustus, E jr. **'Logistics Automation Support'** in *Army Logistician.* (Jul. - Aug. 1993). pp. 36 - 37.

Bacon, P. H. **'Repair Parts to the Rescue'** in *Army Logistician* (Jan. - Feb. 1993). pp. 11 - 13.

Ballantyne, R A. **'The Aerospace Industry 2: Supporting the Military - Operations Granby and Desert Storm'** in *RUSI* Journal 137 (4) (Aug. 1992). pp. 50 - 52.

Beaton, D J. **'Operation Granby 23 Engineer Regiment'** in *The Royal Engineers* Journal 105 (3) (Dec. 1991). pp. 268 - 271.

Bernard, D C. **'Support to Strike - Tactical Supply Wing Support During Operation Granby'** in *Air Clues* 45 (11) (Nov. 1991). pp. 410 - 412.

Blackwell, J. **'An Initial Impression of the Logistics of Operation 'Desert Shield'** in *Military Technology* 14 (12) (Dec. 1990). pp. 57 - 61.

Bosco, L. **'Making things happen in the Persian Gulf'** in *All Hands* 891 (Jun. 1991). pp. 32 - 35.

Bottoms, A. M. **'Desert Shield and the Acquisition Manager'** in *Program Manager* 20 (1) (Jan. - Feb. 1991). pp. 8 - 13.

Brame, W. L. **'From garrison to desert offensive in 97 days'** in *Army* 42 (2) (Feb. 1992). pp. 28 - 35.

————. **'Planning Desert Storm Logistics'** in *Army Logistician* (May - Jun. 1992). pp. 16 - 21.

Bryant, B. **'City in the Sand'** in *The Military Engineer* 83 (546) (Nov. - Dec. 1991). pp. 10 - 14.

Byther, D. **'Desert Shield Contingency Contracting'** in *Army Logistician* (Mar. - Apr. 1991). pp. 22 - 25.

Cain, F. M. III. **'Building Desert Storm Force Structure'** in *Military Review* 73 (3) (Jul. 1993). pp. 21 - 30.

Carbonneaux. **'Daguet: Logistics in broad daylight'** in *Military Technology* 15 (8) (Aug. 1991). pp. 33 - 35.

Collar, L. L. **'Desert Storm and Its Effect on U.S. Maritime Policy'** in *Defense Transportation* Journal 47 (3) (Jun. 1991). pp. 67 - 68.

Defense Transportation Journal. **'Desert Shield/Desert Storm: USTRANSCOM's First Great Challenge'** 47 (3) (Jun. 1991). pp. 14 - 19.

————. *'The Railroaders' Finest Hour: Desert Shield/Storm'* 47 (3) (Jun. 1991). pp. 27 - 30.

————. *'U.S. Ports Rise to the Challenge of Desert Shield/Desert Storm'* 47 (3) (Jun. 1991). pp. 36 - 41.

Donohue, T. J. **'Trucking Industry Played Key Role in Desert Shield/Storm'** in *Defense Transportation* Journal 47 (3) (Jun. 1991). p. 26.

Donovan, F. R. **'Test of Sealift Planning for MSC'** in *Defense Transportation* Journal 47 (3) (Jun. 1991). pp. 60 - 62.

Donovan, F. R. (et. al.). **'Sealift - The Steel Bridge'** in *Defense Transportation* Journal 47 (6) (Dec. 1991). pp. 30 - 32.

Driscoll, E. J. **'They Also Serve'** in *Defense Transportation* Journal 47 (3) (Jun. 1991). pp. 58 - 59.

Eanes, J. T. **'USCENTCOM As Focal Point of Mobility Effort'** in *Defense Transportation* Journal 47 (3) (Jun. 1991). pp. 72 - 73.

Edgar, E. **'249th Engineers Storm the Desert'** in *The Military Engineer* 83 (546) (Nov. - Dec. 1991). pp. 4 - 9.

Elder, J. P. **'Survey Operations'** in *The Royal Engineers* Journal 106 (2) (Aug. 1991). pp. 125 - 131.

Evans, J. D. **'Air Force Buys Support'** in *The Military Engineer* 83 (546) (Nov. - Dec. 1991). pp. 28 - 29.

Fellows, R. **'Computers Aid in Medical Supply'** in *Army Logistician* (Sept. - Oct. 1992). pp. 12 - 13.

Felton, M. W. **'Managing ODS Supply : The Untold Story'** in *Army Logisitician* (Jul. - Aug. 1993). pp. 13 - 15.

Fortner, J. A., Doux, J. T. & Peterson, M. A. **'Bring on the HETs ! Operational and tactical relocation of heavy manoeuvre forces'** in *Military Review* LXXII (1) (Jan. 1992). pp. 36 - 39.

Fuchs, W. F. **'Sealift ships play major role in Desert Storm'** in *Defense Transportation Journal* 47 (2) (Apr. 1991). pp. 10 - 14.

Galway, L. A. *Management Adaptations in Jet Engine Repair at a Naval Aviation Depot in Support of Operation Desert Shield/Storm* (CA: RAND 1992). N-3436-A/USN.

Gourley, S. R. **'US trucks in Desert Storm'** in *International Defence Review* 25 (1) (Jan. 1992). pp. 34 - 38.

Ham, J. **'Desert Air Terminal Operations'** in *Defense Transportation* Journal 47 (3) (Jun. 1991). p. 74.

Hammick, M. **'Lost in the pipeline speed stretched logistics to the limit'** in *International Defence Review* 24 (9) (Sept. 1991). pp. 998 - 1001.

Hayashi, G. **'Intermodalism Pays Off in the Gulf War'** in *Defense Transportation Journal* 47 (3) (Jun. 1991). pp. 63 - 66.

Hayr, K. **'Logistics in the Gulf war'** in *RUSI* Journal 136 (3) (Autumn 1991). pp. 14 - 18.

Heron, J. M. **'Sanitation in the sand and other tales'** in *The Royal Engineers* Journal 105 (1) (Apr. 1991). pp. 12 - 18.

Hill, R. D. **'Depot operations supporting Desert Shield'** in *Military Review* LXXI (4) (Apr. 1991). pp. 17 - 28.

Hoover, W W. **'The Desert Shield Airlift: A Great Success That Holds Some Valuable Lessons for the Future'** in *Defense Transportation* Journal 47 (3) (Jun. 1991). pp. 54 - 55.

James, G. K. **'Aviation logistics mobility: lessons learned'** in *U.S. Army Aviation Digest* 1-92-2 (Mar. - Apr. 1992). pp. 32 - 36.

Jane's Defence Weekly. **'Achilles heel of an army'**, 18 (13) (Sept. 1992). pp. 19 - 23.

————. *'Success behind the 'Storm' front'* 15 (19) (11/05/91). pp. 783 - 787.

Johnson, L. M. & Rozman, T. R. **'The armour force and heavy equipment transporters: A force multiplier?'** in *Armor* C (4) (Jul. - Aug. 1991). pp. 13 - 16.

Kirsch, R. S. & Magness I.V., **'Iron Sappers'** in *The Military Engineer* 83 (546) (Nov. - Dec. 1991). pp. 25 - 27.

Kondra, V. (et. al.). **'Airlift - First on the Scene'** in *Defense Transportation* Journal 47 (6) (Dec. 1991). pp. 26 - 29.

Langenus, P. C. **'Moving an army: Movement control for Desert Storm'** in *Military Review* LXXI (9) (Sept. 1991). pp. 40 - 51.

Laposata, J. S. & Hatley, C. D. **'Conventional Forces Europe Combat Equipment Retrograde: A Dress Rehearsal For Desert Storm'** in *Defense Transportation* Journal 47 (4) (Aug. 1991). pp. 10 - 11.

Leonhart, L. **'Challenges for tomorrow'** in *Defense Transportation* Journal 47 (6) (Dec. 1991). pp. 33 - 38.

Los Angeles Times, **'Ground assault could clog U.S. supply lines'** (25/01/91).

Luizet, M. **'La logistique des matériels - La Direction centrale du matériel de l'arm ee de terre et la guerre du Golfe'** in *Armees d'Aujourd'hui* 158 (Mar. 1991). pp. 58 - 61.

MacGinnis, G. R. W. **'Waterbeach at war - 39 Engineering Regiment's part in the Gulf War'** in *The Royal Engineers* Journal 105 (3) (Dec. 1991). pp. 272 - 277.

Magness, J. **'Rapid reload operations'** in *U.S. Army Aviation Digest* 1-91-2 (Mar. - Apr. 1991). pp. 50 - 51.

McDonald, W. S. **'Providing engineering support to operation desert shield'** in *U.S. Army Aviation Digest* 1-91-1 (Jan. - Feb. 1991). pp. 8 - 11.

McGill, I. D. T. **'British Forces Kuwait in the Aftermath of the Gulf War'** in *The Royal Engineers* Journal 105 (3) (Dec. 1991). pp. 212 - 221.

Menarchik, D. *Powerlift - Getting to Desert Storm: Strategic transportation and strategy in the New World Order* (London: Praeger 1993).

Meyers, J. E. **'Building the desert logistics force'** in *Military Review* LXXI (4) (Apr. 1991). pp. 13 - 16.

Military Review. **'Chronology'** (9) (Sept. 1991). pp. 65 - 78.

Miles, D. **'Sustaining The Force'** in *Soldiers* 46 (5) (May 1991). pp. 20 - 23.

Moore-Bick, J. D. **'Operation Granby Preparation and Deployment for War'** in *The Royal Engineers* Journal 105 (3) (Dec. 1991). pp. 260 - 267.

Munro, S. H. R. H. **'The armoured delivery group 1st Armoured Division Operation GRANBY'** in *British Army Review* (102) (Dec. 1992). pp. 53 - 61.

Oler, R. **'Desert Shield and the army aviation national maintenance point'** in *U.S. Army Aviation Digest* 1-91-1 (Jan. - Feb. 1991). pp. 2 - 5.

Pagonis, W. G. **'Good logistics is combat power: The logistic sustainment of Operation Desert Storm'** in *Military Review* LXXI (9) (Sept. 1991). pp. 28 - 39.

—————. *Moving mountains: Lessons in leadership and logistics from the Gulf War* (USA: Harvard Business School Press 1992).

Pagonis, W. G. & Krause, M. D. **'Theater Logistics in the Gulf War'** in *Army Logistician* (Jul. - Aug. 1992). pp. 2 - 8.

—————. **'Observations on Gulf War Logistics'** in *Army Logistician* (Sept. - Oct. 1992). pp. 5 - 11.

Phillips, R. S. **'Logistics automation support for desert storm'** in *Military Review* LXXI (4) (Apr. 1991). pp. 9 - 13.

Piatak, J. R. **'MTMC Key to Strategic Mobility'** in *Defense Transportation* Journal 47 (3) (Jun. 1991). pp. 20 - 24.

Piatak, J. R. (et. al.). **'Surface Transportation - Linchpin to Projection'** in *Defense Transportation* Journal 47 (6) (Dec. 1991). pp. 19 - 24.

Porter, K. K. & LePore, H. P. *Legacy in the sand: The US Army armament and chemical command in operations Desert Shield and Desert Storm* (USA: Historical office US Army AMCCOM 1991).

Pyles, R., Shulman, H. L. *United States Air Force Fighter Support in Operation Desert Storm* (CA: RAND 1995). MR-468-AF.

Ranken, M. **'The Gulf war - logistics support and merchant shipping'** in *Naval Review* 79 (1991). pp. 198 - 206.

Ross, J. D. **'Victory: The Logistics Story'** in *Army* 41(10) (Oct. 1991). pp. 128 - 138.

————. **'Combat Power Where Needed'** in *Army* 41 (10) (Oct. 1991). pp. 138 - 140.

Rutherford, W. R. III & Brame, W. L. **'Brute Force Logistics'** in *Military Review* 73 (3) (Mar. 1993). pp. 61 - 69.

Salomon, L. E. & Bankirer, H. **'Total army CSS: providing the means for victory'** in *Military Review* LXXI (4) (Apr. 1991). pp. 3 - 8.

Sanchez, E. **'La logistica gano la guerra del Golfo'** in *Revista Espanola de Defensa* 38 (4) (Apr. 1991). pp. 56 - 59.

Schnell, W. **'Anytime, Anyplace'** in *U.S. Army Aviation Digest* 1-91-1 (Jan. - Feb. 1991). pp. 6 - 7.

Schuster, C. R. **'A Real-Life Test of Flexible Readiness'** in *Army Logistician* (Nov. - Dec. 1991). pp. 38 - 40.

Shrader, C. R. **'Gulf War Logistics'** in *Parameters* 25 (4) (Winter 1995 - 1996). pp. 142 - 148.

Southworth, M. S. **'When War Wouldn't Wait'** in *Army Logistician* (Jan. - Feb. 1992). pp. 8 - 11.

Springett, R. **'Supply Support for the Support Helicopter Force in the Gulf War'** in *Air Clues* 47 (7) (Jul. 1993). pp. 248 - 254.

The Royal Engineers Journal. **'Report by Engineer in Chief'** 106 (2) (Aug. 1991). pp. 102 - 111.

Tow, S. L. **'Airlift-Delivered Victory'** in *Defense Transportation* Journal 47 (3) (Jun. 1991). pp. 47 - 48.

Tuttle, G. T. jr. **'Operation Desert Storm Demonstrates AMC's Mission Framework'** in *Army* 41 (10) (Oct. 1991). pp. 80 - 85.

Vuono, C. E. **'Desert Storm and Future Logistics Challenges'** in *Army Logistician* (Jul. - Aug. 1991). pp. 28 - 31.

Waller, C. A. H. **'Waller's Keynote Address: A Celebration of Victory'** in *Defense Transportation* Journal 47 (6) (Dec. 1991). pp. 14 - 19.

Walton-Knight, M. P. **'Supplying Water to the British Army During the Gulf War'** in *Royal Engineers* Journal 108 (2) (Aug. 1994). pp. 154 - 159.

Wells, G. W. jr. **'Aerial Delivery in Desert Storm'** in *Army Logistician* (Mar. - Apr. 1993). pp. 35 - 37.

White, M. S. *Gulf Logistics: Blackadder's war.* (London: Brassey's 1995).

Wilson, A. A. **'A Theatre Overview of Engineer Aspects Of the Gulf War'** in *The Royal Engineers* Journal 106 (2) (Aug. 1991). pp. 116 - 124.

Wilson, B. F. **'Forward-Area Support in Desert Storm'** in *Army Logistician* (Sept. - Oct. 1992). pp. 2 - 4.

Williams, B. J. **'Tyre changing facilities for Operation GRANBY'** in *Journal of the Royal Electrical and Mechanical Engineers* 41 (May 1991). pp. 19 - 23.

Williams, J. C. **'Quality Control for the Patriot'** in *Army Logistician* (Jan. - Feb. 1992). pp. 6 - 7.

Zorpette, G. **'From factory floor to desert war'** in *IEEE Spectrum* 28 (9) (Sept. 1991). pp. 40 - 42.

Part Two // Defence Procurement

Case 2.1 // EM-2

A Rifle Ahead of its Time: The EM-2 (Rifle No. 9 Mk 1)

Peter D. Antill[1]

Centre for Defence Acquisition, Cranfield University, Defence Academy of the UK.

Introduction

Few items that have come into service with the British Army have caused more controversy over their operating lives than that of the SA80 (Small Arms for the 1980s) series of weapons. This family of small arms, chambered for NATO's 5.56x45mm ammunition, consists primarily of the L85 Individual Weapon (IW) and the L86 Light Support Weapon (LSW), but also includes the L22 Carbine and L98 Cadet Rifle. The L85 and L86 replaced the L1A1 Self-Loading Rifle (SLR), L4A9 Bren Light Machine Gun (LMG), L7A2 General Purpose Machine Gun (GPMG) and L2A3 Sterling Sub-Machine Gun (SMG). Officially handed over on 2 October 1985, production had reached over 323,920 when it finished in 1994.[2] This however was not the first bullpup-style[3] military rifle to have been adopted for the British Army, for in August 1951 "the British unilaterally adopted the EM-2 as the short-lived "Rifle No. 9 Mk 1" and the .280/30 cartridge as "Cartridge, SA, Ball, 7mm Mk1Z" (the "Z" indicating that the cartridge was loaded with nitrocellulose propellant)."[4] This case study is the story of that rifle; its cartridge and its demise due to post-World War II political decisions to bow to the wishes of the United States.

Background

At the end of World War II, the British remained one of the few major powers that had yet to introduce a self-loading rifle into service.[5] The Soviets had fielded the Simonov AVS36 in 1936 but had failed to trial it properly with the result that all sorts of problems in reliability occurred in the field. Further development resulted in the Tokarev SVT38 and SVT40, both of which continued the 7.62x54R calibre. In 1943, they also developed the 7.62x39mm intermediate cartridge and the following year brought out the Simonov SKS carbine to use it. The USA had adopted the M1 Garand in 1936 and during the war, the Germans had developed the FG42 paratroop rifle, the G41 and G43 semi-automatic rifles as well as the revolutionary StG44 (also known as the MP44). The British Army meanwhile, were still using the No. 1 and No. 4 Short Magazine Lee Enfield bolt-action rifles in .303 calibre, a rifle which, tried and tested though they were, traced their roots back to the late 19th Century.

Experience of infantry combat during the two world wars, questioned the need for infantry to be armed with weapons that were chambered with cartridges that were accurate out to 2,000 yards. "In fact, despite the evidence that most shooting during WW1 was at short range, armies continued to show an interest in full-power rifle/MG rounds."[6] This experience from World War I was reinforced by events during World War II, where "it was obvious that the modern warfare will require the infantry to be armed with light, selective fire weapon with effective range of fire much longer than of submachine gun, but shorter than of conventional semi-automatic or bolt-action rifles."[7]

A New Calibre

The concept of a bullpup service rifle for the British Army actually goes back to a decision taken in late 1945, at the very end of the war. The Ministry of Supply set up a "Small Arms Ideal Calibre" Panel to determine the optimum cartridge for a lightweight self-loading rifle. A great deal of experimentation and testing took place, primarily of calibres between .250 and .270 (approximately 6.35 to 6.8mm), undertaken at the Armament Design Establishment, Enfield under Dr Richard Beeching, the Deputy Chief Engineer.[8] The panel reported in March 1947, recommending the further development of two different designs. The first, appearing in November 1947, was in .270 calibre (6.8x46mm) with a steel-cored 100gn bullet travelling at between 2,750 and 2,800fps (approximately 840 to 850m/s).[9] This round retained 81 ft lbs of energy (109j) at 2,000 yards (1,830m) with 60 ft lbs of energy (80j) reportedly being necessary to injure an unprotected human being. The second was in .276 calibre (7x43mm), later re-designated as .280 to avoid any confusion with earlier ammunition, such as the American .276 Pedersen and .276 Enfield P13, both of which had been considered as potential new rifle calibres for their respective countries, in 1932 and 1913 respectively.

Indeed, the .276 Pedersen round may well have been the cartridge used in the M1 Garand, had it not been for the intervention of General Douglas MacArthur.[10] This round was tested with a number of bullets weighing between 130 and 140 grains (8.4 to 9g), with velocities between 2,330 and 2,450fps (710 to 747 m/s). The 130gn/2,450fps combination had a retained energy of 100 ft lbs (135j) at 2,000 yards (1,830m). In time, a combination entailing a Belgian-designed 140gn (9g) bullet fired at a velocity of 2,415fps (736m/s) was chosen for further development[11] and in November 1948, the .270 calibre round was dropped and attention focused on the .280 cartridge. The .280 calibre was slightly larger than originally intended but was selected in order to try and meet the American desire for good long-range performance. In line with this, and with an eye to the possibility of it being standardised within NATO,[12] the original case rim diameter was enlarged slightly to more closely match that of the .30-06 round in order that it be easier to re-barrel existing weapons, with the result that the designation was changed to .280/30.[13]

A New Rifle, A New Enfield

World War II impacted on British small arms design in two ways. Firstly, in terms of actual small arms design, there was very little in terms of an indigenous knowledge/skill base for up until the outbreak of war, "There was very little 'know how' in Great Britain before the war. Practically all small arms were purchased abroad and as the Germans over-ran Europe, so a number of European gun designers managed to escape and come to England and join the Ministry."[14]

Secondly, the war had seen the capture of a large number of German weapons, many of whose features were incorporated into British designs. It was first thought that these new designs were 'ersatz' or cheap items due to the large amount of sheet metal stampings used, but closer examination proved that this was not the case. In the latter stages of the war, the Germans had suffered shortages of expensive alloys and so weapon designers had been forced to come up with equipment that used normal carbon steels wherever possible. The Germans had eventually mastered the art of mass production during wartime that meant that they could produce items at a price far cheaper than most of the rest of the world was paying (who were using the time-consuming, expensive method of extensively machined metal forgings) but without any loss in quality.[15]

Two different weapons influenced the post-war bullpup designs from the ADE. The first was a sniper rifle, also designed at the ADE, which was intended to be a solution to the problem of snipers being visually spotted when operating the bolt on the No. 4 Lee Enfield rifle. Three designs were produced but only one prototype was built, in 7.92x57mm calibre. The second weapon became known as the EM-3 or Hall rifle and was developed from a design solution put forward to a problem set on the 8th SAT(War) Course at RMCS Shrivenham in 1944. The design solution, put forward by Major J E M Hall of the Australian Army, was favourably received by the staff at RMCS and additional development was undertaken to the point where a patent, No. 589394, was granted to Major Hall by the British Patent Office.[16] It is interesting as it was even shorter than the Enfield designs under development, was completely sealed against the external environment and featured over-the-shoulder ejection, thus it was able to be fired either left or right-handed. Both of these had an impact on the subsequent designs coming out of the ADE.

Four design teams were formed at Enfield, under the overall project leadership of Noel Kent-Lemon. The first of the bullpup designs was the EM-1 or Korsac rifle which originated with the Polish design team based at Cheshunt, Hertfordshire, headed by Roman Korsac[17] until his retirement when Captain Januszewski (who later changed his name to Stefan Kenneth Janson) took over. This weapon was mainly used to investigate the feasibility of the bullpup layout in this sort of weapon and owed its design to German weapons produced during World War II, most notably the FG42 paratroop rifle. The initial FG42s were made by Heinrich Krieghoff Waffenfabrik of Suhl and first used in action during the rescue of Mussolini by German commandos led by Otto Skorzeny in September 1943. The design was such that the rifle used the same 7.92x57mm round as the regular Mauser Kar98k but was shorter and lighter. This was done by moving the magazine up to the left-hand side of the receiver and onto the same vertical axis as the trigger and pistol grip, along with moving a hollow, recoil-cushioned butt forward so that it surrounded the rear of the receiver.

In addition to the Korsac rifle, which would replace the 9mm Sten SMG and .303 Lee Enfield rifle, an experimental GPMG known as the Taden gun was developed, which would replace the Bren LMG and the Vickers heavy machinegun. "It was belt-fed and air-cooled, could be fitted with a butt and fired from a bipod or fitted with spade grips and fired from a tripod and generally appeared to be just the answer to the machinegun question."[18] The Korsac rifle design was moved to museum status on 25 September 1947.[19]

It was succeeded by a new design, from a team led by Stanley Thorpe that ran in parallel with a second design from the team led by Stefan Janson. This new EM-1 was therefore also known as the Thorpe rifle. It was a light automatic rifle that shared characteristics with the Mauser Gerät 06 No. 2 and a construction based upon the sheet metal stampings method pioneered by the Germans.[20] The Mauser Gerät 06 No. 2 was a prototype weapon, itself the forerunner of the Sturmgewehr 45 (StG45), a replacement for the StG44 (also known as the MP44) which had come into service with the Wehrmacht in late 1943. The StG45 was intended to be even simpler than the StG44 and therefore easier and cheaper to manufacture (45 Reichsmarks compared to 70 Reichsmarks).[21]

The first Thorpe rifle was proofed at Enfield on the 19 and 20 December 1949 to a pressure of 23 tons with a 130 grain .280 calibre bullet with a mild steel core and a charge weight of 30.1 grains. The design was eventually dropped due to problems with the stamped sheet steel used in the weapon's construction, the complexity of the weapon, the difficulties of field stripping it and the advanced state of the EM-2, also known as the

Janson rifle. The EM-2 employed the standard method of machining and finishing forged steel as used in the FG42 and Korsac rifle. It also used characteristics found in the Gewehr 43 (or Karabiner 43), a German semi-automatic rifle that was itself inspired by the Tokarev SVT40 semi-automatic rifle, including a gas operating system and pivoting locking lugs. The Janson rifle kept all these features as well as lessons from the Korsac rifle to include an identical butt plate retainer and release, the same basic bullpup layout, a similar receiver design and methods of manufacture and an identical cocking handle location.

Towards the end of World War II, the United States had also been considering the replacement of its venerable .30-06 cartridge and the weapons that used it, looking to develop what was in essence, a selective-fire, lightweight rifle. They were therefore after a rifle that was small, handy and capable of selective fire just like the .30 M2 carbine but with the stopping power of the M1 Garand and having a weight of around 7lbs (3.2kg). Perhaps optimistically, the new rifle was required to replace both the M1 Garand and the .30 Browning Automatic Rifle in .30-06 calibre (7.62x63mm), both the M1 and M2 carbines in .30 calibre (7.62x33mm) and the M3 SMG in .45 ACP calibre (11.5x23mm). The new lightweight rifle was expected to fire a round that was shorter than the .30-06 cartridge but still retain satisfactory performance at longer range so it could entirely replace the older cartridge – "a stopping and wounding power which shall not be less than that of the standard calibre .30 ammunition [7.62x63] fired from the M1 at ranges of 400, 800, 1,200 and 2,000 yards [up to 1,830m]."[22]

Once again, a great deal of experimentation followed but against all logic, the Americans decided to stay with the .30 calibre. Ultimately, they reached the point where they had a length of 51mm instead of 63mm but with very similar performance to the .30-06 round (made possible by modern propellants) and therefore very similar recoil. This meant that the new selective fire, lightweight rifle would prove uncontrollable in full automatic fire, something proved years before the introduction of the M14. However it must be noted that American opinion on this matter was far from homogenous, as will be explained later.

The Trials

The EM-2 and FN lightweight rifle in .280 calibre (the Belgians were enthusiastic supporters of the British concept, along with the Canadians) were tested in the UK from late 1948 to late 1949 and then taken to the USA to begin trials, in competition with the American T25 experimental rifle. First to commence were the engineering trials at Aberdeen Proving Ground between March and May 1950, followed by troop trials at Fort Benning, Georgia in front of the US Army Infantry Board between May and November 1950. After firing almost 57,000 rounds of .280 ammunition, the EM-2 had suffered stoppage rates of under 5 rounds per 1,000 for automatic fire and only 3.4 rounds per 1,000 for semi-automatic fire.

The M1 Garand, used as a control weapon, suffered stoppages of 3.8 rounds per 1,000 firing on semi-automatic only.[23] The T25 was chambered for the original T65 cartridge (7.62x47mm) which was still in the early stages of development but eventually became the 7.62x51mm NATO round. At Fort Benning, the Trials Board reported that "the T65 Cal .30 is not satisfactory because of its excessive recoil, blast, flash and smoke. That the Cal .280 is not satisfactory because of its relatively high trajectory. That of the two basic types of rounds submitted for the test the British calibre .280 is preferred."[24]

The actual findings of the tests showed that while the trajectory of the T65 was somewhat flatter and produced greater wounding at ranges of up to 1,000 yards (900m), the British .280 cartridge became more effective than the T65 at the longer distances due to its superior ballistic coefficient. At 1,000 yards, the British round could penetrate body armour seventy percent of the time, compared to the T65 that achieved sixty percent. The British round was also superior in terms of the collateral effects of shooting, that is, blast, recoil, flash and smoke. In addition, the T25 was found to be slightly more accurate and achieved more hits per minute when fired from a bipod, but the EM-2 was vastly superior when fired from the shoulder.[25]

The Results

Overall, the British design team had achieved all that they had hoped to. In spite of being the most accurate and reliable rifle in the competition and the Trials Board's recommendations to focus development on the .280 round, the decision was ultimately rejected by the US Army Chief of Staff, General J Lawton Collins. This was because the clear preference of the US Ordnance Department (who considered the .280 cartridge underpowered) as well as the wider American political establishment and military-industrial complex, was in favour of a full-power .30 calibre cartridge of American origin, and so having a domestically designed contender meant that the US Ordnance Board considered all other designs, especially when they were foreign ones, to be virtually irrelevant.

It didn't matter how good the .280/30 cartridge might or might not have been, or indeed the rifle and GMPG designed to fire it as "the 7mm cartridge was Not Invented Here and could not therefore be accepted by the US forces."[26] The Americans accepted "the concept of standardisation, but only if their round was the one to be standardised. There was thus little hope of persuading them to re-tool to a new calibre, with all the attendant costs that would be incurred, especially if it was of foreign design."[27] The British were aghast at the decision and felt the trials at the Aberdeen Proving Ground and Fort Benning should have settled the matter decisively and so were determined not to give up that easily.

The British focused their efforts on overcoming the American objections by producing more powerful versions of the .280 round. With the support of both Belgium and Canada, they first increased the charge in the 43mm case and added a 140gn bullet to produce a velocity of 2,550fps (777m/s) to overcome the trajectory problem and the complaints that arctic conditions reduced performance to an unacceptable level.[28] This increased the energy available at 2,000 yards to 126 ft lbs (170j). The Americans would not consider however, any movement from their preferred calibre of .30 and so the British unilaterally adopted the EM-2 rifle and the .280/30 cartridge in August 1951 following the decision in April of that year by the UK's Minister of Defence, Emanuel Shinwell.[29]

A Political Storm

This decision provoked immediate alarm in the rest of NATO. The issue was discussed at length at a meeting convened by the Canadian Defence Minister, Brooke Claxton, along with the defence ministers of Britain, France and the USA at the Pentagon on 2 August 1951. By this time, a General Election in the UK had led to a change of government, with the Conservatives under Winston Churchill replacing Labour under Clement Atlee. At the meeting in the Pentagon it was made clear that the .280/30 (7mm) cartridge for which the EM-2 was chambered would never be accepted by the other members of NATO and so in the interests of the majority, it was agreed that only weapons then currently in production, would remain so until NATO decided on the new round to be adopted. However, it was clear that "the US was in a position to dictate to an almost bankrupt Europe the path that must be taken"[30] and so despite the vision Shinwell had for the British Army to have a shorter, lighter and more accurate replacement for the 9mm SMG and .303 rifle, Churchill, on being returned to power, understood what was being hinted at by the USA. Shinwell proved that he was ready to promote new technology but Churchill was a realist.

The West's relations with the Soviet Union and its Eastern Bloc allies were becoming more strained by the day and the UK could not take such a big risk as to unilaterally introduce a new service rifle, especially one whose cartridge was not going to be adopted by any other member of the Alliance. Any serious crisis and breakdown in relations with the USSR could mean that British troops would be deployed to reinforce those already in Europe, armed with an unproven weapon and firing a non-standard round, making re-supply from friendly forces difficult if not impossible, meaning that any major shortfalls in the availability of ammunition could be disastrous.[31]

In a move that was seen by many at the time and since, as bowing to American pressure to maintain good relations with the Atlantic Alliance, Churchill agreed in early 1952, to rescind Shinwell's decision,[32] reportedly after a meeting between Churchill and Truman on 9 January.[33] Despite this, Britain, Belgium and Canada came together to form what became known as the 'BBC' Committee to have a try at producing a 7mm cartridge that would be acceptable to NATO (i.e. the Americans). A number of different cartridge lengths were tried, with such names as 'Optimum' (the bullet seated less deeply), 'High Velocity' (7x49.5mm), 'Compromise' (7x51mm) and 'Second Optimum' (7x49.15mm). The final variant was a 7.62x51mm round necked-down to 7mm. All this was to no avail and by the end of 1953, the BBC Committee caved in to American pressure and the 7.62x51mm cartridge was formally adopted as the standard NATO round on 1 February 1954.[34]

The EM-2 was briefly re-chambered for the new round but as with the enlarged .280 cartridges, the recoil and blast effects had increased to such a level as to be completely contrary to the original concept and it was almost impossible to control when firing in full automatic. The .280 cartridge and EM-2 concept was shelved and the design team disbanded, the only practical result being a cartridge called '7mm Medium' which saw service in an FN FAL variant that was sold to Venezuela.

Further Developments

That however, is not quite the whole story. As mentioned earlier, American opinion was not completely behind the idea of keeping a .30 calibre cartridge. Even as the trials were starting, two American research projects were begun, that would come to fundamentally different conclusions than that of the US Ordnance Board. The first originated with the Operations Research Office (ORO – formerly the General Research Office) and called Project BALANCE. It was a spin off from Project ALCLAD which looked at how to improve personal body armour. The head of the division, Norman Hitchman, reasoned that in order to improve body armour, one needed to look at how wounds were created and where they tended to occur.

A mathematical analysis of over three million casualty reports from both world wars was entered into ORO's computers along with reports from Korea, which led to the creation of Project BALANCE, a theoretical study of the use of infantry rifles in combat.[35] The second, started in early 1950 when the Ballistic Research Laboratory (BRL), based at the Aberdeen Proving Grounds was asked by Colonel René R Studler, the Chief of Small Arms Research and Development in the US Army Ordnance Department, to look at the effectiveness of the infantry combat rifle. Studler was attempting to buttress his own views in supporting the adoption of a full-power rifle cartridge and distrusted ORO's civilians who he thought was starting to infringe upon his area of responsibility.[36]

Both reports were completed in 1952. In March, BRL's Donald Hall published *An Effectiveness Study of the Infantry Rifle*, outlining that a high-velocity, smaller calibre bullet could give just as good or better hit probability than a slower, larger calibre bullet owing to its flatter trajectory and that at short range, the smaller calibre might even outperform the larger one.[37] In addition, due to the decrease in individual cartridge weight, an infantryman could carry a larger number of small calibre rounds and therefore in theory, could inflict a greater number of casualties than an infantryman carrying larger calibre rounds.[38] In June, Hitchman's report entitled *Operational Requirements for an Infantry Hand Weapon* was published, much to the discomfort of many in the Ordnance Department. The report analysed data from combat records in World War II and the conflict in Korea (interviews were conducted with over 600 personnel) where it found that it was rare for rifles to be used against targets at ranges greater than 300 yards, and that marksmanship dropped quickly beyond 100 yards due to such factors as eyesight, the terrain and visibility.

The average distance for bullets hitting from aimed shots was between seventy-five and 100 yards, while eighty percent of effective rifle and LMG fire occurred at ranges that were less than 200 yards and ninety percent at ranges under 300 yards. Hits at ranges greater than 300 yards were found to be so infrequent that they were considered to be on a par with the chances of being hit by shrapnel from an artillery shell. As such, a small calibre weapon that provided good single-shot characteristics up to 300 yards and controllable 'pattern-dispersion' at longer ranges might double the chances of hitting over a single shot from an M1 Garand.[39]

On top of that, the test results for the treasured .30 calibre Lightweight Rifle, duly reported by the ORO, proved to be a major disappointment for the Ordnance Department. The tests consisted of firing controlled, five-round bursts at silhouette targets held in a frame that was 6ft (1.8m) square. In repeated tests, conducted at no more than 100 yards (90m), only a single bullet from each burst even hit the frame, let alone the target while at 50 yards (45m), only one bullet hit the target, that being the first bullet fired, with the remainder firing high as the muzzle rose under the force of the recoil. "The inevitable conclusions of the Hitchman Report were that the .30 cal Light Rifle cartridge was vastly overpowered for its purpose and that fully-automatic fire with a powerful rifle was a waste of time and ammunition."[40]

Conclusion

The demise of the .280 chambered EM-2 rifle was ultimately due, not to technical reasons, for it was a "well balanced and laid out rifle, with comfortable controls, accurate and reliable...had it been put into service, the British troops could have a first class assault rifle prior to 1960"[41] but political ones, both national and international. It fell victim, despite the results of the trials themselves, the recommendations of the Trials Board overseeing the tests at Aberdeen Proving Ground and Fort Benning, as well as two official reports, firstly to American domestic sensitivities from the military, industrial and political establishments and their preference for a domestic .30 calibre cartridge and rifle,[42] quickly followed by the United States' ability to put pressure on a bankrupt Western Europe to accept their solution, the overpowered 7.62x51mm round. "The high-handed approach to the idea by the Americans, the insistence on a round of their design, and a .30-calbre full-powered one, did not sit well with the other countries. But many of them were already armed with American weapons from World War II, and ... their economies were very dependent on the billions of dollars that the United States was spending as part of the Marshall Plan to rebuild Europe."[43]

Although the 7.62x51mm round was seen as a compromise between the British .280/30 cartridge (a true intermediate round) and its most direct predecessor, the .30-06 round (a full power rifle cartridge, measuring 7.62x63mm), it was in truth, merely a slightly shortened .30-06 round that maintained similar ballistic qualities to its parent and as such, represented no sort of advance on a fifty-year old design.[44] "As a result, it falls between two stools, being less powerful than the rounds it replaced but too powerful to be a practical assault rifle round."[45] It also caused a political storm in the UK and had an impact on the outlook of the UK Armed Forces. "Psychologically, the frame of mind of the British officer class had been damaged by "the EM-2 experience", wherein this audacious "bullpup" rifle (which was, not only in their opinion, an excellent rifle; embodying optimum weapon characteristics and firing a thoroughly well-researched "Ideal Calibre" cartridge), was betrayed by politicians who were seen as bowing to pressure from the US Army Ordnance."[46]

Frustratingly, for the proponents of the smaller calibre, continuing trials in the USA suggested that the new AR-15 assault rifle developed by Eugene Stoner and chambered for 5.56x45mm (.223) ammunition was superior to the M-14, the rifle chosen to chamber the 7.62x51mm cartridge. The Ordnance Department recommended that development should be pursued with a view to replacing the 7.62mm rifle, which had only been formally adopted for service two years before.[47] With the US Air Force ordering the rifle in 1960 to replace the M2 Carbine and adopting it in 1964, while the US Army finally followed suit in

February 1967, the British .280 cartridge proved to be far ahead of its time with the USA replacing the M14 with the M16A1 in most combat units by the end of the decade.[48]

The M16A1 was designed to use the M193 cartridge but this has been the subject of considerable controversy over its lack of stopping power, reported even then by troops in Vietnam.[49] These problems influenced the next set of NATO Standardisation Trials as, to few people's surprise, the 5.56x45mm cartridge was adopted, but it was decided to adopt the Belgian SS109 cartridge as this had a 62gn bullet, contained a hard steel element near the tip and had better long-range performance. The USA too adopted this variant, as the M855. However, these problems of lethality have never really disappeared, for if "the 7.62mm was overpowered, the 5.56 was distinctly underpowered for military use, being essentially a small game ("varmint") cartridge. In fact, the 5.56x45mm is illegal for deer hunting in many areas of the US".[50]

This lack of lethality has been evidenced in the conflicts in Iraq and Afghanistan, especially for those troops using the M4 and M4A1 carbines, which in this case primarily equates to US Special Forces. The problem seems to relate to the difference in velocity that the SS109 round leaves the M4 (792m/s) as compared to the standard M16 (884m/s). The SS109 is most lethal when it strikes a target at a velocity of 731m/s or greater, after which the bullet starts to tumble and fragments along the cannelure causing a miniature explosion, enhancing the size of the wound cavity. For a standard M16, the range at which the SS109 round drops below 731m/s is around 200m, as opposed to the M4, where this occurs at a range of less than 100m.[51] This has led the USA to procure a new 5.56x45mm cartridge, the Mk 262, which fires a 77gn (5g) bullet at 832m/s from a 16in barrel, while fragmenting at velocities as low as 610m/s, meaning it will have enhanced lethality at about 300m for a 16in barrel and 250m for the 14.5in (M4) barrel.

The problem with the lack of lethality of 5.56x45mm ammunition at ranges over 300m has also plagued the British Army, which has also seen combat operations in both Iraq and Afghanistan (where some 50% of engagements occur over 300m).[52] The problem has been partly tackled at least, by firstly the re-introduction of large numbers of 7.62mm L7A2 GPMGs which are deemed to be "essential" for operations and provide a "battle-winning capability" at platoon or even section-level.[53] Secondly, has been the procurement under an Urgent Operational Requirement (UOR) of 440 7.62x51mm LM7 semi-automatic rifles from Law Enforcement International (LEI) who beat competition from Heckler & Koch, FN Herstal and Sabre Defence Industries. The rifles will be brought into service under the designation L129A1 and will be issued to 'sharpshooters', who are service personnel who complete a designated course of instruction and are regarded as being a grade below that of 'sniper'. 'Snipers' will continue to be issued with the Accuracy International .338-calibre

L115A3 rifle but the new L129A1 will enable soldiers to effectively engage targets between 300m and 800m alongside the GMPG, filling a capability gap.[54]

Even with the introduction of the Mk 262 cartridge, some Special Forces were sceptical as to whether any 5.56x45mm could fulfil what they were looking for so began to look at other options. Two such calibres were 6.8x43mm Remington SPC and 6.5x38mm Grendel both of which had ballistic properties broadly similar to the British .280/30.[55] While these programmes were shelved due to the introduction of the Mk 242, the USA is again looking at its future requirements in the Lightweight Small Arms Technologies (LSAT) development programme which aims to develop a next generation LMG to replace the 5.56mm M249 (FN Hertsal Minimi) with a weapon that is much lighter. The initial calibre and ballistics have been chosen to match the 5.56x45mm SS109 round for comparison purposes but it is also hoped that the programme may lead to the development of both a rifle and machinegun of an intermediate calibre round that could replace both the 5.56x45mm and 7.62x51mm. One candidate is a 6.5mm cartridge firing a 120 grain bullet which approximates the effectiveness of 7.62x51mm at 1,000 yards. "Is it possible to achieve a suitable common cartridge? The evidence suggests strongly that it is. The British aimed to do this with the 7x43 cartridge half a century ago, and by all accounts succeeded admirably."[56]

Bibliography

Articles

Cutshaw, Charles Q. **'Barrett's M468 special-purpose carbine'** in *Jane's International Defence Review*, May 2004, posted on http://www.janes.com, 02 April 2004.
Ezell, Edward C. **'Cracks in the Post-War Anglo-American Alliance: The Great Rifle Controversy, 1947-1957'** in *Military Affairs*, Vol. 38, No. 4 (Dec. 1974), pp. 138-141.
Gelbart, Marsh. **'The Story of Britain's Lost Bullpups'** in *Small Arms Review*, Volume 6, Number 2 (November 2002), pp. 41 – 48.

Books

Dockery, Kevin. *Future Weapons*. Berkley Caliber, New York, 2007 (Reprint).
Dugelby, Thomas B. *EM-2 Concept and Design: A Rifle Ahead of its Time*. Collector Grade Publications, Toronto, 1980.
Dugelby, Thomas B. *Modern Military Bullpup Rifles*. Arms and Armour Press, London, 1984.

Hogg, Ian V. *Machine Guns: A Detailed History of the Rapid Fire Gun 1300 – 2001*. KP Books, London, 2002.

Popenker, Maxim and Willliams, Anthony G. *Assault Rifle: The Development of the Modern Military Rifle and its Ammunition*. Crowood Press, Ramsbury, 2004.

Raw, Steve. *The Last Enfield: SA80 – The Reluctant Rifle*. Collector Grade Publications, Cobourg, Ontario, 2003.

Internet Resources

Drury, Ian. *'British troops get new Sharpshooter weapon to blast Taliban ... because their weapons have a longer range than ours'* posted on 18 January 2010, located at <http://www.dailymail.co.uk/news/article-1244085/British-troops-new-Sharpshooter-rifle-blast-Taliban-half-mile-away.html>

Jane's Ammunition Handbook. *'7.62 x 51 mm cartridge'*, dated 25 January 2005, located on the Jane's Information Group website <http://www.janes.com>

Jane's Infantry Weapons. *'L85A1/L85A2 5.56mm Individual Weapon'*, dated 12 May 2009, located on the Jane's Website <http://www.janes.com>

Mail on Sunday. *'British troops to get U.S. rifles to tackle Taliban'* posted on 17 January 2010, located at <http://www.dailymail.co.uk/news/article-1243851/British-troops-U-S-rifles-tackle-Taliban.html>

Popenker, Max R. *'Enfield EM-2/Rifle, Automatic, caliber .280, Number 9 Mark 1 (Great Britain)'*, currently located at <http://world.guns.ru/assault/as59-e.htm>

Watters, Daniel E. *'Special Purpose Individual Weapons'*, currently located at <http://www.thegunzone.com/spiw.html>

White, Andrew. *'UK forces take on small-arms lessons from Afghanistan'*, posted on 02 September 2009 and located on the Jane's Website <http://www.janes.com>

White, Andrew. *'UK selects 7.62mm Sharpshooter weapon for Afghan ops'*, posted on 24 December 2009 and located on the Jane's Website <http://www.janes.com>

Wikipedia. *'.280 British'*, located at <http://en.wikipedia.org/wiki/.280_British>

Wikipedia. *'EM-2'*, located at <http://en.wikipedia.org/wiki/EM-2_rifle>

Wikipedia. *'SA80'* located at <http://en.wikipedia.org/wiki/SA80>

Wikipedia. *'StG 45(M)'*, located at <http://en.wikipedia.org/wiki/StG_45(M)>

Wikipedia. *'Taden Gun'*, located at <http://en.wikipedia.org/wiki/Taden_gun>

Williams, Anthony G. *'Assault Rifles and their Ammunition: History and Prospects'* webpage, located at <http://www.quarry.nildram.co.uk/Assault.htm>

Williams, Anthony G. *'Why Bullpups?'*, located at <http://www.quarry.nildram.co.uk/bullpups.htm>

Monographs

Drummond, Nicholas & Williams, Anthony G. *Biting the Bullet*, October 2009, located at <http://www.quarry.nildram.co.uk/btb.pdf>

Hall, Donald L. *An Effectiveness Study of the Infantry Rifle*, Memorandum Report No. 593, Ordnance Corps, Ballistic Research Laboratories, Aberdeen Proving Ground, MD, March 1952.

Hitchman, Norman A et al. *Operational Requirements for an Infantry Hand Weapon, Operations Research Office*, Johns Hopkins University, Chevy Chase, MD, June 1952.

Kern, Danford A. *The Influence of Organisational Culture on the Acquisition of the M16 Rifle, Master of Military Art and Science Thesis*, US Army Command and General Staff College, Fort Leavenworth, Kansas, 2006.

Notes

[1] The author would like to thank Jonathan Davies for his help in putting this case study together.

[2] Jane's Infantry Weapons. *'L85A1/L85A2 5.56mm Individual Weapon'* webpage, currently located on the Jane's website <http://www.janes.com>

[3] A bullpup rifle is one which sees the action, magazine and barrel pulled backwards into the stock with the pistol grip and trigger mechanism moving forward, resulting in a weapon that has a much shorter overall length than a traditional rifle, but retains the same barrel length. Such a configuration has both its advocates and critics - see Williams, Anthony G. *'Why Bullpups?'*, dated 19 May 2009, located at <http://www.quarry.nildram.co.uk/bullpups.htm>

[4] Raw, Steve. *The Last Enfield: SA80 – The Reluctant Rifle*, Collector Grade Publications. Cobourg, Ontario, 2003, p. 224.

[5.] Ibid. p. 5.

[6.] Williams, Anthony G. *'Assault Rifles and their Ammunition: History and Prospects'* webpage, dated 11 June 2009, located at <http://www.quarry.nildram.co.uk/Assault.htm>

[7.] Popenker, Max R. *'Enfield EM-2/Rifle, Automatic, caliber .280, Number 9 Mark 1 (Great Britain)'* webpage, located at <http://world.guns.ru/assault/as59-e.htm>

[8.] Ezell, Edward C. **'Cracks in the Post-War Anglo-American Alliance: The Great Rifle Controversy, 1947-1957'** in *Military Affairs*, Vol. 38, No. 4 (Dec., 1974), p. 138.

[9.] Popenker, Maxim and Willliams, Anthony G. *Assault Rifle: The Development of the Modern Military Rifle and its Ammunition*, Crowood Press, Ramsbury, 2004, pp. 53 – 54.

[10.] Dockery, Kevin. Future Weapons, Berkley Caliber, New York, 2007 (Reprint), p. 43; Kern, Danford A. *The Influence of Organisational Culture on the Acquisition of the M16 Rifle, Master of Military Art and Science Thesis*, US Army Command and General Staff College, Fort Leavenworth, Kansas, 2006, p. 32.

[11.] Op Cit. Popenker and Williams, 2004, p. 53 – 4.

[12.] Op Cit. Raw, 2003, p. 224.

[13.] A note on terminology and measurements: in – inch; fps – feet per second; m/s – metres per second; gn – grain with 15.43 grains equalling 1 gram; g – gram; calibre tends to be measured in millimetres on the continent but in inches in the UK and USA, for example, 7.62mm is also known as .308 and 5.56mm is also known as .223.

[14.] Recollections from Lt Col Noel Kent-Lemon MBE, TD – reproduced in Dugelby, Thomas B. *EM-2 Concept and Design: A Rifle Ahead of its Time*, Collector Grade Publications, Toronto, 1980, p. 15.

[15.] Op Cit. Dugelby, 1980, pp. 22 – 23.

[16.] Ibid. p. 8.

[17.] Op Cit. Williams, 11 June 2009.

[18.] Hogg, Ian V. *Machine Guns: A Detailed History of the Rapid Fire Gun 1300 – 2001*, KP Books, London, 2002, p. 172.

[19.] Op Cit. Dugelby, 1980, p. 17.

[20.] Ibid. p. 43.

[21.] Wikipedia. *'StG 45(M)'*, located at <http://en.wikipedia.org/wiki/StG_45(M)>

[22.] Quoted in Op Cit. Popenker and Williams, 2004, p. 54.

[23.] Op Cit. Raw, 2003, p. 7.

[24.] Quoted in Op Cit. Popenker and Williams, 2004, p. 54.

[25.] Op Cit. Popenker and Williams, 2004, p. 54.

[26.] Op Cit. Hogg, 2002, p. 172.

[27.] Op Cit. Raw, 2003, p. 8.

28. Op Cit. Popenker and Williams, 2004, p. 54.

29. Op Cit. Raw, 2003, p. 8.

30. Ibid.

31. Op Cit. Raw, 2003, p. 9.

32. Op Cit. Popenker and Williams, 2004, p. 55.

33. Op Cit. Dugelby, 1980, p. 166.

34. Ibid.

35. Watters, Daniel E. *'Special Purpose Individual Weapons'* webpage, located at <http://www.thegunzone.com/spiw.html>

36. Ibid.

37. Hall, Donald L. *An Effectiveness Study of the Infantry Rifle*, Memorandum Report No. 593, Ordnance Corps, Ballistic Research Laboratories, Aberdeen Proving Ground, MD, March 1952, p. 11.

38. Ibid..., p. 9.

39. Hitchman, Norman A et al. *Operational Requirements for an Infantry Hand Weapon*, Operations Research Office, Johns Hopkins University, Chevy Chase, MD, June 1952, pp. 40 – 1.

40. Op Cit. Popenker and Williams, 2004, p. 56.

41. Op Cit. Popenker, *'Enfield EM-2/Rifle, Automatic, caliber .280, Number 9 Mark 1 (Great Britain)'*.

42. Op Cit. Kern, 2006, p. 36.

43. Op Cit. Dockery, 2007, p. 44.

44. Cutshaw, Charles Q. **'Barrett's M468 special-purpose carbine'**, *Jane's International Defence Review*, May 2004, posted on <http://www.janes.com>

45. Jane's Ammunition Handbook. *'7.62 x 51 mm cartridge'* webpage, dated 25 January 2005, located on the Jane's website <http://www.janes.com>

46. From *'Editor's Foreword'* – Op Cit. Raw, 2003, p. xxv.

47. Op Cit. Popenker and Williams, 2004, p. 60.

48. Wikipedia. *'.280 British' webpage'*, located at <http://en.wikipedia.org/wiki/.280_British>

49. Op Cit. Williams, 11 June 2009.

50. Op Cit. Cutshaw, <http://www.janes.com>, 02 April 2004.

51. Ibid.

52. Drummond, Nicholas & Williams, Anthony G. *Biting the Bullet*, October 2009, located at <http://www.quarry.nildram.co.uk/btb.pdf>

53. White, Andrew. *'UK forces take on small-arms lessons from Afghanistan'*, posted on 02

September 2009 and located on the Jane's website <http://www.janes.com>

[54.] White, Andrew. *'UK selects 7.62mm Sharpshooter weapon for Afghan ops'*, posted on 24 December 2009 and located on the Jane's website <http://www.janes.com>; *Mail on Sunday. 'British troops to get U.S. rifles to tackle Taliban'* posted on 17 January 2010, located at <http://www.dailymail.co.uk/news/article-1243851/British-troops-U-S-rifles-tackle-Taliban.html>; Drury, Ian. *'British troops get new Sharpshooter weapon to blast Taliban ... because their weapons have a longer range than ours'* posted on 18 January 2010, located at <http://www.dailymail.co.uk/news/article-1244085/British-troops-new-Sharpshooter-rifle-blast-Taliban-half-mile-away.html>

[55.] Op Cit. Wikipedia, *'.280 British'*; Op Cit. Cutshaw, <http://www.janes.com>, 02 April 2004; Op Cit. Williams, 11 June 2009.

[56.] Op Cit. Williams, 11 June 2009.

Case 2.2 // TSR-2

The Plane That Barely Flew – A Case Study in Single Service Procurement

Peter D. Antill

Centre for Defence Acquisition, Cranfield University, Defence Academy of the UK.

Introduction

The TSR-2 project ran from 1957 until its cancellation on the 6th April 1965 and during this time it became a topic of much heated political debate. This was just at the point in history when the mechanisms for controlling UK aerospace research, development and procurement were in a state of flux and when the UK was giving up her role in South Asia and moving from Great Power status to a major regional power with a defence policy focused on Europe. It had become apparent that while the UK still had a largely self-sufficient weapons base and a technological research and development capability, both were in decline (particularly with greater numbers of science places remaining unfilled at university).

While many darker reasons have been suggested for the project's cancellation (none of which can be dismissed entirely) one of the most important aspects is that the TSR-2 programme was taking place right at the point where the country had found itself at a turning point. As the project was getting under way, the UK still had responsibility for a large section of South Asia, the Persian Gulf and Africa. By the time it had been cancelled, the UK was in the process of imperial decline and divesting itself of many of its former colonies. There was a growing realisation that major changes had to occur in the defence priorities of the country and the way in which it procured its defence equipment. Not only that, but defence inflation (the rise in equipment and manpower costs) had begun to spiral upwards (a factor behind the various procurement initiatives since that time) and the UK could no longer afford to maintain such a large standing armed forces, equip them and keep them in operation over such a large area of the globe. The move towards more effective project management had begun.

Even the UK could no longer afford to buy all that science had to offer for its armed forces and would be forced to effectively prioritise on what it wanted its forces to have. The management (or mismanagement) of expensive and complex programmes had cost the taxpayer around £430 million pounds[1] between 1952 and 1966 but it has been the enduring idea that financial and technical resources wasted on cancelled defence programmes are somehow worse than such resources being wasted in other areas of public expenditure that has led to such an interest in defence procurement both then and today.

The TSR-2 project has a place in the historical development of UK defence procurement as a spin-off of the programme was the Multi-Role Combat Aircraft (MRCA) which eventually became the Tornado programme that included West Germany and Italy in the development and production consortium, Panavia. The project also had repercussions for the system of control over research and development and UK defence procurement decision-making, which are with us up to the present day.

The Defence Environment

Throughout the 1950s a debate took place within the UK about the philosophy of deterrence and the role of the strategic nuclear forces. The public debate centred around the morality of such weapons but the debate in defence circles concerned the rightful place of the manned aircraft and the rapid pace of technological change in the fields of aerospace, ballistic missiles and guided weapons. Such developments had an impact on the RAF which had initiated a programme in1947 to procure bombers (such as the Vulcan) to act as delivery vehicles for nuclear weapons and fighters (such as the Hunter) to act as an air defence force. This was a continuation of the belief that strategic air power could independently win or deter wars separate from land and sea forces. However, developments in Germany regarding ballistic missile and rocket research seemed to indicate that the days of the manned aircraft were numbered, and when Germany surrendered at the end of World War II, the UK (along with the USA and USSR) captured this research and initiated a research programme into the basics of this technology.

From 1954 onwards the Air Staff began to look at the relationship between manned aircraft and unmanned missiles in the period from 1965 onwards. There were two lines of thought on this. The first was that up until 1962 guided weapons should not be given priority over manned aircraft and after 1965 manned aircraft would be of little use in a total war situation.[2] While in many ways they were right, it is important to note that there have been many lesser conflicts fought across the world where the flexibility of the manned aircraft has been demonstrated time and again. It was within the context of total war that Duncan Sandys wrote his White Paper of 1957.

The 1957 White Paper set out that henceforth, the priority in procurement policy would be the defence of the UK through nuclear deterrence and that deterrence being through ballistic missiles, while the air defence of the UK would be centred on guided missiles. As a result there was a cancellation of a number of projects including the Avro 730 supersonic bomber (Operational Requirement (OR) 330) and a replacement interceptor (OR 329).

These decisions received a mixed reception in the RAF. While none of their primary missions had been taken away, they feared a move away from a 'pilots' air force to one concentrated on the ground with missile technicians. While reference was made to developing the Vulcan bombers to cover the period up until the end of the 1960s, there was no mention of any other aircraft except transports. This was taken to mean that within Europe, the bombers would be replaced with tactical nuclear missiles. Outside Europe, the naval carrier task forces were assigned to providing air power in small-scale emergencies or limited hostilities, but in the event of a serious confrontation would be augmented by land based air power, the type and numbers of which was not properly defined.[3]

Thus it was assumed, it would be appropriate to develop new aircraft in this role, as, even within Europe, "there are some jobs that which missiles cannot do. They cannot reconnoitre enemy positions and bring back accurate photographs. They cannot be rapidly moved from one theatre of operations to another. Nor can they be switched from one target to another. Only a manned vehicle can provide such flexibility."[4] It was therefore logical to deduce that it would be legitimate to develop a manned system to help pinpoint the targets for the tactical nuclear missiles but which also had the capability to attack them if necessary. Such a system would prove invaluable as it could be moved to dispersal airfields quickly and make a contribution to the continuation of conflict after a nuclear exchange ('broken-backed' warfare) as there was "a possibility that the nuclear battle might not prove immediately decisive".[5]

Defence procurement philosophy in the period up until 1957 had been examined in the White Paper on the Supply of Military Aircraft, Cmnd 9388. In the immediate post-war period procurement was governed by the assumption that no immediate re-equipment of the RAF was necessary as the threat of another war in the immediate future was very unlikely and the UK's economic situation made such a programme extremely difficult to implement. Attempts were therefore made to draw up a series of operational requirements that would still be valid in 1957 (also when the UK was expected to have a reasonable stockpile of nuclear weapons). Out of these came the Vulcan, Victor, Hunter and Javelin but only two aircraft were developed as a stopgap measure, the Canberra and Valiant.

After 1950 and the outbreak of the Korean War had shattered this optimistic assumption, the public alarm over the conflict in the Far East had enabled the Government to increase defence spending considerably. Many of the 1957 projects were relabelled as high priority, and despite the cancellation of the Swift experimental fighter, were successful.[6]

This seemed to encourage the planning for a large and well-equipped air force for the 1960s. The programme began to come under pressure in the mid-1950s however, as the fear of war receded once it was seen that Korea had been a localised conflict, the cost of the programme had escalated greatly, and the aircraft industry was overloaded with new projects and production orders.

By 1955 the future for the aircraft industry in the UK was looking rather bleak. Firstly, there was the overloading as mentioned previously. Secondly, with the increasing emphasis being placed on the likelihood of a short and devastating nuclear exchange, it was questionable whether there was actually a need for a large number of aircraft manufacturers that presumably would form the basis of an expanded wartime industrial base. Thirdly, there seemed to be an indication that there would be a cutback in the number of aircraft projects being considered by the Government as there was "evidence that the industry is being asked to attempt too much".[7] Fourthly, while the last of the Korean War rearmament orders were being processed, there seemed to be a switch towards ballistic missile and guided missile development in defence thinking. The Select Committee recommended a re-evaluation of the aircraft programme as they felt that as aircraft grew more complicated, expensive and difficult to maintain there would be fewer of them and modern warfare was evolving in a way that it would be unlikely that manned fighters would be unable to provide complete protection.[8] They did however suggest that a measure of coalescence (that is, industrial consolidation and mergers) be stimulated by the Air Ministry and Ministry of Supply but with a view to maintaining a reasonable amount of competition.[9] Given the state of the industry and the defence environment at the time it is likely that such consolidation of the industry was inevitable anyway and that the use of selectively awarding Government contracts merely hastened the process.[10]

The procurement process was coming under scrutiny as well, and some aspects do bear relevance to the TSR-2 programme. Firstly, the procedures for the close examination of the proposals for both service aircraft and research projects were far from adequate within the procurement process.[11] The procurement process was started by a particular service through the issuing of an Operational Requirement (OR) which then consulted the Ministry of Supply whether such a project could be reasonably undertaken in the time-scale that was specified. The OR then went to the Defence Research Policy Committee (DPRC) which then graded it in priority with reference to the other projects that were going on at the time. The problem with this however was that the DPRC had no technical assistance attached to it and neither was the Treasury represented. The procedures for liaison and co-operation between the Ministry of Supply, Ministry of Defence (MoD) and industry were defective as well and it wasn't until the White Paper on the Supply of Military Aircraft was being prepared in 1955 that the MoD "became aware that the number of military aircraft projects was having an adverse affect on the resources of the industry."[12] Additionally, the Ministry of Supply felt it had little control over the starting and progress of projects so long as the services firmly stated a demand for such an item and those in Government agreed. While remedies to this lack of co-ordination were discussed, one of the main solutions put forward, that the RAF adopt the naval system of direct procurement and the intermediary role of the Ministry of Supply be abolished, was rejected on the grounds that the present system aided weapons standardisation among the services.[13] There was also the question of the continual demands for modification that often impeded the developmental process. Many were suggested by the industry themselves and the service would often accept them as they feared rejection might lead to the equipment being obsolete as it entered service. Many of the changes in requirement stemmed, it was claimed, by the turnover in service personnel and the continual movement and the promotion and posting of officers acted against the demands of continuity within a project. While the Select Committee felt unable to recommend the setting up of a procurement branch within the RAF, it did recommend that officers in these positions should serve a longer term than was normal and have the technical qualifications to cope.[14] There were three other recommendations in the White Paper: [15]

- That aircraft be treated as a complete weapon system and that responsibility for the co-ordination of all the components lie with the designer.
- Future programmes adopt the developmental batch system (instead of two or three prototypes, around a dozen are ordered to avoid delays due to a lack of prototypes).
- That future programmes proceed by a series of short steps leading to smaller but more frequent advances. This would ease technical problems and in the event of an emergency a relatively up-to-date and advanced aircraft could be put into production quicker, and "the overall result would be economy of the nation's resources and an increase in its preparedness at any point."[16]

The Beginnings of the TSR-2 Project

The English Electric Canberra entered squadron service in May 1951 and was considered an excellent light jet bomber, which would be able to deliver ordnance to most parts of Eastern Europe and survive for the foreseeable future. "Only a year later, the picture had changed. MiG-15s in large numbers were equipping the fighter squadrons of Russia and her allies, and there was an unvoiced but growing feeling that, if the Canberra had to go to war – especially in daylight – the tragedy of May 1940, when the RAF's Fairey Battle light bombers were shot out of the sky over France by the Luftwaffe's fighters, might be repeated."[17]

The Ministry of Supply had issued Specification B 126T that called for a series of design studies for a bomber that could carry a six-ton nuclear payload for a combat radius of 1,500 nautical miles at not less than 0.85M. Although a number firms responded, the technology wasn't available at the time and so the specification was shelved, but the contest for a low-level strike bomber for the Royal Navy remained open. This was covered under Specification M 148T and written around Naval Air Staff Target (NAST) 39 and the contest was won in 1955 by Blackburn with their B 103 aircraft which became the Buccaneer. While the Buccaneer might have seemed the ideal aircraft initially to fulfil the RAF's need for a low-level strike aircraft, the aircraft's systems were not fully up to the task, and it was considered too slow.[18] Despite this, and rumours to the contrary, the RAF looked seriously at the Buccaneer [19] but the Air Staff finally modified B 126T to include an over-the-target speed at low-level of 1.3M and the incorporation of an inertial navigation and attack system so that it could deliver conventional weapons accurately. The one thing that was becoming apparent for the Air Staff was that not only did the estimated costs have to fit in with the anticipated budget the RAF was going to receive but that additional funds for research and development would have to be fought for alongside the Army and Royal Navy as the RAF's ability to divert resources to such projects by cutting procurement programmes and support facilities was coming to an end. The selection of the airframe would depend on what the Royal Aircraft Establishment thought was a feasible design, and the actual choice lay with the Ministry of Supply and its view on the different capabilities of the aircraft industry. It was also likely that the Government would try and encourage industry restructuring with the promise of the contract for the new aircraft. "Thus it was clear at the outset that the TSR-2 project was to be used as a means of imposing an arbitrary and radically altered basic structure upon the industry. The structural policy may have been sound in itself, but the use of a massive and extraordinarily complicated contract to achieve it was another matter."[20]

The 1957 White Paper had implications for both the RAF and Royal Navy. In the event of total war, the Navy's role was uncertain, but the carrier task forces would be of use in the event of limited conflicts outside the European theatre. There was thus a debate between the two services about who was best suited to perform airborne strike and reconnaissance 'East of Suez' and this wasn't helped by the feeling both services had that the future of each service was at stake. In the aftermath of the White Paper, the RAF's requirement for a new aircraft was formalised as General Operations Requirement (GOR) 339 in September 1957. Eight firms were invited to submit proposals. These included Vickers, who produced two submissions in conjunction with Bristol Aircraft (both fixed wing, one with a single engine, the other with two Rolls Royce engines known as the Type 571), and English Electric in combination with Shorts with their P 17/P 17D combination which comprised a strike aircraft (P 17) and a Vertical Take-Off and Landing (VTOL) platform to recover it (P 17D). The P 17 was based upon English Electric's experience with the Canberra and the P 1 and called for an aircraft with a take-off weight of between 61,000 and 70,000 lbs. It would be powered by twin Rolls Royce or Bristol Olympus engines and be capable of speeds in excess of Mach 2. Hawker-Siddeley already had a submission in the form of its P 1121 interceptor/strike aircraft (developed from the P 1103 design submitted in response to Operational Requirement (OR) 329 of March 1954) which was not exactly what GOR 339 wanted, but what Hawker-Siddeley thought it should contain. A variant was produced however, which contained two crew and had a greater range and payload and which came closer to the requirements of GOR 339.

The Air Staff were coming down in favour of the P 17 but were sufficiently impressed by the Vickers submission that they included certain features of it in the form of a refined operational requirement, OR 343 in which the VTOL requirement was dropped, thus ending Shorts' participation. It also more or less demanded the amalgamation of English Electric and Vickers, who along with the Bristol Aeroplane Company formed the British Aircraft Corporation (BAC) in February 1960.[21] The experience of the companies was complimentary in meeting the requirement in that English Electric had concentrated on the low-level design work, had experience of Mach performance up to 2 and feasibility studies had shown that their design of wing was the better one. Vickers had conducted studies into Short Take-Off and Landing (STOL) had done a lot of work in electronics and air-borne equipment. The project was formally announced in the House of Commons in December 1958 by Mr George Ward and the new aircraft was to be capable of very high performance at all levels from 250 feet to above 50,000 feet and have an automatic terrain-following radar, and automatic navigation system based upon dead reckoning using Doppler and have inertial platform and air data computer information sources. In the reconnaissance role it was to be equipped with photographic cameras and a sideways looking radar, and be capable of operation from small airfields and rudimentary surfaces. "Here perhaps is the basic weakness of the TSR-2 concept, the attempt to meet too many new and complex specifications at the same time."[22]

On the 1st January 1959, the Ministry of Supply announced that "Vickers-Armstrong and English Electric had been awarded the contract to develop a new tactical strike and reconnaissance aircraft, known as TSR-2, to replace the Canberra."[23] The airframe would be developed from that of the P 17 from English Electric and would be powered by two afterburning Olympus engines (a variant of the 320 known as the 22R) from Bristol-Siddeley, an amalgamation of Armstrong Siddeley Motors and Bristol Aircraft Engines. There was concern from the design team as regards using the Bristol-Siddeley engine (the design team wanted a Rolls Royce engine) but the decision was pushed through so that the Government could achieve another industry consolidation. While there were other factors that pointed towards the desirability of Armstrong Siddeley Motors and Bristol Aircraft Engines merging (such as competing with Rolls Royce), "the scale was tipped by the Government making it clear that unless the merger took place the supersonic engine for TSR 2 would not be the Bristol Olympus 320 but an engine of R.R. origin."[24] This led to a situation that "should be avoided in aircraft development – a new design of aeroplane with a new design of engine right from flight one."[25] Given the problems that were to occur, it is interesting to speculate on the failure to choose a Rolls Royce powerplant, as it was "undoubtedly the development costs of the Olympus [that] were a big factor in the escalating costs"[26] and "the engines were to prove probably the biggest technical worry in the entire programme."[27] But the Bristol engine was the only one available at the time for immediate development.

At the time, Government departments were continually reviewing the procedures they were using for the procurement of equipment. As such, these procedures were in a state of flux, and the TSR-2 project could be seen as a sort of experimental project where new methods were being tried out. TSR-2 was to be procured under the 'weapons system' procedure where a 'prime contractor' is chosen, agrees an overall price and then sub-contracts as much of the work as it deems necessary. BAC was directly responsible for the airframe and had some 1,800 factories under direct sub-contract to them.[28] They were also directly responsible for some of the electronics suite on the aircraft and jointly responsible with the Ministry of Aviation for much of the remainder. It retained a close eye on what its sub-contractors were doing to ensure production standards and was responsible for the co-ordination of the whole project as specified in the main and all sub-contracts. Any firm that was not under a direct contract with the prime contractor (BAC) could appeal to the Ministry of Aviation if it didn't like what was happening. The Ministry of Aviation and Technology was responsible for the engine (with which it placed a contract with Bristol Siddeley), a number of electronic components, with which it placed contracts with firms such as EMI, and any work carried out at the Royal Aircraft Establishment and the Royal Radar Establishment and the use of existing RAF equipment. Taking the figure of £125 million that was spent up until cancellation,[29] BAC was responsible for only 30 percent of it[30] of which 17 percent was 'in-house' expenditure.[31] Thus the 'prime' contractor (BAC) did not really have the contractual authority to control the whole project. "Many of the sub-contractors were working directly with the Ministry and not under the control of the central management organisation, that is, BAC."[32] It could also be argued that the Ministry failed to specify accurately what equipment was actually required and to keep a tight enough watch on how their contracts were being handled.

The Construction of TSR-2

The basis that was laid for Ministry control of the TSR-2 project was in a detailed development cost plan that was submitted in March 1960. This provided the basis for the TSR-2 Management Document "which showed in detail a phased development programme for each component part, the interrelation between each and the correspondence between cost and technical progress."[33] As specified in the contract, the cost estimates were to be revised on an annual basis, expenditure returns were to be put in quarterly and the development plan was to be kept up to date by regular progress reports from the contractor.

The first real cost estimate of £137 million was not available until March 1962 and so the early cost-plans must have been rough estimates only. In January 1963, the Ministry decided better control was needed and so appointed a single project manager from BAC. For their part BAC introduced the PERT (Programme Evaluation Review Technique) system, set up a new team to keep a check on costs and appointed a manager in charge of value engineering. Implementation of control for the project was by means of a number of specialist committees with the TSR-2 Steering Committee (and below that a Management Committee) overseeing the whole affair with representatives from the RAF, Ministry of Aviation and industry sitting on it.

There were even two committees on the financial side of the project.[34] The specialist committees could decide minor issues, but if they involved cost changes they had to be referred to a higher level. "It seems clear that there was a failure of communication over the working of this machinery"[35] and that "throughout its development, TSR-2 was to be bedevilled by the Board's decisions and compromises. In effect, it was the first time in the history of British aviation that decisions affecting the design of an aircraft were taken away from the design team involved and placed in the hands of a committee."[36] This sort of management structure proved difficult to co-ordinate and was very time consuming.[37]

As a project, TSR-2 never went beyond the development stage due, in many ways to the use of the development batch procedure, which meant that the prototypes were built on an assembly line with production line jigs. The procedure was meant to reduce the development time in the aircraft's life cycle through ordering a number of prototypes. This procedure leads to rising research and development costs, which was undesirable with the British system of Treasury control.[38] It also takes a long time for a prototype to get into the air, which can help in terms of the politics of defence procurement as it enables officials to show that public money is producing something that actually works. Under the prototype procedure, a machine is literally constructed from scratch and then displayed to encourage further investment. If the TSR-2 been constructed under that procedure and not intended to have the avionics immediately available for squadron service, Mark 1 TSR-2s might have been produced with COTS (commercial off-the-shelf) equipment and been in squadron service around 1966 – 7.[39]

The aircraft was designed to fly fast at both high and low altitudes. This meant "changes in design"[40] that affected the aircraft's development as the aircraft was required to fly supersonically near to the ground. The TSR-2's wing was relatively small and used large blown flaps over the full span of the wing combined with a high thrust-to-weight ratio so that it could achieve good STOL performance. It was also reported to be completely stable while landing and in the clean configuration, good control with the transition to supersonic flight and little response to turbulence. The aircraft's electronics were complex and varied but almost completely new. The navigation system was based on a doppler/inertial system that was updated by the radar. The nose radar was for weapon delivery and terrain following, while the sideways-looking radars could be matched against maps to aid navigation. All data output was fed into the central computer which governed the flight of the aircraft via the automatic control system.

The TSR-2 could also be equipped with a comprehensive reconnaissance set-up which could be aided by the sideways-looking radar. The engines were a pair of Olympus Siddeley 22R turbojets, each having a development potential of 33,000lbs of static thrust. It would be armed with air-to-ground missiles (such as the Matra HSD AS37/AJ, which was being developed for both the TSR-2 and Mirage IVA), conventional bombs or nuclear weapons. It may also have acted as a patrol fighter, and so could have been given air-to-air missiles. The aircraft as a whole had significant redundancy built-in in order to survive system failure and battle damage.

The Cancellation of TSR-2

The TSR-2 first flew at the end of September 1964 (Prototype XR219) and by early 1965 the flight testing had picked up rapidly with a second prototype joining the testing programme and the first supersonic flight was on 21 February. None of this seemed to matter however, as the Labour Government cancelled the project on 6 April 1965. The Government did not even allow the BAC Management to tell his staff the news that they were going to be made redundant as it was considered a budget secret.[41] After this announcement, the decision was taken to destroy the production line and the aircraft still awaiting assembly. The two complete and one almost complete prototypes were taken to the gunnery ranges at Shoeburyness and used to test the effects of gunfire. "The assassination was to be complete; no trace of the project was to survive."[42] Why was the project cancelled? The main reasons that have been cited are:

Cost[43] – Cost is one of the main reasons cited for the cancellation of the project as the overall cost of the programme was seen to overshadow the military utility of the aircraft. There were several reasons for this. Firstly, there was the problem of sub-contractors putting in low figures for work and those figures being passed on from the Ministry of Aviation to the Treasury. Secondly, the plane was built in two different places, by two different partners who had not worked together before and control was invested in a series of ad hoc Ministry committees whose exact responsibilities and powers were not clearly defined. Third, the unprecedented level of reliability and redundancy that the RAF had asked for. Fourth, the difficulties and delays as already mentioned over the engines. Fifth, the tendency for the RAF to overload the project with optimum requirements. It was difficult to see whether the Air Staff could fully comprehend the scale of costs this was adding to the programme. Sixth, the delay in getting official decisions and the uncertainty this caused as well as the effect on morale of the production and design teams. Even so, by American standards the project had suffered only minor delays and technical hitches.

The F-111 had suffered in many ways, such as the fuselage having to be lengthened to cut down drag, the engine intakes had been wrongly designed and the all-up weight was over the original limit. Any of the faults suffered with the F-111 would have got the TSR-2 cancelled. The Government also contended that it would be cheaper to buy the F-111 but they were proved wrong when they were forced to eventually cancel the order. The overall cost of the project and the purchase of fifty TSR-2s has been quoted as being around £400 million.

The purchase of fifty F-111A aircraft (assuming £2.5 million per aircraft) would come to around £325 million (cost of TSR-2 cancellation: £200 million; cost of F-111s: 125 million excluding spares and servicing). While there seems to be a saving of £75 million, one must not forget the charge for servicing and spares, the loss to the exchequer of the money going abroad and not back into the economy as would have happened with the purchase of TSR-2.[44] There are however, additional factors to take into account, such as the UK's foreign exchange crisis and the desire to obtain a loan from the International Monetary Fund for which we needed American support, the burden of increased costs that were passed onto the Concorde project, the loss of foreign military sales of the TSR-2, the loss to UK technological research and development as a whole, and the damage done to the UK's ability to compete in the field of advanced military aviation. "The TSR-2 was virtually our last chance to move back into a lead position in military aviation and this was thrown away, ostensibly due to costs; but this excuse, if it was in fact the true motivation for cancelling which is doubtful, has never stood close examination."[45] Even so, "axing the TSR-2, HS681and P1154 did not lead to a better future along the lines Plowden recommended."[46]

The lack of foreign orders – the decision to cancel the TSR-2 programme was not complicated by the existence of foreign orders for the aircraft. Had such orders been placed, the decision to cancel would have had international repercussions and perhaps staved off cancellation. The only real customer that had expressed interest was the Royal Australian Air force, which had begun to think about replacing their ageing Canberra's and had looked at what the RAF was going to do.

Despite an official mission under Air Marshall Sir Valston Hancock, Chief of the Air Staff, which looked at both the TFX (the F-111) and TSR-2, and recommended TSR-2, the Australian Government eventually, chose to favour the F-111 and ordered twenty-four aircraft. The reasons for this are several, and include closer Australia – USA relations (the Australians had opted to buy the 'Charles F Adams' guided missile destroyer from the United States), comparative costs of the two aircraft (the United States may have offered a more 'cut-price' deal) and the effects of the official and unofficial 'anti-TSR-2' lobby in the UK.[47] This included many in the Labour Party (then in opposition but likely to return to power at the 1964 General Election) and the press, but the project also had its opponents in Whitehall itself including the Chief of the Defence Staff, Lord Mountbatten who had fought long and hard to protect the interests of the Navy and to unify the three services resulting in the reorganisation of the Ministry of Defence.

From the outset of the project the Chief of the Defence Staff had favoured the RAF acquiring the Buccaneer, despite it unable to meet the requirements of OR343. The Chief must have been aware of the increasing competition between the three services for funding and was determined that the Navy obtain a new series of aircraft carriers. The situation wasn't helped by the decision at Nassau to go for Polaris as the replacement deterrent for the V bombers, despite the potential of the TSR-2 to handle the strategic nuclear role, amongst other things. This decision alarmed the Air Staff but the Admiralty had little reason to welcome it either, as to secure both the carrier programme and the strategic deterrent would involve a major clash of interests with the RAF. It was unlikely that the strategic deterrent would go, and Mountbatten viewed the carrier replacements as absolutely vital, so in an era of budgetary pressure something would have to give. In the end it was TSR-2.[48] While it is natural that the First Sea Lord would want a naval solution to the RAF's problem, the continual undermining of the TSR-2 project and his advocacy of the Blackburn Buccaneer couldn't have helped but be noticed by the Australians, particularly after Sir Frederick Scherger had visited the UK in April 1963. "There are strong reasons to believe that Scherger left this country with his confidence in TSR-2 virtually destroyed."[49] The second major opponent (who also saw Scherger) was Sir Solly Zuckerman, Scientific Advisor to the Chief of the Defence Staff since 1960. Zuckerman had never taken it upon himself to visit or obtain information on the British aviation industry and its place in defence and the economy as a whole, but had looked at the US aviation industry quite closely. He recommended for many years that the UK should seek to procure its military aircraft from the United States. He was a willing ally for the Chief of the Defence Staff and saw quite a lot of Harold Wilson while he was in opposition, probably giving comfort in that the UK could buy American if TSR-2 was cancelled.

The changing strategic environment – how far was the decision to cancel affected by our changing strategic position? The 1966 Defence Review foresaw that British forces would remain in an 'East of Suez' role for at least a decade. However, while the cancellation of TSR-2 was not the immediate prelude to a massive change in the UK's strategic role (that happened after the 1967 Stirling crisis which caused the cancellation of the CVA-01 as well), it must also be remembered that there were also other cuts (such as the P1154 and HS681) and the UK had entered a period of retreat from many of its imperial commitments after the Suez crisis of 1956.

The political symbolism of TSR-2 – the TSR-2 project was considered to be a Conservative 'prestige project' by the Labour Party, one of the major causes of the UK's economic difficulties. This was unfair as the project had only cost some £125 million up until cancellation, and had in fact, quite a slow rate of spend for a major weapon system of this type.

Conclusion

The decision to cancel was controversial in many quarters and remains so today, including the way it was done, particularly the destruction of the prototypes, as that did not even allow the flight team to continue testing the aircraft for the benefit of other programmes. The decision came as a major blow to the UK aerospace industry as there is "no doubt that TSR-2 would have been an outstanding strike and reconnaissance aircraft, with the potential of filling other roles".[50]

It also damaged defence and aerospace research and development that, along with other decisions British Governments had made, meant that the UK "stood eight years behind America in the continued process of development and production of new supersonic military aircraft, through no technical fault of its own but due to inept decisions by succeeding Governments which utterly failed to understand the problems involved and took the wrong decisions. The TSR-2 was virtually our last chance to move back into a lead position in military aviation and this was thrown away".[51]

Many of the project team who were made redundant ended up going to the United States. The RAF in the end, did not even receive the F-111 which ran into considerable technical difficulties and cost-overruns and which the Government eventually cancelled the order at great cost.[52] While the Buccaneer was adapted to RAF service, for which it performed an admirable job, it wasn't until the arrival of the Tornado in 1982 that the RAF acquired an aeroplane, which approached the TSR-2 in terms of capability.[53] Much has been said about this decision and its effects. Some examples are:

"I think it's the most shameful aspect of this sad story. There is no sort of reason whatsoever, except a, what you could really describe as, a selfish determination to ensure that that aeroplane would never be built under any other circumstances or in the future."[54]

"I never heard anybody connected with this project who was other than shocked by this decision."[55]

"I personally think it was a criminal act to get rid of it as it was done."[56]

"I called my book 'The Murder of TSR-2' and I believe that is exactly what happened."[57]

"An example of ... British Governments not seeming to value the importance of a thriving aerospace industry."[58]

"This was fundamentally, lack of faith in the ability of our aircraft industry to produce those sort of goods – we were in fact leading the world at the time, but the actions of the politicians made sure we would never do it again."[59]

Bibliography and Further Reading

Amery, Julian. **'Real Lessons of TSR-2'** in *The Sunday Telegraph*, 11 April 1965, p. 13.

Beamont, Roland. *Phoenix into Ashes*. William Kimber & Co, 1968.

Beesly, L. R. **'Military Aircraft Procurement'** in *Flight International*, 26 May 1966, pp. 871 – 872 & 2 June 1966, pp. 924 – 927.

'Britain's VG Team' in *Flight International*, 5 October 1967, pp. 557 – 558.

'British Aircraft Corporation TSR-2' in *Aircraft Engineering*, November 1964, pp. 338 – 353 & 361.

Comptroller and Auditor General. *Civil Appropriation Accounts,* (Classes I – V) 1964 – 65, House of Commons Paper No. 28, Session 1965 – 66, HMSO, London.

Fishlock, David. **'From the Ashes of TSR-2'** in *New Scientist*, 8 April 1965, p. 88.

Gunston, Bill. **'TSR-2: What Went Wrong?'** in *Aeroplane Monthly*, September 1973, Volume 1, Number 5, pp. 216 – 220.

Hastings, Stephen. *The Murder of TSR-2*. MacDonald, 1966, London.

Horsfield, W. D. **TSR-2 – A Comparison of Actual Handling Qualities with Estimates**, Report 534, North Atlantic Treaty Organisation Advisory Group for Aerospace Research and Development, presented at the 28th meeting of the AGARD Flight Mechanics Panel held in Paris, France, 10 – 11 May 1966.

Impact Image. *TSR-2: The Untold Story*, DD Video, 1995, DD1092.

Jackson, Robert. *Combat Aircraft Prototypes since 1945*. Airlife Publishing, 1985, Shrewsbury.

Lord Plowden. *Report of the Committee of Enquiry into the Aircraft Industry*, Cmnd 2853, December 1965, HMSO, London.

Ministry of Defence. *Defence: Outline of Future Policy*, Cmnd 124, February 1957, HMSO, London.

Ministry of Defence. *The Supply of Military Aircraft*, Cmnd 9388, 1955, HMSO, London.

2nd Special Report from the Committee of Public Accounts. Bristol Siddeley Engines Ltd, House of Commons Paper No. 571, Session 1966 – 67, HMSO, London.

2nd Report of the Select Committee on Estimates. *The Supply of Military Aircraft*, House of Commons Paper No. 34, Session 1956 – 57, HMSO, London.

'TSR.2 – Britain's New Weapon: an Assessment by the Technical Editor' in *Flight International*, 31 October 1963, pp. 710 – 711 & 738 – 739.

'TSR-2: Integrated Weapons System' in *Aircraft Engineering*, December 1963, pp. 358 – 362 & 371.

Williams, Dr G., Gregory, F. & Simpson, J. *Crisis in Procurement: A Case Study of the TSR-2*. Royal United Services Institution, 1969, London.

Notes

[1] Dr G. Williams, F. Gregory & J. Simpson. *Crisis in Procurement: A Case Study of the TSR-2*, Royal United Services Institution, 1969, London, and Stephen Hastings. *The Murder of TSR-2*. MacDonald, 1966, London.

[2] Ibid. pp. 10 – 11.

[3] Ibid. p. 12.

[4] Amery, Julian. **'Real Lessons of TSR-2'** in *The Sunday Telegraph*, 11 April 1965, p. 13.

[5] Ministry of Defence. *Defence: Outline of Future Policy*, Cmnd 124, February 1957, HMSO, London. Also known as the *Sandys White Paper*.

[6] Ministry of Defence. *The Supply of Military Aircraft*, Cmnd 9388, 1955, HMSO, London, p. 8.

[7] 2nd Report of the Select Committee on Estimates. *The Supply of Military Aircraft*, House of Commons Paper No. 34, Session 1956 – 57, HMSO, London, p. v.

[8] Ibid. p. xviii.

[9] Ibid. p. xxix.

[10] Op Cit. Williams, Gregory & Simpson. p. 14.

[11] Op Cit. 2nd Report of the Select Committee on Estimates. *The Supply of Military Aircraft*, p. vxiii.

[12] Ibid. p. xix.

[13] Ibid. pp. xix – xx.

[14] Ibid. p. xxv.

[15] Op Cit. Ministry of Defence. *The Supply of Military Aircraft*. pp. 9 – 12.

[16] Ibid. p. 12.

[17] Jackson, Robert. *Combat Aircraft Prototypes since 1945*. Airlife Publishing, 1985, p. 107.

[18] Ibid. p. 107.

[19] Air Chief Marshal Sir Neil Wheeler, Dep Dir Op Req 1, 1953 – 1957 interviewed on *TSR-2: The Untold Story*.

[20] Hastings, Stephen. *The Murder of TSR-2*. MacDonald, 1966, London, p. 29.

[21] Op Cit. Jackson. p. 107.

[22] Op Cit. Williams, Gregory & Simpson. p. 20.

[23] Op Cit. Jackson. p. 107.

24. *2nd Special Report from the Committee of Public Accounts.* Bristol Siddeley Engines Ltd, House of Commons Paper No. 571, Session 1966 – 67, HMSO, London, p. 116.

25. Roland Beamont, interviewed in *TSR-2: The Untold Story.*

26. Ibid.

27. Gunston, Bill. **'TSR-2: What Went Wrong?'** in *Aeroplane Monthly*, September 1973, Volume 1, Number 5, p. 217.

28. Op Cit. Williams, Gregory & Simpson. p. 22.

29. Comptroller and Auditor General. *Civil Appropriation Accounts* (Classes I – V) 1964 – 65, House of Commons Paper No. 28, Session 1965 – 66, HMSO, London, p. xviii.

30. Op Cit. Williams, Gregory & Simpson. p. 22.

31. Dr A. Hall (BAC's Chief Project Engineer) in **'Britain's VG Team'** in *Flight International*, 5 October 1967, p. 558.

32. Roland Beamont, interviewed in *TSR-2: The Untold Story.*

33. Op Cit. Comptroller and Auditor General. *Civil Appropriation Accounts* (Classes I – V) 1964 – 65, p. xx.

34. Op Cit. Hastings, pp. 36 – 37.

35. Ibid. p. 37.

36. Op Cit. Jackson. p. 107.

37. Beamont, Roland. *Phoenix into Ashes.* William Kimber & Co, 1968, pp. 37 – 38.

38. Op Cit. Williams, Gregory & Simpson. p. 25.

39. Ibid.

40. Op Cit. Comptroller and Auditor General. *Civil Appropriation Accounts* (Classes I – V) 1964 – 65, p. xviii.

41. Stephen Hastings, interviewed in *TSR-2: The Untold Story.*

42. Op Cit. Jackson. p. 111.

43. Op Cit. Hastings, pp. 57 – 64.

44. Ibid. p. 61.

45. Op Cit. Beamont, p. 165.

46. Op Cit. Williams, Gregory & Simpson. pp. 32 – 33.

47. Ibid. p. 33.

48. Op Cit. Hastings, pp. 68 – 71.

49. Ibid. p. 91.

50. Op Cit. Gunston, p. 220.

51. Op Cit. Beamont, p. 165.

52. Op Cit. Jackson. p. 111.

53. Ibid.

[54.] Stephen Hastings, interviewed in *TSR-2: The Untold Story.*

[55.] Ibid.

[56.] Peter Arnold, Defence Research Establishment, Shoeburyness, interviewed in *TSR-2: The Untold Story.*

[57.] Stephen Hastings, interviewed in *TSR-2: The Untold Story.*

[58.] Air Chief Marshall Sir John Baraclough, Director of Public Relations, RAF 1961 – 64, interviewed in *TSR-2: The Untold Story.*

[59.] Roland Beamont, interviewed in *TSR-2: The Untold Story.*

Case 2.3 // Rapier

The Procurement of an Air Defence Missile System for the British Army

Peter Tatham, Stuart Young and Prof. Trevor Taylor
Centre for Defence Acquisition, Cranfield University, Defence Academy of the UK.

Background

The Rapier family of Surface-to-Air missile systems began its development some fifty years ago with the original design studies being carried out by the then British Aircraft Corporation (BAC) and the Royal Signals and Radar Establishment (RSRE) in the late 1950s. Although grounded in the Cold War, the requirement for air portable defence against aircraft is enduring – albeit the UK's current military deployments are not taking place in the context of a significant threat from this quarter.

As a result, and as will be discussed later in this case study, the requirement for this capability is currently at a lower priority, and the reductions in operational firing platforms reflect this. Nevertheless, in other ways, such as the novel support arrangement, Rapier remains at the forefront of acquisition developments and is one of the last significant complex weapons systems to be developed indigenously within the UK.

Doctrine

Rapier was always perceived as being in support of world wide deployments and as such it was imperative that it was capable of swift strategic deployment into all climates to provide air defence in support of the operations of the main force. Given the potentially fluid nature of the battle, the requirement was for a mobile installation with a short reaction time, compactness and low weight, a high rate of fire and kill potential, and good defensive coverage.

As a result, the initial concept was to be able to intercept aircraft flying at up to Mach 1.5 and 3,000 metres, with a medium range Surface-to-Air missile (Thunderbird) taking on higher level targets. Although these metrics have changed with the advent of increasingly agile aircraft and cruise missiles, and associated challenges such as the need to provide sophisticated electronic counter-measures (ECM) and significantly improved system availability, the underpinning doctrinal imperative remains unchallenged.

Equipment

Although the basic requirement for Surface-to-Airdefence is enduring, the capabilities of the resultant system have had to be modified to reflect the developing threat as discussed above. This section of the case study is, therefore, structured to reflect the three generations of Rapier – Field Standard A (FSA), Field Standard B1, B1M & B2 (FSB) and Field Standard C (FSC).

Rapier Field Standard A (FSA)

The original FSA project began in 1959 but, in the absence of any significant MOD interest, it was subsequently cancelled in 1962. However it was revived in 1965 as a result of studies which showed that the existing radar controlled gun system was not particularly effective and was very expensive in terms of manpower and logistics.

The project continued with live firings in 1967 and it achieved its Initial Operating Capability (IOC) in 1973 – i.e. the initial phases of what would now be described as the CADMID cycle took slightly over ten years which, for such a system incorporating a considerable number of new technologies, processes and doctrine (it is still the only true Shorad system where the missile is a two man lift), was exceptional. The basic FSA system consisted of an Optical Tracker and a Fire Unit with four missiles, to which was added a Radar Tracker from 1977. In addition to the UK market, it achieved respectable export success – it has been estimated that 600 launcher, 350 radar systems and 25,000 missiles (of all variants) have been sold worldwide over the thirty five years in which Rapier family has been in service.

This latter point was demonstrated when, in 1972, an Iranian FSA system achieved its first recorded success by shooting down an Iraqi "Blinder" Supersonic Bomber. Twelve FSA Fire Units (subsequently rising to thirty two) were also deployed during the 1982 Falklands War and it was estimated that these achieved one confirmed and four assisted kills from sixty firings. More importantly, the presence of Rapier forced Argentinean pilots to avoid low level sorties and, thereby, reduced the effectiveness of the air threat.

Notwithstanding its success in the Falklands Conflict, Rapier FSA suffered from four major operational constraints:

- It was necessary for the tracker (optical or radar) to follow the target until impact before engaging a further target.
- It had limited all weather capability.
- The missile warhead was less than fully effective and the system had a limited rate of fire.
- It was difficult to keep it operationally available in a light weight deployment without maintenance support vehicles.

Rapier Field Standard B (FSB)

The result was the development of FSB which, as indicated above, incorporated a series of improvements including the "Blindfire" radar system that provided an all weather capability, upgrades to the system's computers and the introduction of more sophisticated ECCM capability, a six round launcher. FSB also included the MK2 missile that was developed for the FSC, but backwardly compatible with FSB. This missile not only had a 20% increase in range over the Mk 1 but it also had a vastly improved lateral acceleration capability that allowed it to prosecute faster and more agile targets.

The missile came in two variants – the Mk2A that was detonated on contact in the same way as the Mk1, and the Mk2B that contained an infra red proximity fuze optimised against cruise missiles. FSB2 (forty eight systems) was also a risk reduction exercise to demonstrate the Electro-Optic (EO) and Built In Test Equipment (BITE) capabilities for FSC whereas B1MLI went only to the RAF with enhanced availability and countermeasures modifications. A tracked variant (originally developed for the Shah of Iran but subsequently taken over by the UK MOD) was also used to equip two regiments based in Germany. This system, which entered service in 1990, had the benefit of being mobile and, therefore, able to follow the armoured forces. However, it did not include a tracked radar, and so it had a limited all weather capability, although with the advent of the Electro-Optic sight, it achieved a good night time capability.

Rapier Field Standard C (FSC)

In parallel with the interim "quick fixes" of FSB to the shortcomings of FSA identified in the Falklands conflict, a formal feasibility study for an improved system (subsequently designated FSC) took place in the period 1977–1979, and these were followed by a project definition study between 1979 and 1983, the total cost of these two phases being £80M at 1987 prices (some £150M in 2007). It is interesting to note that the British Aerospace-led feasibility study identified technical solutions to all the new requirements and saw no problems in achieving an In-Service Date (ISD) in 1986 (the study even proposed bringing forward the ISD by one year). As will be seen, this was not borne out in practice.

In reality, FSC included a number of major changes to FSB and was at the cutting edge of missile, computing and radar technology. As a result, it was not clear at the time what might be technically achievable (and at what cost). Almost inevitably, such an open-ended approach ran into difficulties, and as part of the subsequent enquiry by the House of Commons Defence Committee (HCDC), the MOD admitted that the project was "very high risk" and that "full development began without the certainty that key aspects of the requirement would be feasible...". The HCDC also noted that "The absence of a [firm and precise] specification before development of Rapier FSC was far from satisfactory".

In practice, once full development began in 1983, it was progressed on a "cost plus" basis. Whilst this enabled control on expenditure, the contractor (BAe) had limited incentive to minimise costs. Furthermore, because of the innovative and technologically risky nature of the programme, there was no formal milestone plan and, as a result, the ISD slipped by some four years. By 1986, however, the technical risks had been sufficiently identified to allow a new contract to be negotiated on a "Target Cost plus Incentive Fee with a Maximum Price" basis. But this was only signed following agreement to increase the original price estimate by some £57m and also to take out some requirements such as reliability proving.

As a result, the overall cost of the system's development (including the cost of slippage caused by the failure of the MOD to supply key components and test range facilities, which in themselves represented a complex project management task) rose from £445M to £733M (a rise of £288M or 65%). Fortuitously, it was possible to accommodate this major price increase by restricting numbers to the fifty seven fire units in the initial order as opposed to the aspiration for a total of two hundred and four units, reflecting the end of the Cold War and the reducing requirement for anti-aircraft area defence. This was complemented by a reduction in the system's operational capability (saving £46m), a reduction in the level of support (saving £150m), and a further slippage of two years in the ISD – in essence, a classic example of the juggling of the Performance, Cost & Time envelope.

Initial deliveries of FSC eventually took place in 1995 (i.e. some twenty years after the start of the feasibility studies and almost ten years after the initial forecast ISD), with the system being declared operational in 1995. Its current out of service date is understood to be around 2020, and so the system will have actually achieved some twenty five years of operational life. It is also important to appreciate that the FSC contract represents an example of the use of a prime contractor to manage the activities of a number of other companies each of which is, in its own right, a major player in the defence sector. Thus MBDA (which is, itself, a joint venture between BAE Systems (37.5%), EADS (37.5%) and Finmeccanica (25%)) acts as the prime to:

- BAE Systems (CS&S): Blindfire (Targeting) Radar.
- BAE Systems (Insyte): Dagger (Surveillance) Radar.
- BAE Systems (Land): Warhead.
- Roxel (A merger of CELERG of France and the former Royal Ordnance Rocket . Motors Division – and 50% owned by MBDA): Missile.

- Raytheon: SIFF (Selective Identification of Friend or Foe).
- Lex Multipart Defence: Logistics.

Interestingly, during the development phase of the FSC variant of Rapier, the complexity of the sub-contractor matrix was rather less complex, and this made the developmental task simpler for both the MOD and Prime Contractor easier to manage.

Organisation

Although 204 FSC Fire Units were originally planned, as discussed above, following the end of the Cold War and the consequential reduction in the need for wide area anti-air defence, this was reduced to fifty seven. This was sufficient to arm two Royal Artillery Air Defence batteries and three RAF Air Defence Squadrons. In 2004, the RAF Air Defence Squadrons were disbanded and only twenty four Fire Units are currently maintained in an operational status.

Training

Training requirements were developed after extensive human factors work which examined the role of the human operator in achieving system effectiveness. Unfortunately, procurement of training systems before operational systems, and subsequent funding restrictions on updates, meant that training systems were not fully representative of in-service systems.

Three types of trainer were used:

- **Part Task Trainer.** Utilised actual Rapier system electronics. Used for individual operator training and enabled smooth transition to real systems. The Trainers were allocated at the Battery/Squadron level.

- **Detachment Engagement Trainer.** This utilised a full replicated system housed within a dome and provided team training for full detachments. Three units procured – two Army, one RAF.

- **Maintainer Trainer.** Full scale mock-up of Rapier FSC used for fault finding. Enables replacement of electronic units.

In addition, live firings were carried out on the Hebrides range. These were of limited training value as they were carried out under technical rather than operational conditions.

Logistics

The logistic support for the FSA, FSB and early stages of the FSC systems followed the standard approach in which the MOD managed and executed the process by purchasing spares to meet defined stock and availability levels, warehousing the spare parts and distributing them as necessary. However, in 2004 the £740K TRADERS (The **RA**pier **D**irect **E**xchange of **R**epairables **S**cheme) initiative was implemented through the prime Contractor (MBDA) and its sub-contractor Lex Multipart Defence (LMD). LMD became responsible for the total management, warehousing, distribution and replenishment of all consumable and repairable stocks. This contract required direct delivery to units in their peacetime or training locations (in UK, Germany or Canada), together with calibration and maintenance of all 2nd Line test equipment. The contract was subsequently externally audited and it had achieved 97% on time in full delivery against the requirement of 85%.

In 2007, the TRADERS contract was replaced by the ADAPT (**A**ir **D**efence **A**vailability **P**roject **T**eam) contract. This is a Contracting for Availability (CfA) arrangement that includes total fleet management, a joint management team and obsolescence management. It is anticipated that ADAPT will lead to savings of £175m over the twelve years until the planned out of service date (2020). The ADAPT contract includes a gain share arrangement that increases towards the end of the contract life as a means of helping to ensure that MBDA continues to search for efficiencies and improvements.

In addition, operational support is provided at two levels:

- Unit Repair (1st Line) – carried out by a Forward Repair Team equipped with a 4-tonne truck with crane, a small workshop and spare Line Replaceable Units (LRUs).

- Field Repair (2nd Line) – for more extensive repairs. Comprises:
  - o Shop Equipment Electronic Repair – containerised on a 14-tonne truck. Provides full facilities for test and repair of LRUs.
  - o Electronic Repair Support Equipment – containerised on a 4-tonne truck with application test packages for more extensive fault finding.
  - o Shop Equipment Repair – containerised on a 4-tonne truck for testing the hydraulic and cooling systems.

Personnel

The ADAPT support contract (see above) also incorporates the use of Sponsored Reserves (SPO RES) to provided deployed 2nd Line support. Under this model, two team of 8 MBDA personnel, together with their associated test, maintenance and spares support can be deployed to an operational theatre at short notice as Reserve personnel (rather than as Contractors on Deployed Operations – CONDO). This arrangement has the support of the Front Line Command who welcome the engagement of staff with specific and detailed knowledge of the operation and support of the Rapier system. In parallel, it has also been welcomed by the REME as it reduces the requirement for skilled craftsmen who are in short supply.

Information

Integration with external systems to achieve a networked capability is currently limited. At an early stage in its development, networked operation of FSC utilising the Clansman radio system in its data-linked mode was investigated but never incorporated. Connectivity with Bowman could be achieved but this requirement has not yet been defined.

Infrastructure

The TRADERS support arrangement incentivised the prime contractor to maintain a support infrastructure and manufacturing jigs. It is yet to be seen whether this will continue under ADAPT. Apart from the original Bedford site used for final integration testing, no further infrastructure is required.

Summary

The RAPIER family offers a classic story of the technical capability of a system developing over time to meet an increasingly challenging operational requirement with similar improvements in both reliability and availability. But, at about the time the RAPIER FSC achieved the necessary technical capability, real world events (ie the end of the Cold War) reduced the priority for the requirement. As a result, it was possible to reduce the number of planned fire units and the consequential reduction in expenditure more than covered the earlier cost over-runs.

In parallel, however, the support arrangements have developed in line with the DE&S strategy and, indeed, present an excellent illustration of the planned support concepts. As a result, the current capability represents a technologically sophisticated system, with high reliability and availability, underpinned by an equally high quality support contract.

Case 2.4 // Horizon

European Collaborative Procurement – The Horizon Common New Generation Frigate

Peter D. Antill

Centre for Defence Acquisition, Cranfield University, Defence Academy of the UK.

Introduction

The vastness and complexity of the oceans seem to be all too often underestimated and it is remarkable that up until very recently, man knew more about the distant Moon than the depths of the oceans, which cover 71 per cent of the planet. The surface and weather conditions on the surface ranges from total calm and absolute clarity, to raging seas and zero visibility. The three dimensional space under the surface is even more hostile, with the depth ranging from less than 130 meters on the continental shelf to around 6,000 meters in the abyssal plains and pressure going up to thirty tonnes per square meter.

The land is characterised by hills, valleys and political boundaries, but the surface of the ocean is uniform and apart from the small amount of 'territorial waters' granted to each country, has no boundaries. Naval forces can thus travel at will, and moving quickly from over the horizon, can alter the balance of power in an area. They thus have a powerful power projection capability.

For much of history, ships have been powered by sail (or by oars at certain times) and the last great sea battle involving sails was the Battle of Trafalgar, fought on 21 October 1805 between the British fleet led by Admiral Lord Nelson and the combined Spanish and French fleet. Although he used new tactics, the equipment used by Nelson had changed little from that of Drake's time in 1588. The pace of technological development quickened through the late 19th Century and early 20th Century. Muzzle loading cannon and sails gave way to beach loading guns and steam propulsion. Better ammunition, gun tubes and propellants gradually lead to an increase in ranges, which were eventually married up with improved fire control for greater accuracy. The First World War saw the emergence of the submarine (as typified by the German U-Boat) and the aircraft carrier, although they did not really come into their own until the Second World War, with the Battles of the Atlantic, Midway and Leyte Gulf, for example. The surface war saw the twilight of the battleship as the primary weapons platform and the rise to prominence of the aircraft carrier, and subsequently the submarine. The carrier carries a flexible air group, which in the case of an American supercarrier, has more aircraft than many air forces. The submarine, with advances in hull, propulsion and weapons technology, has evolved into a significant threat to both land and surface targets.[1]

The collapse of the Warsaw Pact and Soviet Union at the start of the 1990s meant that a significant threat to western navies in the NATO area dissolved. Political instability, from whatever source, be it nationalism, ethnic hatred, or religious intolerance, has increased steadily in the post Cold War world. Armed conflict has often occurred as a result of this instability, with examples being the breakup of the Soviet Union and disintegration of Yugoslavia. In many instances, these conflicts have resulted in multinational humanitarian and peacekeeping operations, often under a UN Security Council mandate. As such, Western naval forces have had to play their part, and such conflicts have brought a number of challenges to bear on the blue-water navies of the West. For much of their time they exercised, trained and were equipped in order to deal with the ships, submarines and aircraft of the Soviet Navy.

But now, blue-water operations have given way to operations in the littoral, an area defined by the US Navy as the 'near land area' of the world (any land or ocean within 650 nautical miles (1046 kilometers) of a coastline). This change in environment has meant that Western naval forces have had to radically re-think their anti-air warfare (AAW) strategy, tactics and procedures in order to cope with possible scenarios far different from those envisaged just a decade ago. In the littoral, the AAW battle is compressed into a much smaller space than a blue-water engagement would be, and thus reaction time is reduced, which is complicated by restrictive rules of engagement (ROE), if they exist. It is thus vital, that warships operating under these conditions have the sensors and weapon systems necessary to enable them to carry out their job, particularly if they are escorting a high value target.[2]

In recent years, the cost of warships has soared, making a hull of about 4000 tonnes the largest many navies can afford to buy in quantity, in terms of initial monetary outlay, dockyard construction facilities and maintenance. The frigate (and destroyer for that matter) are medium sized ships, that tend to fulfill a 'jack-of-all-trades' role within a fleet, having to counter air, surface, missile and submarine threats.[3]

Background to the Project: Royal Navy Frigates in the 1980s

During the 1980s, the Royal Navy has had two major frigate classes in service, the Amazon class (Type 21) and the Broadsword class (Type 22) frigates. The origins of the Amazon class lay in the 1966 decision by the Labour Government to phase out the Royal Navy's aircraft carriers, and the cancellation of CVA-01, and three out of the four Type 82 destroyers.

The Type 42 destroyers were at a very early stage of development, as was their sister ships the Type 22 frigates (to replace the Leander class (Type 12M)), and so a large gap was threatening to open up in the Royal Navy's modern fleet escort surface vessels. As the Royal Navy's ship design departments were at full stretch with working on the Type 42 and Type 22 designs, there was an invitation to tender issued to the commercial sector for a suitable design. Vosper Thornycroft offered a development of its Mk 5 and Mk 7 frigate designs, and this was chosen as the basis for a contract. The contract was awarded to Vosper on 27 February 1968 and Yarrow Ltd was nominated to assist in the design and building stages.

The new Type 21 frigates were the first major warships that were designed to be propelled only with gas turbines. Two Rolls Royce Olympus gas turbines supplied 50,000 shp allowing a speed of up to 32 knots at full load. As with the Tribal class, they were designed as general purpose vessels and so had a mix of armament. They had the new Mk 8 automatic 4.5in gun, developed from the British Army's Abbot self-propelled artillery piece. There was also a quadruple Seacat launcher on the roof of the helicopter hanger, a pair of Knebworth/Corvus chaff launchers forward of the bridge and a pair of single 20mm guns either side of the bridge. There was also a substantial increase in the electronics and computer assistance in running the ship. This meant that the overall crew complement was reduced to 177, as opposed to 250 or so for the Leander class.

The one criticism of the class was that they seemed under-armed for their size and cost, and so the Exocet missile system was installed on all ships after the first three, and those would be fitted with it at their mid-life refits (although *Antelope* was sunk during the Falklands conflict). They were also fitted with additional electronic equipment including a comprehensive ESM aerial array carried around the top of the foremast. Generally, the Type 21s have been popular with their crews and served with distinction during the Falklands conflict. Both *Antelope* and *Ardent* were sunk, and while some criticism has been leveled at the use of aluminium alloys in the superstructure, on the whole they proved reliable and sturdy vessels.[4]

There have been a number of further upgrades and modifications though, starting in around 1983 when large cracks were noticed in the hulls of *Arrow* and *Amazon*, which required a large steel strap to be fitted throughout the class. Additionally, ballast has been added to improve stability, the sonar equipment has been updated and all ships have had four light 20mm anti-aircraft guns fitted, although it was decided not to fit Seawolf in place of Seacat.[5] In 1993-5 the ships were transferred to the Pakistani Navy and renamed the Tariq class.[6]

The development of the successor to the Leander class frigates proved to be a long drawn out affair, and eventually became the Type 82 fleet escort destroyer which was designed to act as an escort for the CVA-01 fast carrier. With the cancellation of this project and with only one out of the four Type 82 vessels (HMS *Bristol*) being completed, and the decision to phase out the remaining aircraft carriers, by the mid-1970s, the Royal Navy would have no effective air defence vessels.

Top priority was given to the design of an air defence escort vessel, and the result was the Type 42 destroyer, the first of which was HMS *Sheffield*. With the priority given to the Type 42, the Type 21 was brought in as a stopgap measure, and work on the Type 22 proceeded slowly. The first of class was eventually laid down in February 1975, with the others proceeding at yearly intervals. The specification called for the ability to conduct sustained anti-submarine warfare, but have a general capability as well. An early decision was made to have the same engines as the Type 42 destroyers, which meant that the Royal Navy would benefit from economies of scale, and operate a common spares holding.[7]

The main anti-submarine weapon was the Lynx helicopter (of which there were two with a double hanger) and together with the Stingray torpedo, formed a capable system. The design of the hull closely followed that of the Leander class, which was renowned for its sea-keeping ability. The hull is flush-decked with the forecastle being raised, as in the Leander ships. A superstructure deck stretches nearly the full length of the ship and provides a generous amount of internal space for its electronic systems. The upper superstructure was kept as clean as possible with only the necessary systems visible, to help with cleaning down nuclear fallout, and to minimise the surfaces a radar signal could lock onto.

The ship was armed with four container Exocet launchers, two sextuple Seawolf missile launchers, two 40mm Bofors guns and the Knebworth/Corvus multiple rocket launcher. There is an extensive range of electronic and radar systems, including the Type 967/968 radar, providing information for the Seawolf tracking radar and Exocet guidance system, and the Type 910 tracking and guidance radar systems, a comprehensive ECM outfit, and a Type 1006 navigation and short range surveillance radar.

The first of class, HMS *Broadsword*, was laid down in February 1975, and commissioned May 1979. From the fifth of class onwards, it was decided that the hull would be lengthened by 41 feet to provide more internal space, for munitions and additional sonar equipment. *Boxer* was the first of the Batch II ships, and as well as the additional space, had a Computer Assisted Command System (CACS-1) fitted, which was intended to replace the Computer

Assisted Action Information System (CAAIS), but needed a great deal of work just to get it up and running.

Although the original intention was to build a large number of Type 22s, the 1981 Defence White Paper cut the programme back as they were deemed too expensive, and only one more was to be added to the six already approved by the time of the White Paper. However, the Falklands conflict saw *Brilliant* and *Broadsword* assigned as close escorts to the carriers Hermes and Invincible. The ships emerged from the campaign with excellent reputations, and the Seawolf missile system had proven a great success. The Seawolf was completely automated, with no human operators, and on many occasions "the first warning of impending attack was the 'whoosh' as a Seawolf left its launcher."[8]

As a result of the conflict, it was announced that a further five Type 22s would be ordered to replace the losses suffered in the campaign. Originally, three of the five ships would be of a new Batch III version. However, as a result of delays to the Type 23 programme, a further two Batch III ships were ordered and earlier details revised, so that overall, four Batch I, six Batch II and four Batch III ships were finally completed. Alterations to the design of the later ships, included the Mk 8 4.5in automatic gun, an enlarged hangar to house an EH101 Merlin or Sea King helicopter, and a new Type 911 tracking radar for use with the Seawolf GWS25 system. The previous Type 910 radar systems had been found to have difficulty tracking targets very close to the sea due to 'second path' reflections. The new radar used higher 'K' band frequencies, thereby reducing the problem.

HMS *Brave* was also used as a test bed for two Rolls Royce Spey SM1 gas turbine engines which needed to be tested in an operational environment before being fitted to the Type 23 frigate. While the remaining Batch II ships kept the Olympus/Tyne configuration, the Batch III ships had the Spey/Tyne arrangement in which both engines could be coupled to the shafts simultaneously. Additionally, the weapons suite of the Batch III ships was enhanced with the 4.5in gun, replacement of the Exocet missiles with the Harpoon, fitting of the 'Goalkeeper' Close In Weapon System, which is designed to engage incoming missiles from 1500 meters down to 350. They also carry the STWS-2 A/S torpedo system, two LSE 30mm guns and have been fitted with two GSA-8 Sea Archer 30 series Electro-Optical directors.[9]

The late Batch II and Batch IIIs were fitted with CACS-5, as fitted to HMS *Boxer* and destined for the Type 23s. Despite the problems, the Type 22s remain some of the most heavily armed ships in the Royal Navy, and demonstrate the increase in the size, capability and cost of modern frigates. The ships ordered in the wake of the Falklands conflict were finally completed by 1989, but there was a gap of several years before reasonable numbers of Type 23s came on line. The Royal Navy should count itself lucky to have been able to secure "so many of these fine ships although it was unfortunate that it took a war in the South Atlantic to provide the incentive to build them."[10]

Into the 1990s

The Type 22 frigates are among the most successful warships built for the Royal Navy since 1945, but their continued evolution has led to a ship that is probably closer to a cruiser or destroyer (in terms of capability and cost) than a simple frigate, particularly in their weapons fit. What was needed was a smaller more modest design that could back up the Type 22s and provide the numbers required for the Royal Navy to maintain their out-of-area commitments.

The shipbuilding industry also welcomed the idea as a more moderate design would be more attractive to foreign navies, and help exports. In an increasingly competitive world, British shipbuilders were losing out to the Italians with their 'Lupo' class frigates and the Germans with their various 'MEKO' designs. Initial design work was started in conjunction with the shipbuilders, and the new design would be known as the Type 23. As a result of the Falklands conflict, the Type 23 programme was reviewed and altered substantially.[11]

The Type 23 was originally envisaged to act as a platform for a towed sonar array, carry light missile system against air attack and have facilities to land and refuel helicopters, although it would not carry a hangar. This was basically to keep the unit cost to around £70 million and in many ways, resembled the philosophy behind the limited capability Type 14 frigates of the 1950s. In order to achieve the lowest acoustic signature to allow the towed array sonar to function successfully, there was detailed research into suitable hull forms, and a novel propulsion system – Combined Diesel Electric And Gas Turbine (CODLAG).

The system had three separate elements – gas turbines for cruising and high speed, and diesel generating sets driving electric motors for low speeds and quiet running. In the end, the Rolls Royce SM1A Spey gas turbine was chosen, and although it only produced 18,775 shp as opposed to the 28,000 shp in the TM3 version of the Olympus engine, great efforts were being put into keeping the ship to a length of 100 meters and displacement to around 2,500 tonnes. The armament initially consisted of one OTO Melera 76mm gun, Exocet MM40 missiles and two STWS-2 triple AS tubes. Although this represented the 'minimum frigate', and may have been built for something fairly near the target price, it would have been limited in its operational applications.[12]

Even before the Falklands campaign the design was recast, lengthening the hull by 15 meters, adding a hangar for helicopter operations, light automatic guns and the Seawolf missile system. The ship now began to approach the Type 22 for cost and complexity (£90 million as opposed to £120 million), and these were not yet the last changes that would be made. As a result of the Falklands conflict, modifications were made to damage control arrangements, and the ship was divided into five self-contained fire control zones, each with their own fire-fighting equipment, escape routes and electrical power supply.

New fireproof materials and non-toxic substances were incorporated into the design, and many areas were armoured against shrapnel damage. The Vickers 4.5in gun replaced the OTO 76mm, and a vertical launch system for the Seawolf missiles was included (GWS26), instead of the GWS25 six-round launcher previously. Two additional fixed torpedo tubes were added, two single 30mm guns were installed on mountings abreast of the funnel and the bow-mounted Type 2050 sonar replaced the hull-mounted Type 2016. Length thus increased to 133 meters overall, and displacement (fully loaded) rose to 3,100 tonnes.

Some of these decisions caused a series of political battles, including selection of the surface-to-surface missile system (the McDonnell-Douglas Harpoon was finally chosen over the Italian Otomat, French Exocet and British Aerospace Sea Eagle) and the choice of the tracking radar for the Seawolf missile system (originally intended to be the Hollandse Signaalapparaten VM40 tracking radar, but the GEC-Marconi 805SW I-band radar won the tender). To this was also added the eventual cancellation of the CACS-4 and the tendering for a new command system, which meant that the early Type 23s were without a computer command system at all.

A British consortium (which included Ferranti) won, and the new system was based on Ferranti's FM2400 computers which were designed to replace the older FM1600s in the Type 21 and 23 frigates. On top of this, the Government was slow to place orders despite the importance of the Type 23 to the Royal Navy. This caused apprehension to among those dockyards that were looking for work and those in political circles seeking a commitment to maintain a 50-ship destroyer and frigate fleet. Despite the various technical and political problems that have surrounded the Type 23 frigate, the final ship is a well balanced design but also an excellent anti-submarine platform.[13]

The NFR90 Project

Despite having worldwide commitments, the Royal Navy's main task in the 1980s and early 1990s was still the defence of Western Europe from Soviet aggression, in partnership with our NATO Allies. One of the main problems that has beset NATO is the lack of standardisation across many of the weapon systems the various Allies deploy. Each country has tended to produce its own design in relatively small numbers. They all reflected each country's desire to maintain its own shipbuilding industry and to demonstrate a design capability that would encourage export orders. It had often been suggested that a standard design could be produced for the Alliance as a whole, particularly with the continuing rise in the cost of modern warships, and that such a programme would benefit from economies of scale. Eight nations were involved with the NFR90 project, the United Kingdom, Canada, France, Germany, Italy, the Netherlands, Spain and the United States. Greece, Turkey and Belgium had also expressed an interest in joining. Feasibility studies were completed at the end of 1985, which showed that it should be possible for such a collaborative project to be undertaken, and that certain national variations in equipment would have been possible to apply, and not risk the whole project.[14]

The biggest source of disharmony was the choice of anti-ship and anti-aircraft missile, and while the McDonnell Douglas Harpoon had become almost standard in NATO, the French were likely to go their own way with the Exocet and the Italians would keep Otomat.[15] The British were also concerned over the lack of a close-in missile defence system, either gun-based, or missile-based.[16]

Following on from this, an official NATO Staff Requirement was drawn up and presented to each member country by mid-1987, after which there was a period of consideration before a Memorandum of Understanding (MoU) could be signed to then proceed to the Project Definition Stage. Although it seemed likely at this stage that the British Government would withdraw from the project (due to cost and incompatibility with Royal Navy Staff Requirements), a major argument against such an action was that British companies would then almost certainly be excluded from bidding for contracts to supply equipment. Therefore it was announced in early 1988, the UK would participate (at a cost of £100 million) in the Project Definition Stage and Admiral Geoffrey Marsh RN would be Project Manager. Unfortunately, late in 1989, the UK declared that it would no longer be continuing in the NFR90 project, and that it would pursue a national replacement for the Type 42 destroyers at the end of the 1990s.

The NFR90 project offered a great opportunity for European defence collaboration and would have been a significant symbol of allied unity.[17] Unfortunately, there was difficulty in reaching agreement on work sharing, which often conflicted with the goal of cost-effectiveness. Also, approval from each participating nation was for one stage only, and all work would have to stop while the following stage is negotiated. A British, nationally designed frigate could take up to eight to ten years; the NFR project could have lasted fifteen or twenty. There was also the problem of designing a ship to accept weapons that do not as yet exist, especially when the participants do not want the same weapons fit. While there was a Project Office in which the representatives from each country were responsible to the Director, as well as seeing that their particular national requirements were met. They also had to organise a consortium of many different firms from the participants, many of whom were not keen on sharing trade secrets.[18]

The Horizon Project

With the eventual collapse of the NFR90 project, the Italians, French and British decided to build on what had been achieved and commence the Common New Generation Frigate (CNGF) – Horizon – programme, followed by Germany, the Netherlands and Spain with their own future frigate programme. The aim of the Horizon Project was to salvage as much as possible from the NFR90 project, and to ensure that the frigate was as 'common' as possible, for economic reasons. The fewer variations there are in the different national specifications, the more money can be saved on development and production.[19] It was envisaged that an International Joint Venture Company (IJVC) would be set up, which would manage the multi-billion dollar scheme. It would be made up of the prime contractors from each nation: GEC-Marconi Naval Systems-Yarrow shipbuilders, BAe Defence Systems and Services and BAeSEMA and Vosper Thornycroft for the UK; Fincantieri and Finmeccanica (with Alenia) for Italy, and DCN International for France. It was intended that after the setting up of the IJVC, it would formally bid for the contract to design the frigates and build the initial three, one for each navy.[20]

Even at that early stage, it was recognised that the allocation of work sharing would be a delicate matter. The problem was seen as the fact that all three partners were capable of building a nationally designed frigate themselves and with no-one being a specialist in one area, or lagging behind the others in another, there was no natural division of labour. It was therefore envisaged that the work would be done in 'packs' by the naval yards and then proceed according to a 'building block' approach. Additionally, a major obstacle to the project proceeding on schedule was the disagreement between the UK and its continental partners over the radar configuration for the frigate's Principle Anti-Air Missile System (PAAMS).

The UK wanted a MESAR variant, while the Italians and the French being happy with the EMPAR system. As a compromise, the three governments asked two consortia, Eurosam (Thomson-CSF/Aerospatiale/Alenia) and UKAMS (BAe Dynamics – GEC Marconi) to conduct studies to see if a single interface could be produced that could handle either type of radar.[21] By the end of October 1994, the question of whether to adopt the EMPAR or MESAR/SAMPSON variant had been effectively left up to the UK Government as the defence industry had guaranteed that both EMPAR or SAMPSON could be integrated in to the PAAMS architecture. One defence industry source described PAAMS as the main hold up in the project – with the UK's insistence on having a MESAR/SAMPSON radar system, this could well lead to a break-up in the programme.[22] Even, the IJVC's managing director commented on the unusually high level of risk, especially given the constraints within which it had to work – the tender would only be accepted once the three governments had signed the Supplement to the July MoU. This would release the money for the definition phase, but up until that point, all the costs would have to be met from the companies' own resources.[23]

Finally, in February 1995, the IJVC was formally established in London, the national prime contractors being DCN International, GEC-Marconi Naval Systems and Orrizonte SpA. The new consortium was supposed to have been formed in the previous October, but a dispute over commercial conditions with the three National Armament Directors delayed its establishment. This strained relations with the Government-level Joint Project Office, which was unable to begin actual procurement without its industry counterpart. The situation was resolved when the IJVC Horizon partners received assurances that non-recurring costs incurred in setting up the joint venture would be considered as a through-life overhead, recoverable at a later stage. By this time, concerns had been voiced over the possible gap between the CNGF requirement and the level of funding available. According to one source, IJVC Horizon would have to spend double what it had originally intended to produce all the deliverables currently agreed.[24]

By the middle of 1995, CNGF was the largest surface warship project in Europe, and exerted influence on procurement policies all over Western Europe. But the complexities of European defence collaboration were starting to have an impact and a report from the National Audit Office thought that the MoD had underestimated the overall procurement cycle by up to four years. It also warned that care would be needed if work share and cost problems were going to be avoided. While being reasonably settled in terms of management and leadership of the industrial partners, the three Governments seemed to have difficulty in agreeing certain aspects of the vessel and the main weapon system.[25] The delays in signing the MoU for the PAAMS meant that the remainder of the programme was delayed, including the start of the definition work on the combat management system (CMS), electronic warfare system (EWS), the integrated communications system (ICS) and the warship platform. The UK effectively put a hold on further progress by refusing to sign the Supplement 1 to the Project Horizon MoU until the three partners had reached agreement over PAAMS in that they should find a cost-effective technical solution that also met national work share arrangements.

The UK also had a desire to derive maximum long-term benefit from the radar for PAAMS and had reservations about the performance of EMPAR. Issues of work share, cost and competition complicated the final decisions over two major PAAMS subsystems – the Long-Range Radar (LRR) and vertical launch system (VLS).[26]

Three MoUs were finally signed in March 1996. They covered the general rules governing the three partners' collaborative effort for overall development and production of the PAAMS programme (PAAMS MoU), the PAAMS Full Scale Engineering Development Initial Production Phase (PAAMS MoU Supplement 1) and a supplement to the Horizon programme covering the design definition phase (CNGF Programme MoU Supplement 1). Despite this, many commentators commented on the continuing delays and arguments, which would slow the programme down, over differing national requirements, the choice of electronic systems and defensive weapons. As 1996 progressed, the PAAMS related delays and the hold up in the award of the SAMPSON development contract meant that the probable in-service date again slipped to sometime in 2006. This meant that there would be growing problems in maintaining the Royal Navy's Type 42 destroyers, as well as the French and Italian Suffren-, Cassard-, Doria- and Audace-class anti-air warfare ships, as the earliest Type 42 destroyers would be approaching thirty years old, HMS *Birmingham* being commissioned in December 1976. The cost of continuing to maintain these ships was regarded as considerable, as was the cost of the Sea Dart missile system.[27]

Charting a Difficult Course

Despite the resolution of the immediate issues, the programme still had to chart a complex course, with the finalization of a full-scale engineering development and initial production contract for PAAMS considered to be the most pressing priority. Another problem was one of affordability, where the costs of collaboration should be outweighed by the shared costs in all phases of the ship's life cycle: development, production and through-life operation. Cost-capability and programme-capability trade-offs are seen as central in achieving this.

Non-Developmental Item (NDI) selection was also difficult, with a number of NDI requirements having already generated national variations in order to placate partisan national industrial interests. This will reduce the economies of common procurement and through-life support. The In-Service Dates (ISD) for the three First-of-Class (FOC) were still troublesome, as the original 2002 ISD was by then clearly unobtainable, due to the misalignment between the warship and PAAMS, with the EMPAR equipped ships looking to enter service in around 2004 – 5 and the SAMPSON equipped ship in 2005 – 6. Clearly the management of the Horizon/PAAMS interface had become an important issue for the JPO. What was a cause for concern was that PAAMS was obviously the primary component of the CNGF combat system, and yet was outside the JPO's responsibility with respect to its stand-alone performance. Despite a formal charter having been drafted by the JPO, PAAMS Project Office (PPO) and an industrial working group, which was to be established between the two contractors (IJVC Horizon and EUROPAAMS) for the transfer of information. There were still however reservations within the industry at the level of cost and risk, chagrin over the national work share, resentment at having to co-operate with companies that are natural competitors in the export market, scepticism as to whether the cost savings sought through collaboration can really be achieved and an unwillingness to commit their own funds to a programme that may not reach fruition. There was therefore this dichotomy: the three governments saw an industry led programme as the best way to gain value-for-money and minimise risk. Industry however, wanted firm government commitments so it could judge how much it would be able to invest and for what return.

As the programme continued, things were reluctant to improve. There were disagreements in early 1997 over the type of vertical launch system to be employed. The French and Italians had favoured the Franco-British-Italian Sylver A50 developed by DCN, Alenia and BAe Dynamics. The UK however, had shifted to support the Mk 41, supplied by Lockheed-Martin, which would allow the installation of US SM-3 based theatre defence missiles.[28] The UK had also been refusing to negotiate a full-scale engineering development and initial production contract for PAAMS on the terms that had been offered by industry to the PPO, and had resisted moves by France and Italy to relax the PAAMS performance specification set out in the original agreement. This continued uncertainty had forced the JPO to shelve a number of key design definition contracts. What was worse, the divergence of performance goals coupled with different national contracting procedures had produced a schism in the programme, which had been made public by a leaked letter from the Chief of Defence Procurement (CDP), Sir Robert Warmsley. The letter was addressed to his French and Italian National Armament Director (NAD) counterparts, and outlined the UK's concerns – that the UK still wanted Column 2 performance, which was the Royal Navy's minimum acceptable performance criteria, meaning a local area capability against simultaneous threats. France and Italy had been ready to accept Column 1 performance, which was really only a replacement for the existing SM-1MR missile. Warmsley had also noted that industry had been unwilling to respond to the joint requests either in the form of an acceptable offer for PAAMS or more information.[29]

Finally, in mid-1997, the NADs of France, Italy and the UK conditionally endorsed the industrial management framework for the collaborative development and production of PAAMS. PAAMS was designed to provide CNGF with area, local and point defence capability, based on Astor 15 and 30 Surface-to-Air missiles. The Astor 15 missile had just been successfully tested in a 'hit-to-kill' engagement with a live MM38 Exocet anti-ship missile, demonstrating that it could carry out the local-area defence mission, against a crossing target, as required by the Royal Navy.[30] Later that year, two consortia were awarded Phase B Project Definition study contracts to develop proposals for the fully integrated EWS. The two groups involved were the ARCEO team (Racal-Thorn Defence with Alenia Difesa and CS Defense) and the JANEWS consortium (Elettronica, Horizon GIE – a joint venture between Thomson-CSF and Dassault Electronique – as well as GEC-Marconi). They were awarded £4.5 million each to conduct an eighteen-month programme of definition work. The EWS was designed to be a fully integrated and autonomous subsystem, with electronic support measures (radar-band), onboard jammers and offboard decoys, coordinated by advanced software. The specifications of the system were ambitious, with high sensitivity, high bearing accuracy, modulation on-pulse analysis and programmable high-radiated-power jamming among those deemed necessary. The initial Phase A Project Definition work had been done in 1994 and identified point defence jamming as a potential problem area. Italy favoured the cross-eye technique (using solid-state transmitters) while France and the UK looked towards the use of offboard decoys for angular seduction.[31] It was also decided that PAAMS would be procured via a 'twin-track' approach due to the differing contracting requirements of the UK and France/Italy.[32]

By the summer of 1998, the industrial partners developing PAAMS were at the point of being ready to sign an agreement on full-scale engineering development and initial production. The PAAMS Programme Office had virtually completed its deliberations concerning the selection of outstanding elements of the PAAMS system, as well as the allocation of work shares, while the UK Government had nearly finished its Strategic Defence Review. Aerospatiale had already begun extolling the virtues of the PAAMS system as a unique multi-mission air-defence system, which uses a single active-homing missile and sensor system to fulfill the self-defence, local-area defence (over seven kilometers) and area defence (up to fifty-five nautical miles radius). Combined with a radar that has a sufficiently high update rate (such as SAMPSON), Aster could be used in the theatre anti-ballistic missile role, but it would need both a new warhead (to produce larger fragments) and greater range.[33]

Royal Ordnance had also unveiled proposals for its medium-calibre gun system and hence went head-to-head with Italy's OTOBreda. The requirement called for a 5 in/127mm weapon, with gun mount, control system, magazine, hoist structure and handling system, a computer system and munitions. OTOBreda had offered its new 127mm/54 Lightweight Gun Mount (in partnership with VSEL Armaments) while Royal Ordnance's offered the Mk 45 CNGF Gun Weapon System (GWS), which was almost identical to the US Navy's new Mk 34 Aegis GWS. It was the only system at that time to be purposely designed to take advantage of the new EX171 extended-range munition being developed by Raytheon TI Systems.[34] Finally, the PAAMS Programme Office chose the Sylver missile launcher (made by DCN) rather then the rival Mk 41 vertical launcher, which is manufactured by Lockheed Martin. This was something as a blow to the Royal Navy as they had expressed interest in acquiring a theatre ballistic missile defence capability based on the Standard Missile Block IV and the Tomahawk Land Attack Missile. The Sylver launcher, as it was originally configured, could not fire either of these missiles.[35] With the impasse over PAAMS seemingly cleared, attention turned to the Project Horizon programme, which covered the ship and all other combat systems.

A second grant was made for further design definition work to the prime contractor, Horizon IJVC, for outline design work on the ship, combat system and subsystems and the selection of marine engineering. Supplement 2 to the Project Horizon MoU was due to be signed late in 1998 and would have paved the way for a detailed design and FOC build contract award to Horizon IJVC sometime in 1999. Adhering to this schedule was seen as vital to stay in touch with the 2004 in-service date. However, budgetary problems had already started to loom, with the partner countries starting to demand a reduction in the size and cost of the CNGF before committing to it. Work nonetheless continued on assessing the rival bids for the Combat Management System and the Fully Integrated Communications System. EUROCOMBAT (BAeSEMA, Thompson-CSF and Alenia) and HEPICS EEIG (Marconi, Matra Defense, Dassault Electronique, Matra-CAP Systemes, and Datamat) were contesting the CMS competition. The FICS competition was being fought by NICCO Communications (a limited company comprising Thompson-CSF, Redifon MEL, Marconi Communications and Elmer) and a rival team comprising Italtel, BAe Defence Systems, Racal Communications, Bull SA Europe and Dassault Electronique. Bids for the Electronic Warfare System had been in by the previous September and the two rival EWS consortia are ARCEO (comprising Racal, Alenia and Matra Defense) and JANEWS (comprising Electronica, Thompson-CSF, Dassault Electronique and Marconi Electronic Systems).[36]

The End of the Horizon Project

By early 1999, attempts at breaking the deadlock had resulted in the Horizon IJVC in developing a package that had the partners in a new framework with Marconi Marine in overall leadership, DCN International acting as combat system prime and Orrizonte as platform prime. Horizon IJVC's new costing proposals took account of the year-long programme of affordability studies which were designed to balance capability against affordability. There was a relaxation of the selection criteria for major non-developmental items with a matrix of options to help satisfy national requirements. There was still concern though over the hardening attitude in Whitehall over the lack of progress on the project in recent times.[37] This was combined with a fear in the UK defence industry that it was being disadvantaged by a lack of coherent Government policy and the recent pull out from the Trimilsat programme and troubles with MRAV/VCBI/GTK programme.[38] Statements from both John Spellar, the Parliamentary Under-Secretary of State for Defence and Sir Robert Warmsley, the Chief of Defence Procurement indicated that the UK was not going to wait forever while the financial, managerial and industrial aspects of collaboration continued to fail to provide a basis for an effective transition to design and production. The Royal Navy continued to have an urgent need for a replacement for its aging Type 42 air defence destroyers after 2005. Thus MoD officials started to formulate a fallback national procurement strategy for a national anti-air warfare escort should the UK decide to pull-out of the Horizon programme.

Finally, in April 1999 the UK decided to pull out of the platform component of the Common New Generation Frigate, although it confirmed it would still continue with the PAAMS component. This effectively marked the end of Project Horizon, with the French and Italians looking at the possibility of bilateral cooperation, and the UK having decided on a national warship programme. The decision to pull out was eventually arrived at due to the UK's continuing dissatisfaction with the industrial and management structure of the collaborative project, despite trying to stimulate the process by refusing to sign the PAAMS agreement. The UK wanted the prime contractor designate, IJVC Horizon, which was owned in equal shares by DCN International, Orrizonte SpA, and GEC's Marconi Marine to reconstitute itself with Marconi Marine taking on the single prime contractor role and full responsibility for price, performance and delivery. Further difficulties arose when France insisted that DCN should take responsibility for the combat system within this new structure. It also insisted that the system should be based around a HEPICS combat management system (CMS), drawing on core technology from DCN's SENIT 8 system. The UK preferred the rival Eurocombat CMS on both price and performance. The Phase One design definition studies for Project Horizon will be used by all three nations to determine their national procurement strategies.[30]

A New Warship

In mid-August 1999, Marconi Electronic Systems, leading a joint team with British Aerospace, began a ten week study to examine the design parameters for the UK's replacement to the Horizon CNGF programme – the Type 45 anti-air warfare destroyer.

The MoD meanwhile, continued work on finalising the Type 45 procurement strategy and produce a new user requirements document.

At the time of writing the programme was expected to cost around £7 billion for twelve ships, each of which would weigh about 6,000 tons and be armed with a variant of the PAAMS. While the MoD had been reluctant to confirm the selection of Marconi/BAe as prime contractor, a fully-integrated prime contracting office was being set up in Bristol with personnel from Marconi Naval Systems and BAE's Defence Systems business. Talks had already begun with possible major subcontractors such as Vosper Thornycroft, Racal and Redifon MEL.

The MoD's sensitivity to making an official announcement could probably be put down to their reluctance to award a non-competitive contract to Marconi Naval Systems due to their performance on the Astute-class nuclear attack submarine, Auxiliary Oiler and Landing Platform Dock programmes, which were running late. It may also have been that the MoD would have preferred to wait until the sale of Marconi from GEC to BAe was finalised. This would have simplified contracting arrangements.

The MoD was probably willing to select the Marconi/BAe team without recourse to a competition due to the fact that Marconi was the UK partner in the IJVC Horizon programme which meant that it had the best chance of pulling through work already done on the Horizon project. The UK had committed over £75 million up to that point and the MoD would have been keen for that money not to go to waste. There were some doubts in industry as to the amount of 'pull-through' that could have occurred between the programmes. The limited development budgets meant that the Type 45 would probably have been equipped with derivatives of current RN systems, rather than 'clean-sheet' systems pursued by the Horizon programme.

At the time of writing, the design was expected to have a sufficient growth potential for there to be technology insertion throughout the life-cycle of the platform. There was also some pressure to re-evaluate the decision to accept the SYLVER launcher, which is limited to using the Aerospatiale Aster missile, and procure the US Mk 41 vertical launcher system that has the flexibility to fire a range of missiles. An integrated project team was due to be set up in the Defence Procurement Agency, its leader being Colonel Keith Prentice who was promoted to one-star level on taking the position.[40]

Conclusion

The differences over the CMS and its requirements had caused stalemate, the final blow being that the UK considered the new proposals from IJVC Horizon to be some 20 percent over-budget and too risky.[41] The failure to resolve these differences in the requirements stems from the different roles that the CNGF was expected to take by the three participants. Even though the crucial role of air defence was shared by all, and the ships were to share the same missile, the Franco-Italian Aster, the navies had somewhat different missions in mind. The Royal Navy wanted to use the ship in an area-defence role, such as the safe-guarding of convoys or amphibious task forces, whereas France and Italy desired the ship to perform as a point defence platform for the immediate defence of an aircraft carrier, for example.

While this caused no essential clash, the missile being capable of both missions, the radar needed to perform the area defence role (as required by the Royal Navy) would have to be of a markedly different nature to that required by the point defence role (France and Italy). It was eventually agreed that the participants could employ the radar of their choice, even though this meant a departure from the principle of utilising common equipment throughout the design. Not only that, but the different navies effectively operate in different environments. The Italian Navy is essentially concerned with the shallower, more confined waters of the Mediterranean and Adriatic Seas, while the French Navy looks to operate around the Bay of Biscay, the littoral Atlantic waters near to Europe and the Mediterranean as well. The Royal Navy however, needs ships that can operate in true deep-water conditions, at mid-Atlantic range. So, while France and Italy would have been happy with a ship of around 3 – 4,000 tons, the UK could not really make do with anything less than 6,000 tons. "A compromise was reached in definition of a common operational requirement but it was only a compromise – with all that word implies."[42]

While the joint project office (JPO) that had been formed in London was representative of the Ministry of Defence of all three nations and the commercial shipbuilders that had come together for the project, it was only really a cipher. The JPO had no real authority, in that it required consent from all three governments to make decisions of any importance. In the end, national interests overrode common goals. There has also been strong criticism of the slow decision-making process and that because no single production line was envisaged, it would have been difficult to realise a reduction in hull production costs and collaborative overheads would probably increase programme costs. This negated the entire reason for a collaborative project.

There was additional concern being voiced from British industrialists in that although the UK was paying one-third of the development costs, it would actually end up receiving a low share of the work involved. While Rolls Royce was very likely to end up the winner of the contract to supply propulsion plants, because of juste retour considerations, there would be little left over to go to the electronics industry. The French and Italians were seen as suppliers of sixty-six percent of the equipment on each of the British ships and since the equipment was to be the basis for other UK warships, British industry envisaged itself being excluded from other UK programmes while French and Italian competitors won contracts at their expense. Horizon's death knell should sound a warning to those who would sacrifice national defence requirements in the interest of international collaboration. That is not to say that collaboration cannot be advantageous, for example in allowing some nations to implement programmes that they could not afford alone, via the spreading of costs and encourages the sharing of technology, cross-border industrial alliances, joint ventures and partnerships and adds closeness to partner nations.

Collaboration is an ideal way forward in an era of ever tightening defence budgets, but at times, idealism must make way for realism. Also, some benefits have emerged from the Horizon project, such as the PAAMS programme, and the electronic warfare and integrated communications systems that could still be applied to other warship programmes. The lesson that must be learnt for future collaborative projects is that common requirements are laid out and commercial practices are brought to bear from the outset.[43]

Notes

[1] Walmer, Max. *'Modern Naval Warfare'*, pp. 8 – 13.

[2] Harrison, Peter. *'Anti-air warfare in the littoral'*, p. 16.

[3] Ibid. p. 22.

[4] Marriott, Leo. *'Royal Navy Frigates since 1945'*, pp. 105 – 112.

[5] Ibid. p. 113.

[6] Jane's Fighting Ships 1998 – 99. *'Tariq (Amazon) Class (Type 21) (DDG/DD/FFG/FF)'*, 22 May 1998.

[7] Marriott, Leo. pp. 114 – 115.

[8] Ibid. p. 122.

[9] Ibid. pp. 125 – 6.

[10] Ibid..., p. 126.

[11] Ibid. pp. 128 – 9.

[12] Ibid. pp. 129 – 30.

[13] Ibid. pp. 130 – 32.

[14] Ibid. pp. 141 – 42.

[15] Paloczi-Horvath, George. *'Gunboat Diplomacy'*, p. 25.

[16] Dickey, Alan. *'MoD's fears delay orders for 50 NATO frigates'*, p. 8.

[17] Marriott, Leo. p 142.

[18.] Brown, D. K. *'The Future British Surface Fleet'*, pp. 167 – 68.

[19.] Lewis, J. A. C. *'Sharing out the CNGF load'*, p. 22.

[20.] Ibid.

[21.] Ibid.

[22.] Jane's Information Group. *'Building Europe's future frigates'*, p. 41.

[23.] Jane's Information Group. *'Euronaval: First News - Horizon IJVC unhappy at high level of risk'*, p. 9.

[24.] Scott, Richard. *'IJVC Horizon sets sail'*, p. 36.

[25.] Bickers, Charles. *'UK report highlights problems facing collaborative projects'*, p. 17.

[26.] Jane's Information Group. *'PAAMS partners set sights on November MoU'*, p. 8.

[27.] Foxwell, David. *'Trouble over the Horizon. Europe's common frigate programme beset by delays'*, p. 38.

[28.] Jane's Information Group. *'Horizon's PAAMS industries to submit work share proposals'*, p. 14.

[29.] Tusa, Francis & Scott, Richard. *'PAAMS partners paper the cracks'*, p. 5.

[30.] Jane's Information Group. *'PAAMS paves way for Horizon'*, p. 11.

[31.] Jane's Information Group. *'Project Horizon: Funding released for EWS studies'*, p. 4.

[32.] Scott, Richard. *'PAAMS partners plot twin-track target'*, p. 28.

[33.] Pengelley, Rupert. *'PAAMS ready for full-scale development and initial production'*, p. 13.

[34.] Jane's Information Group. *'Opening shots fired in CNGF gun battle'*, p. 11.

[35.] Jane's Information Group. *'PAAMS launcher set to arouse controversy'*, p. 57.

[36.] Jane's Information Group. *'Key decisions loom for CNGF programme'*, p. 35.

[37.] Scott, Richard. *'Latest talks aim to break Horizon deadlock'*, p. 67.

[38.] Jane's Information Group. *'Is Whitehall wavering?'*, p. 7.

[39.] Scott, Richard. *'Horizon warship project sinks as UK pulls out'*, p. 3.

[40.] Scott, Richard. *'Marconi/BAe team start work on UK navy's Type 45 destroyer'*, p. 13.

[41.] Corless, Josh. *'Horizon sinks', PAAMS flies*, p. 8.

[42.] Witt, Mike. *'Lost Horizon – an obituary'*, p. 90.

[43.] Ibid. pp. 89 – 91.

Bibliography

Bassett, Richard. **'Vosper Thornycroft's destroyer prospects hike its share price'**, *Jane's Defence Weekly*, 26 May 1999, p. 21.

Beaver, Paul and Scott, Richard. **'UK considers Horizon alternative'**, *Jane's Defence Weekly*, 3 March 1999.

Bickers, Charles. **'Partners say Horizon in service date could slip'**, *Jane's Defence Weekly*, 12 August 1995, p. 5.

Bickers, Charles. **'UK report highlights problems facing collaborative projects'**, *Jane's Defence Weekly*, 26 August 1995, p. 17.

Brown, D. K. *The Future British Surface Fleet: Options for Medium-Sized Navies.* Conway Maritime Press, London, 1991.

Corless, Josh. **'Horizon sinks, PAAMS flies'**, *Jane's Navy International*, June 1999, p.8.

Corless, Josh. **'Horizon stalemate continues'**, *Jane's Navy International*, 1 March 1999, p. 8.

Dickey, Alan. **'MoD's fears delay orders for 50 NATO frigates'**, *The Engineer*, 21 May 1987, p. 8.

Foxwell, David. **'Trouble over the Horizon. Europe's common frigate program beset by delays'**, *International Defence Review*, June 1996, p.38.

Gibbons, T. and Miller, D. *The New Illustrated Guide to Modern Warships.* Salamander, London, 1992.

Harrison, Peter. **'Anti-air warfare in the littoral'**, *Jane's Navy International*, August 1995, p. 16.

Hooten, Ted. **'NFR90: NATO's Next Generation Frigate?'**, *International Defense Review*, April 1988, pp. 409 – 411.

House of Commons Defence Committee. *Oral Evidence from Sir Robert Walmsley, John Cox and Mark Hutchinson: About the UK's pull-out of the Horizon (Common New Generation Frigate)* programme, heard on the 16 June 1999.

Jane's Information Group. **Building Europe's Future Frigates** in *Jane's Defence Weekly*, 15 October 1994, p. 41.

Jane's Information Group. **'Euronaval: First News – Horizon IJVC unhappy with high level of risk'** in *Jane's Defence Weekly*, 29 October 1994, p. 9.

Jane's Information Group. **'Horizon's PAAMS industries to submit work share proposals'**, *Jane's Defence Weekly*, 16 April 1997, p. 14.

Jane's Information Group. **'Is Whitehall wavering?'** in *Jane's Defence Industry*, 1 March 1999, p. 7.

Jane's Information Group. **'Key decisions loom for CNGF programme'** in *Jane's Defence Weekly*, 21 October 1998, pp. 32 – 35.

Jane's Information Group. **'Opening shots fired in CNGF gun battle'** in *Jane's Navy International*, 1 July 1998, p. 11.

Jane's Information Group. **'PAAMS launcher set to arouse controversy'** in *Jane's Navy International*, October 1998, p. 57.

Jane's Information Group. **'PAAMS partners set sights on November MoU'** in *Jane's Navy International*, October 1995, p. 8.

Jane's Information Group. **'PAAMS paves the way for Horizon'** in *International Defense Review*, August 1997, p. 11.

Jane's Information Group. **'Project Horizon: Funding released for EWS studies'** in *Jane's Navy International*, September 1997, p. 4.

Lewis, J. A. C. **'Sharing out the CNGF load'** in *Jane's Defence Weekly*, 12 March 1994, p. 22.

Lewis, J. A. C. & Scott, Richard. **'Three-nation PAAMS deal is finally sealed'** in *Jane's Defence Weekly*, 18 August 1999, p. 3.

Lok, Joris J. **'Common ground sought on future frigates'** in *Jane's Defence Weekly*, 12 February 1994, p. 21.

Marriott, Leo. *Royal Navy Frigates since 1945*. 2nd Edition, Ian Allan Ltd, London, 1990.

Paloczi-Hovarth, George. **'Gunboat Diplomacy'** in *The Engineer*, 26 January 1989, pp. 24 – 25.

Pengelley, Rupert. **'PAAMS ready for full-scale development and initial production'** in *Jane's Missiles & Rockets*, 1 May 1998, p. 13.

Scott, Richard. **'Horizon back on an even keel?'** in *Jane's Navy International*, August 1996, p. 17.

Scott, Richard. **'Horizon warship project sinks as UK pulls out'** and **'UK sets sights on national solution'** in *Jane's Defence Weekly*, 5 May 1999, p. 3.

Scott, Richard. **'IJVC Horizon sets sail'** in *Jane's Navy International*, March 1995, p. 36.

Scott, Richard. **'Latest talks aim to break Horizon deadlock'** in *Jane's Defence Weekly*, 13 January 1999, p. 67.

Scott, Richard. **'Marconi/BAe team start work on UK navy's Type 45 destroyer'**, in *Jane's Defence Weekly*, 18 August 1999, p. 13.

Scott, Richard. **'PAAMS partners agree framework'** in *Jane's Navy International*, June 1997, p. 59.

Scott, Richard. **'PAAMS partners plot twin-track target'** in *Jane's Defence Weekly*, 1 October 1997, p. 28.

Scott, Richard. **'Three more years of birth pangs'** in *Jane's Navy International*, October 1995, p. 3.

Scott, Richard. **'UK MoD disputes VT frigate claim'** in *Jane's Defence Weekly*, 26 May 1999, p. 6.

Scott, Richard & Lewis, J. A. C. **'Affordability clouds Horizon outlook'** in *Jane's Defence Weekly*, 23 September 1998, p. 6.

Tusa, Francis & Scott, Richard. **'PAAMS partners paper the cracks'** in *Jane's Navy International*, May 1997, p. 5.

Walmer, Max. *An Illustrated Guide to Modern Naval Warfare*. Salamander, London, 1989.

Witt, Mike. **'Lost Horizon – an obituary'** in *Defence Procurement Analysis*, Summer 1999, pp. 89 – 91.

Case 2.5 // JSF

Multi-National Collaborative Procurement – The Joint Strike Fighter Programme

Peter D. Antill and Pete Ito[1]

Centre for Defence Acquisition, Cranfield University, Defence Academy of the UK.

Introduction

"The joint strike fighter will be the world's premier strike platform beginning in 2008 and lasting until 2040."[2]

The last twenty years has seen a radical shift in the nature of the aerospace market and in particular its defence sector. The civilian aerospace market has become dominated by two giant rivals, the US-based Boeing Corporation and the European consortium of Airbus. With the end of the Cold War, the defence market has seen rationalisation similar to that in the civilian aerospace market but conducted at a slower pace. Boeing merged with McDonald Douglas in August 1997 after acquiring Rockwell's aerospace and defence concerns the year before. Lockheed and Martin Marietta merged in March 1995 and then acquired the defence electronics and systems integration businesses of the Loral Corporation in 1996 as well as elements of General Dynamics, while Northrop Aircraft merged with Grumman Aerospace in April 1994 and acquired a number of smaller companies such as Teledyne Ryan, Litton Industries, Newport News Shipbuilding, Federal Data Corporation and Sterling Software Inc.[3]

Today, the defence market is still dominated by companies such as Boeing, Lockheed Martin, General Dynamics, Northrop Grumman, Raytheon, Thales and BAE Systems.[4] In conjunction with this slow but steady rationalisation, the end of the Cold War has meant that most countries have sought some form of 'peace dividend' and diverted resources away from the defence budget to other areas of public expenditure.[5] In addition to this, weapon systems have been incorporating ever increasing amounts of cutting edge technology which itself requires additional expenditure on software development, systems integration and supporting technology such as satellites and real-time global communications. This has meant that the proportion of fixed costs to variable costs has been rising and that, combined with the decline in defence spending, has resulted in a decrease in the number of actual systems procured from one generation to the next, so that the unit production costs have risen as well.[6]

All this has forced companies on both sides of the Atlantic to look for partnering opportunities with their opposite numbers,[7] although challenges have remained – "it is currently harder for Lockheed Martin to do business in Europe than in Russia."[8] Indeed, despite the USA still having a total national defence budget which is twice that of the rest of NATO put together ($625.9bn[9] compared to $311bn in 2007)[10], even the USA is finding it increasingly difficult to adequately fund all its modernisation programmes. For example, cancelling the RAH-66 Comanche helicopter programme in 2004[11] and reducing the planned number of F-22 Raptor air-superiority fighters from its original requirement of 750, down to 381 by November 2008 – a figure that the US Air Force is, even now, unlikely to receive.[12]

As a result, not only are companies looking increasingly at opportunities for international cooperation, countries are as well. This case study looks the history, development and procurement of the Joint Strike Fighter, now known as the F-35 Lightning II, an aircraft that is "the Department of Defense's (DOD) largest weapon procurement program in terms of total estimated acquisition cost."[13] This programme promises to become the largest aerospace programme in history, aimed at reversing the spiralling cost of combat aircraft procurement while still including new 5th generation technology for both domestic and export aircraft. For the first time, the USA invited allied countries to participate in the programme at different levels and contribute (depending on the level of participation) towards the cost of the development.[14]

Background

The Pentagon's Joint Strike Fighter (JSF) programme "will set the world fighter market stage for the next 50 years".[15]

NATO studies conducted in the 1950s and 1960s predicted that the airfields in Western Europe would be among the first targets for any direct attacks by Warsaw Pact forces during the opening phase of an invasion, making it very difficult for conventional aircraft to take-off or land without substantial repair and clearance work being undertaken. Thus began a quest to develop aircraft that were capable of landing and taking off using very short distances or even doing so vertically in order to alleviate the vulnerabilities of conventional aircraft. Producing aircraft to fulfil such a mission has been challenging for aircraft designers for the last five decades, particularly one that does it at an acceptable cost.[16] The scale of the challenge can be seen by the fact that only the first and second generation Harrier aircraft as well as the Yak-38 Forger have achieved operational service out of a whole series of programmes dating from the 1950s.[17] Such programmes have included the Lockheed XFV-1 Salmon, the Mirage Balzac, and the EWR (a consortium formed by Heinkel, Messerschmitt and Bolkow) VJ 101C. Of the two successful designs, the Harrier has been the most successful commercially, with McDonnell Douglas helping to export the aircraft and adapt it for the US Marine Corps, which operates it as the AV-8 Harrier.

While early critics pointed to its potential vulnerability against supersonic fighters, the aircraft acquitted itself well in the Falklands War against Dassault Mirage III aircraft flown by the Argentinean Air Force. The Harrier was developed from a design called the Kestrel, itself evolving out of a design called the P.1127. A supersonic version was actually on the drawing board at about the same time, incorporating the BE.100 engine from Bristol Engines and was known as the P.1154 project. Unfortunately, it was cancelled on the 2nd February 1965 only four months before TSR-2 was cancelled[18] along with CVA-01.

Early History: Programme Rationalisation

"Affordability and cost is going to drive this programme."[19]

The Joint Strike Fighter programme has its immediate origins in a number of different projects that were being undertaken between the mid-1980s and mid-1990s. These programmes were designed to provide the US Air Force, Navy and Marine Corps (and by extension the USA's close allies) with new warfighting capabilities and to replace a number of legacy systems procured in the 1960s and 1970s. These systems had been introduced with varying degrees of success and included the McDonnell Douglas F-4 Phantom II which while designed as a naval fighter, evolved into a successful multi-role fighter in service with the US Air Force and Marine Corp as well as a significant number of overseas militaries, including the Royal Air Force and Royal Navy.

A less successful programme, the General Dynamics F-111, saw service as a strike aircraft with the USAF and while initially held up as an alternative to the RAF's ill-fated TSR-2, suffered from considerable cost and time overruns meaning that the only export customer proved to be the Royal Australian Air Force which bought twenty-four aircraft, while the navalised fighter variant, the F-111B, was cancelled – the US Navy instead going for the Grumman F-14 Tomcat.[20] These different projects included:[21]

- **Advanced Short Take-Off/Vertical Landing (ASTOVL 1983 – 1994):** In 1983, the Defense Advanced Research Projects Agency (DARPA) initiated a programme to examine the technologies available to design and manufacture a supersonic replacement for the Hawker Siddeley (then British Aerospace) Harrier, in service with the US Marine Corps as the AV-8A.

This led to increased US-UK collaboration but the results of the study made it clear that the technology available was not mature enough to produce a result both countries would have been satisfied with. DARPA then approached Lockheed who ran highly secret R&D projects at their 'Skunk Works' facility in Burbank, California to see if they could help. Lockheed had a number of ideas it could pursue and so DARPA continued with the second phase of ASTOVL to cover Lockheed's involvement.

- **STOVL Strike Fighter (1987 – 1994):** Lockheed, in conjunction with NASA, looked at the feasibility of producing a stealthy supersonic STOVL fighter utilising NASA's facilities (wind tunnels, personnel, super computers etc.) and Lockheed's expertise in designing stealthy aircraft. The research showed that such an aircraft was feasible and Lockheed was convinced it could be sold to both the US Air Force and US Navy (who also procures equipment for the Marines). The services agreed and signed a Memorandum of Understanding (MoU) with the project gradually moving away from its 'black' status.

- **Common Affordable Lightweight Fighter (CALF 1993 – 1994):** While the ASTOVL/SSF programme was initially designed to provide a replacement for the Harrier jump jet in service with both the USA and UK, once it gained support from multiple services and suggested there would be several different variants, it was re-christened the Common Affordable Lightweight Fighter (CALF, but was also known as the Joint Attack Fighter - JAF), its aim being to develop the technologies and concepts to support the ASTOVL aircraft for the US Marine Corps, Royal Navy and Royal Air Force as well as a conventional flight variant sharing a high commonality of parts, for the US Air Force.

- **Multi-Role Fighter (MRF 1990 – 1993):** At the beginning of the 1990s, the US Air Force was looking into designing an F-16 replacement, similar in size and design (single seat, single engine) to the F-16 with a unit flyaway cost of between $35m and $50m. The programme was managed by the Aeronautical Systems Center (ASC) at Wright-Patterson Air Force Base, Ohio. A Request for Information (RFI) was issued in October 1991 with the start of the programme expected in 1994 due to the large number of F-16s reaching the end of their service lives, the end of the Cold War meant that the total number of US Air Force wings were being reduced along with flying time, so the F-16 service life issue became less critical. Budgetary pressures, the Air Force's commitment to the F-22 and continued production of Block 50 F-16s meant that the programme was cancelled in 1993.

•   **Advanced Tactical Aircraft (ATA 1983 – 1991):** This programme began in 1983 when the US Navy proposed a long-range, very low observable, high payload medium attack aircraft to replace the Grumman A-6 Intruder in service with the carrier fleet. A combined General Dynamics/McDonnell Douglas team was selected on 13 January 1988 to develop their ATA concept – a unique 'flying wing' concept that was to be subsonic but with a large internal weapons capacity. However, due to severe cost and time overruns as well as technical issues, the programme was cancelled on 7 January 1991.

•   **Naval Advanced Tactical Fighter (NATF 1990 – 1991):** Following Congressional intervention, the US Navy agreed to evaluate a navalised version of the US Air Force's Advanced Tactical Fighter (which became the F-22 Raptor) as a possible replacement for their aging F-14 Tomcats. In return, the Air Force would evaluate a variant of the ATA as a replacement for the F-111s. A Naval Programme Office was set up in late 1988 at Wright-Patterson Air Force Base and the existing programme altered to include studies of possible naval ATF variants.

The Major Aircraft Review reduced the overall numbers and peak production rates of both the ATF and NATF, substantially increasing the cost of the programme. In August 1990, Admiral Richard Dunleavy, Head of the US Navy's Aircraft Requirements stated he could not see how the NATF could fit into any affordable procurement plan for naval aviation. In early 1991, the NATF was dropped as the Navy decided to go with a series of upgrades to the existing F-14 fleet that would see the aircraft through to at least 2015.

•   **Advanced-Attack/Advanced/Fighter-Attack (A-X/A/F-X 1992 – 1993):** With the cancellation of the ATA and NATF, the Secretary of the Navy, Henry L. Garrett III, asked that a new A-6 replacement programme be started. This programme became known as A-X, an advanced, 'high-end', carrier-based multi-role aircraft with low observability, long-range, two crew, two engines and an advanced, integrated avionics and defensive suite with all-day, all-weather capabilities.

The Air Force decided to participate in the new programme from the start, as it was still seeking to replace their F-111s, and in the longer term, the F-117A and F-15E. Contracts worth $20m were awarded to five contractor teams on 30 December 1991, which included teams from Grumman/Lockheed/Boeing, Lockheed/Boeing/General Dynamics, McDonnell Douglas/Vought, Rockwell/Lockheed, General Dynamics/McDonnell Douglas/Northrop (the prime contractor being listed first in each case). The original concept evaluation and development work was supposed to be completed in September 1992 with Demonstration/ Validation proposals expected towards the end of the year with work starting in 1994. Under the original plan, the short Demonstration/Validation phase would consist of design refinements and various risk reduction activities but in late 1992, Congress directed that it include competitive prototyping.

The duration of this phase therefore increased from two to five years. Also, additional air-to-air requirements were added to the programme as a result of the cancellation of the NATF project so the name was changed from Advanced Attack (A-X) to Advanced Attack/ Fighter (A/F-X). A Defense Acquisition Board (DAB) Milestone I Review was expected in early 1993 but the programme was put on hold with the announcement of the Bottom-Up Review and as a result, cancelled when the Review was published, finally being wound up by 31 December 1993.

• **Joint Advanced Strike Technology (JAST) Programme:** With the winding up of the A/F-X programme, the majority of personnel, skills and knowledge transferred over to what had become known as the JAST programme. JAST was not really about the development of a new aircraft, rather it set out to mature the technologies that a new strike/fighter aircraft would incorporate as well as develop requirements and demonstrate concepts. As it took shape, it became clear that the programme would be funding one or more concept demonstrator aircraft sometime in 1996, which happened to coincide with when the ASTOVL programme planned to enter its third phase (full-flight demonstrators).

The ASTOVL programme, which as an advanced concept for a future joint service strike/fighter, appeared to be fully consistent with the original idea behind JAST. In October 1994, legislation related to the FY95 budget passed by the US Congress mandated that the two programmes be merged immediately.

The United Kingdom remained involved with the project – eventually becoming a Level 1 fully collaborative partner – in order to find an eventual replacement for its Royal Navy Sea Harrier and Royal Air Force Harrier GR7 aircraft. Despite the number of STOVL (Short Take-Off/Vertical Landing) fighters required by both it and the US Marine Corps, these numbers are expected to be dwarfed by the sheer number of CTOL (Conventional Take-Off and Landing) fighters required by the US Air Force as well as other export customers.[22]

The Programme Takes Off

"The JSF's overall mission is unique: to solve the budgetary problems in all services with one programme."[23]

The ASTOVL/CALF and JAST programmes were therefore merged in October 1994 to become the Joint Strike Fighter (JSF) programme. The Department of Defense (DoD), on addressing its future requirements, was faced with major concerns over the affordability of replacing over 3,000 aircraft, with discussions on how to resolve these dilemmas evident in 1993.[24] In order to fund the replacements needed to keep force numbers at an effective level, some innovative thinking was called for, especially in terms of its procurement and support strategy. At the same time, the costs associated with the F-22 Raptor programme were escalating and it was becoming increasingly difficult to see how the respective service budgets could cope with the procurement of new aircraft using the old methods (i.e. each service doing their own thing). The JSF programme was set up with a tough mandate, to develop a highly capable, supportable and affordable aircraft that would be used by all three services and therefore nominally replace the General Dynamics F-16 Fighting Falcon, Fairchild-Republic A-10 Thunderbolt II, Lockheed F-117A Nighthawk, McDonnell Douglas F-15E Strike Eagle, Grumman A-6 Intruder, early models of the McDonnell Douglas F/A-18 Hornet as well as the McDonnell Douglas AV-8B Harrier.[25]

The move from technology demonstration programme to a service mission capabilities acquisition programme led to Requests for Proposals (RfP) for the Concept Demonstration Phase (CDP) being issued to three of the four teams that had been involved with the ASTOVL/CALF studies, on 22 March 1996, these Teams being led by Boeing, Lockheed Martin and McDonnell Douglas. The Boeing Team included Pratt & Whitney as well as Rolls Royce, the Lockheed Martin Team included Pratt & Whitney, Rolls Royce and Allison while the McDonnell Douglas Team included Northrop Grumman, British Aerospace, Pratt & Whitney, Rolls Royce and Allison.[26] All three teams included both Pratt & Whitney and Rolls-Royce as a decision in early 1995 had meant that the preferred engine for the project was the Pratt & Whitney F119 but General Electric was given the contract to develop proposals for an alternative engine, the F120, in case development problems arose, while Rolls Royce had designed the propulsion system to be used in the STOVL variant.[27] In April, the JAST Programme Office assigned the designations X-32A, B and C, as well as X-35A, B and C[28] to the two competing concepts that would be chosen towards the end of the year.[29] It was also viewed by some members of Congress as an attempt by the Pentagon to translate a technology demonstration project into a major procurement programme by the 'back door' and launch a $100bn project without going through the usual channels. In what might have been something of a warning shot, the House National Security Committee's Research & Development Panel recommended that none of the $589m funding for FY97 be spent on the ASTOVL version for the US Marines and Royal Navy.[30]

At this stage it was planned that the JSF would supplement the US Air Force's reduced procurement of just over 400 F-22 Raptor fighters, as an affordable next-generation F-16 replacement with high technology and low-observable characteristics, being a close air support and strike 'workhorse' but with decent air-to-air capabilities. As originally conceived, it was a role that was shared with the US Navy and US Marine Corps variants and would supplement the planned 1,000 F/A-18 Super Hornets. Rear Admiral Dennis McGinn, at the time Air Warfare Director for the US Navy, confirmed that the JSF would be a 'Day One' strike aircraft, attacking high-value targets and enemy air defences due to its stealthy nature, clearing the way for the F/A-18. The US Marine Corps (USMC) would use the JSF as 'flying artillery' in the close support role, replacing the AV-8B Harrier, the variant that the UK Ministry of Defence (MoD) was interested in. This was to replace the Royal Navy's Sea Harriers and potentially the Royal Air Force's Harrier GR7/GR9 at a later stage, with the CTOL variant possibly replacing the Tornado GR4s after 2015. Indeed, the US Marine Corps was so focused on having the JSF that it bowed out of the F/A-18 Super Hornet programme, making itself literally a hostage to the programme's fortunes.

While the US Marine Corps managed to make do without a STOVL capability during the Vietnam War, this was mainly due to the availability of land-based airfields and the US Navy's carriers waiting offshore. Pressures within the US Congress almost got the STOVL variant cancelled on the grounds of cost but it would have forced the Marines back into procuring the Super Hornet and eliminated participation of firstly, the Royal Navy and Royal Air Force who would be looking for the STOVL variant to replace their Harriers and secondly, other countries who also operate the Harrier, such as Spain, Italy, Thailand and India and would be looking for replacements eventually too.

There was also the possibility that, if the commonality, technology, cost and time targets could be achieved, that the JSF could have a major impact on the World's fighter export market. The JSF was originally designed to come in between $28m and $32.5m and so for a similar price to a Sukhoi Su-27 Flanker air superiority fighter, an export customer could have an aircraft with 5th generation, multi-role capabilities with a wide choice of model options, lower life-cycle costs and agility only matched by aircraft with similar thrust vectoring technology, such as the F-22 Raptor or Su-35M.[31]

The three companies put forward three different, innovative designs. McDonnell

Douglas proposed a two-engine design without vertical fins, using stabilizers dihedralled to 23 degrees, pitch/yaw thrust vectoring and active fly-by-wire controls to maintain stability. The proposed aircraft had a highly contoured forebody, a blended mid-wing design and raked rectangular cheek intakes, claiming a maximum turn performance up to 40 percent better than conventional aircraft along with a lower radar cross-section, weight and drag. Lockheed Martin's 'Configuration 200' entry featured delta wings with identical leading and trailing-edge sweep angles, large F-22 stabilizers (which replaced large canard surfaces for pitch control), reverse-raked low-observable rectangular cheek intakes made from fibre composites, a chined forward fuselage and twin-canted fins.

Boeing's entry had a modified delta wing, with a large chin engine inlet and builds on the single-power plant vectored-thrust direct-lift concept from the Harrier with a rear-mounted F-22-style pitch-vectoring nozzle for use during conventional flight and which then closes to divert fan and exhaust flows to the other nozzle systems in SOVL mode.[32] The three teams submitted bids for the $2.2bn CDP in June 1996, with two teams expected to go forward after the decision in November and one of those teams going forward to the Engineering and Manufacturing Development (EMD) Phase in late 2001. [33] Later in the programme, this phase was renamed the System Development and Demonstration (SDD) Phase.[34]

And Then There Were Two

"...operational considerations determine most of the differences. Affordability and manufacturing considerations account for most of the similarities."[35]

On 16 November 1996, the US Department of Defense selected the teams led by Boeing and Lockheed Martin to proceed with the CDP, eliminating McDonnell Douglas from the competition. Naval Air Systems Command (NASC) awarded two 51-month contracts worth US$661 million and US$718 million respectively to Boeing and Lockheed Martin, each being tasked to build two full-scale flying demonstrators. The total cost of this phase was estimated at $2.2bn, including related propulsion developments that were due to be funded separately.[36]

The Lockheed Martin design was the lower-risk, more conservative approach and so made it through to the next stage, while McDonnell Douglas and Boeing were more revolutionary in terms of technology. In choosing the Boeing design over that of McDonnell Douglas, Pentagon officials eliminated the McDonnell Douglas design partly because its two-engine design had greater complexity, thus having greater potential for technical and operational problems, but also because the overall proposal did not include many ideas on manufacturing or acquisition innovation, as compared to the other two. In addition, the design would have increased Life-Cycle Costs over the others in regard to fuel, spare parts and maintenance.

Boeing on the other hand, was selected to ensure a competitor with a greater revolutionary approach to technology to Lockheed Martin as well as having an advantage in being able to adapt commercial practices and having more innovative ideas about manufacturing and design. Their design was also perceived to have the best performance of the three, mainly in terms of up-and-away flight, range and manoeuvring but had problems with heat and vibration during VSTOL operations. While having experience on a large number of military aircraft programmes, Boeing had not built a successful military fighter in over fifty years, and although the company had built sizable commercial airliners on an industrial scale, the production and integration of avionics and weapons systems in a military fighter was considered by some as a very different proposition.

Some analysts were concerned about the single-wing design's ease of repair, including stress fractures and battle damage. The Lockheed Martin design was adjudged to have the advantage in stealth characteristics, possibly approaching that of the F-22 and an improvement on the F-117A, and experience in building these two platforms was considered to give the company an advantage in low-cost stealth manufacturing and repair. It was judged they also had greater experience in the field of international cooperation and commonality of parts between variants. However, would the design offer enough of an improved performance to justify the investment in a new fighter programme?[37]

The American Helicopter Society organised a symposium on the Joint Strike Fighter in April 1997. Speakers from both the US Navy and US Air Force expressed interest in a STOVL version, the Air Force as a possible replacement for the A-10 Thunderbolt II but it was considered unlikely that such an aircraft would fulfil the Navy's requirement for an aircraft having a range of 600 nautical miles. As the project moved into the CDP, Northrop Grumman announced its intention to join the Lockheed Martin team on 8 May 1997. British Aerospace was expected to make its mind up very soon, but in the meantime, the Quadrennial Defense Review (QDR) was published on 15 May 1997. It recommended that the scale of future Pentagon procurement be cut back and the JSF Programme was one that underwent changes. The number of aircraft for the USAF dropped to 1,763 (a reduction of 273 aircraft), the number of USMC aircraft dropped to 609 (a reduction of thirty-three aircraft) while the number of US Navy aircraft went up from 300 to 480, although some 230 purchases would be competed for by both the JSF and F/A-18 E/F. The F-22 Raptor Programme was reduced by a third. On 18 June 1997, during the Paris Air Show, British

Aerospace announced it would join the Lockheed Martin team, although it was represented on the Boeing team as well.[38]

Down to the Wire

"Commercial processes and design skills were important in Boeing's win and this fact should resonate throughout the industry."[39]

With McDonnell Douglas out of the running on the JSF programme, the company was under pressure to consider its future options. With a strong balance sheet and order book, it had plenty of room for manoeuvre but could no longer sit on the sidelines while consolidation was occurring throughout the defence and aerospace industry sector. Grumman, General Dynamics, Lockheed, Northrop and Martin Marietta had all come to a decision to move away from the business-as-usual approach and boldly follow a very different course, in order to ensure the company's *Survival*. McDonnell Douglas found itself at such a crossroads[40] and following talks with Boeing, the companies merged on 4 August 1997.[41] McDonnell Douglas' two former partners, BAe Systems and Northrop Grumman eventually joined the Lockheed Martin team, although BAE was represented on the Boeing team as well.

Meanwhile, under the terms of the contracts awarded by NASC, Boeing and Lockheed Martin received $661m and $718m respectively to each build two full-scale flying demonstrators over a period of fifty-one months, while the JSF Program Office awarded a $96m multi-year contract to the manufacturing team undertaking the JSF Alternate Engine Program. This team comprised GE Aircraft Engines, Allison Advanced Development Company and Rolls Royce Military Aero Engines Ltd who started a four-year core development and technology maturation project, the basis being the GE F-120 engine.[42] Like the Pratt & Whitney engines, the F-120 was developed for the cancelled ATF programme. However, this project was an international one with several companies taking part (the Dutch company, Philips Machinefabriecken (PMF) also joined the team in June 2000). The JSF engine was not designed as a variable cycle engine like the original design for the ATF but retains the same massive combustor and core, managing to propel the YF-23 to mach speeds without the need for an afterburner.[43]

In FY1996, Congress expressed its concern over the lack of competition over the engine for JSF and directed the US DoD to correct this state of affairs.[44] At the time, the project was still in its JAST phase but in FY1998, Congress directed the US DoD to ensure that sufficient funding was available to carry out an alternative engine development programme that included flight qualification of an alternate engine in the JSF airframe.

Since then, Congress has provided over $2.5bn worth of funding for the alternate engine programme and requires something in the region of an extra $900m to see it through to 2013 and the completion of development of the F136 engine. The US DoD's budgets for FY2007, FY2008, FY2009 and FY2010 have all included proposals for the cancellation of the project, a project that was authorised by Congress in the FY1996 Defense Authorisation Act and has received consistent Congressional support since it started.

Congress rejected each of these proposals and provided additional funding to the US DoD's budget request each year for the F136 engine programme. Opinions on this matter are split, with some criticising the US DoD and Air Force for being short-sighted in their attempts to have the programme cancelled – a decision even those in the Air Force and the Office of the Secretary of Defense (OSD) have indicated was due to immediate budgetary pressures rather than any long-term pros or cons associated with the project itself. Indeed, the US Secretary of Defense at the time, Donald Rumsfeld, stated on 16 February 2006 that the merits of terminating the F-136 engine were "clearly debatable".[45]

Others applaud this decision and say that single source engine production contracts have really been the norm rather than the exception and claim that long-term engine affordability is achieved through multi-year contracts with a single supplier. Congress' interest in establishing and funding an alternative programme stem from what has become known as 'The Great Engine War' that run from approximately 1984 to 1994 revolving around the competition between Pratt & Whitney and General Electric to produce engines (the F100 and F110 respectively) for the F-16 Fighting Falcon with the competition continuing after 1994 for business from those foreign air forces that had purchased the F-16 and F-15 Eagle. At the time, this acquisition strategy was unprecedented and controversial. The roots of this war go back to the development of the F100 engine for the F-15 in the 1970s and the USAF frustrations with the management of the programme by Pratt & Whitney and their concerns over using a sole source of supply.

The engine's rushed development (to meet the F-15 fielding deadline) prevented problems from being fully examined and the mounting frustration over Pratt & Whitney's reluctance to tackle these problems without additional funding led to the Air Force, Navy and Congress all working together to find an alternative.[46] After several contentious Congressional hearings, Congress providing funding through the Engine Model Derivative Program (EMDP), a directed programme stemming from the 1960s, for General Electric to develop the F110 so that it could compete with the F100.

The course of the CDP can be charted in these milestones:[47]

- 14 September 1995 – Boeing initiates testing on its Large-Scale Powered Model (LSPM) at Tulalip near Seattle.

- 11 July 1996 – Boeing submits its proposal for the JSF design based on 11,700 hours of tests with small and full-scale models, as well as the LSPM.
- 16 November 1996 – Boeing awarded a $660m, fifty-one month CDP contract.
- 20 January 1997 – Boeing signs MoU with McDonnell Douglas.
- June 1997 – Lockheed Martin passes the Initial Design Review (IDR).
- 10 September 1997 – Boeing passes the IDR.
- 5 February 1998 – Boeing announces that the production of its two demonstrator aircraft will take place at Palmdale, California, near to Edwards Air Force Base (AFB).
- 8 July 1998 – Structural assembly of Boeing's first prototype begins at the St Louis Phantom Works (formerly McDonnell Douglas)
- 23 September 1998 – Structural assembly begins of Boeing's second prototype.
- 4 February 1999 – Boeing presents a new Preferred Weapon System Concept (PWSC), with the design gaining an aft horizontal tail.
- 26 March 1999 – Boeing begins testing avionics in its Flying Test Bed (FTB), a modified Boeing 737-200.
- 1 April 1999 – Final assembly of Boeing's first prototype (X-32A) begins.
- 8 May 1999 – Lockheed Martin's first prototype, the X-35A, completes its major structure assembly.
- 20 September 1999 – Final assembly of Boeing's second prototype (X-32B) begins.
- 14 December 1999 – To the surprise of everyone, Boeing rolls out both demonstrators on the same day. Construction of the second prototype had begun almost six months after the first but finished only six weeks later. Boeing had trialled lean manufacturing processes on a black project known as 'Bird of Prey' (revealed to the public in October 2002) and applied it to their second JSF prototype.
- 7 March 2000 – X-32B completes its maximum power engine ground tests.
- 23 May 2000 – X-32A finishes high and low-speed taxi tests.
- 24 August 2000 – X-35A tests the F119-611 engine on full afterburner for the first time.
- 18 September 2000 – X-32A takes off for its maiden flight.
- 21 October 2000 – X-35A completes its final taxi tests.
- 24 October 2000 – X-32A is grounded for several days after suffering software problems. X-35A achieves its first flight under the control of Tom Morganfeld and transfers to Edwards AFB.

- 18 November 2000 – The test pilot from BAE Systems, Simon Hargreaves, flies the X-35A, as does RAF Sqn Ldr Justin Paines.
- 20 November 2000 – X-32A and X-35A meet at Edwards AFB for a 'meet the press' day.
- 21 November 2000 – X-35A achieves supersonic flight.
- 22 November 2000 – X-35A flies back to Palmdale, California to be converted to the X-35B.
- 2 December 2000 – X-32A completes the low-speed approach flight testing (necessary for the carrier variant).
- 16 December 2000 – X-35C takes off for its first flight under the control of Joe Sweeney and transfers to Edwards AFB.
- 19 December 2000 – X-32A performs its first air-to-air refuelling behind a KC-10 tanker. The refuelling was aborted after the fuel started leaking everywhere.
- 21 December 2000 – X-32A performs a supersonic flight.
- 29 December 2000 – X-35B has its lift-fan installed.
- 8 January 2001 – X-32B finishes low and medium-speed taxi run tests.
- 23 January 2001 – X-35C undertakes its eighteenth flight and its first tanker qualification flight (a KC-10).
- 31 January 2001 – X-35C achieves supersonic flight.
- 3 February 2001 – X-32A flies for the sixty-sixth and final time, back to Boeing's Palmdale facility.
- 9 February 2001 – X-35C flies to Lockheed Martin's Fort Worth facility.
- 10 February 2001 – X-35C achieves a cross-continental flight, the first for any X-plane, when it transfers to the Naval Air Station at Patuxent River.
- 23 February 2001 – X-35B starts hover pit testing.
- 10 March 2001 – X-35C makes its final flight but stays at NAS Patuxent River where it goes on display in April 2003.
- 29 March 2001 – X-32B takes off for its first flight under the control of Dennis O'Donoghue.
- 13 April 2001 – X-32B transitions from normal flight, to vertical flight and then back to normal flight.
- 17 April 2001 – X-35B completes the engine ground tests.
- 12 May 2001 – Lockheed Martin completes installation of the complete STOVL propulsion system.
- 4 May 2001 – X-32B transfers from Edwards AFB to NAS Patuxent River.
- 23 June 2001 – X-35B takes off vertically for the first time.

- 24 June 2001 – X-32B performs the first Outside Ground Effect (OGE) hover during the forty-fourth flight.
- 29 June 2001 – X-35B achieves vertical flight for the first time.
- 3 July 2001 – X-35B transfers to Edwards AFB.
- 9 July 2001 – X-35B engages its lift fan, transitions from normal flight to hover, and then back to conventional flight and achieves supersonic speed.
- 16 July 2001 – X-35B combines a short take-off with a vertical landing.
- 20 July 2001 – X-35B makes a short take-off with the lift fan, achieves supersonic flight and then lands vertically.
- 28 July 2001 – X-32B performed a short take-off, transitioned into conventional flight, went supersonic, transitioned back into STOVL mode, finally making a low-speed landing.
- 6 August 2001 – X-35B makes its final flight and transfers to the Smithsonian Air and Space Museum in Washington DC.
- 26 October 2001 – Winner of the CDP announced.
- 9 April 2005 – X-32A arrives at the US Air Force National Museum in Dayton, Ohio. X-32B stays at NAS Patuxent River.

The Last Man Standing

"Like the elimination of McDonnell Douglas in 1996, the choice of Lockheed Martin had a seismic impact on the aerospace industry."[48]

Finally, on 26 October 2001, Ed Aldridge, Under-Secretary of State of Defense for Acquisition, announced that the team led by Lockheed Martin (which included Northrop Grumman and BAE Systems) had won and acknowledged that "both teams met or exceed the performance objective established for the aircraft and had met the established criteria and technical maturity for entering the next phase of the program".[49] The decision naturally, was greeted with disappointment in St Louis (home of Boeing) and jubilation in Fort Worth.

Lockheed Martin's JSF program manager Tom Burbage, who had smoothly succeeded Cappuccio in late 2000, recalls that he only heard the first syllable of the company's name at Fort Worth: the rest was drowned out as hundreds of workers cheered as they watched on big screen TV.[50] The decision would mean that Lockheed Martin will become the premier aircraft manufacturer in the United States, ensure the growth potential of its aerospace business and create thousands of jobs at its Forth Worth plant, as well as with subcontractors around the country and abroad.

The US GAO however had warned the US Congress that eight critical technology areas (such as radar, mission systems integration and integrated flight-propulsion control) had not yet matured to a point where there could be confidence that the SDD Phase could be considered low risk and be spared schedule slippage and cost overruns.[51] While some hoped that Boeing might be able to continue in the programme as a possible subcontractor, in reality, any continued Boeing involvement would mean Lockheed Martin renegotiating the contract with the US DoD. Any Boeing involvement "would have to bring a specific benefit", in other words, Boeing was out of the picture.[52] Why had Lockheed Martin won? It was due to a number of factors:[53]

• **High-risk option** – Boeing had beaten McDonnell Douglas in the November 1996 down-select to go forward to the CDP because its design was very different and offered both high risks and high rewards. The design had evolved out of the F-16 Fighting Falcon study done in 1992 and was simpler, particularly in the STOVL configuration, and with greater wing sweep, only two tails and a two-dimensional low-observable (LO) nozzle, it was potentially stealthier.

• **Performance** – Lockheed Martin, Northrop Grumman and McDonnell Douglas had all concluded independently that a direct-lift solution would be hard-pressed to meet the JSF requirement and some form of lift boost was critical. Boeing's solution was to build a light-weight, unconventional and compact airframe but this also had inherent risks. Firstly, only one tail-less delta had be used on a carrier (the Douglas F4D Skyray) and the inlet design proved a difficulty due to the need for stealth (having to have a blocker in front of the compressor) and efficient hovering. Lockheed Martin's design drew on the experience gained with the aerodynamics and stealth capabilities of the F-22 Raptor.

• **Pressure on Boeing** – twice in the recent past, had the USAF chosen an unconventional solution from an inexperienced supplier. These were General Dynamics' victory over Northrop with the F-16 and Northrop's winning out over Lockheed/Rockwell with its B-2 Spirit stealth bomber. In both of these cases, the winning design was more than a match for the competition. In order for Boeing to win JSF, its design had to beat Lockheed Martin decisively or for Lockheed Martin to fail spectacularly. Between 1999 and 2000, Lockheed Martin managed to prepare the shaft-driven lift-fan system for testing. When they had problems with that system, that was potentially Fort Worth's darkest hour – when they managed to solve those problems, Boeing's hopes were virtually swept away.

- **LM's team was formed faster** – After Lockheed Martin and Boeing were announced to be the winners of the CDP, McDonnell Douglas' former partners (Northrop Grumman and BAE Systems) quickly joined Lockheed Martin. Boeing set about acquiring McDonnell Douglas. Thus Lockheed Martin acquired two partners who had already been involved in some form of partnership role within the JSF project and included BAE System's experience in STOVL technology and all the US experience in operational stealth aircraft. Boeing finally acquired McDonnell Douglas in August 1997 and only then could start forming a team between the sites in St Louis, Seattle and the former Rockwell company.

- **Requirement changes** – by the middle of 1998, both teams had to alter their designs due to changes in the requirements. In this case, a greater bring-back weight and specific maximum approach speed for the CV variant. Lockheed responded by increasing the wing area to one approaching that of the F-15 while Boeing added horizontal tails and revised its inlet design. These changes could not be reflected on the X-32 prototypes as they were already being built but Boeing was adamant the designs would be so similar that the risk would not increase and that flying the X-32 would validate the design, simulation and ground-testing tools that would be used.

- **Vertical lift** – Boeing could not increase the size of its design and add lift – the fan diameter and inlet size were as large as they could be and pushing the throttle would increase jet velocity but temperature as well, which was already at the limit imposed by the ground environment.

- **Test goals** – when the X-32 prototypes were unveiled, programme manager Frank Statkus said that the company were planning a demonstration – not called for in the actual contract – where the X-32B would undertake a short takeoff, accelerate to supersonic speed and then land vertically. The final demonstration however ended with the aircraft using a short landing. It never made a vertical landing with the moveable inlet lip in place, a piece of equipment that was needed for high-speed flight. Given that vertical lift had been a known area of high risk all through the programme, this event was more than just symbolic. It showed that the X-32 had a small negative margin of vertical thrust, something contrary to Boeing's own predictions.

- **Source selection authority** – Boeing managed to persuade the JSF Program Office that their design changes would not increase the risk, but that view was by no means universally accepted.

- **Higher risk/cost advantage** – When the winner of the CDP was announced, DoD officials made it clear that Lockheed Martin's solution offered the 'best value', not the 'lowest cost'. They said that four times. Boeing's bid may have been lower, but the DoD had every right to reject it if they felt that it was a higher risk option. A logical choice given that, at the end of the day, JSF contracts are not fixed price – and a fixed-price contract would be worthless as in five years time, there would be no alternative to JSF.

The announcement of the CDP winner also affected the project in an entirely unforeseen way – it heralded the wrong designation being given to the JSF. The previous fighter in the US designation system had been Northrop's YF-23A, so the JSF should have been the F-24 or even F/A-24 given its strong bias towards the strike role. However, at the conference, a reporter asked Aldridge what the designation would be, but he didn't know the answer. As it stood, there was no answer as a decision had not yet been taken as to what should be assigned. The Pentagon moderator pitched the question to Program Director General Mike Hough. Caught on the hop and slightly confused, Hough said "X-35". Aldridge misheard him and stated the designation as "F-35". This placed the US DoD in a dilemma – admit a mistake had been made or permit a designation that was in fact out of sequence.

The JSF Program Office made its decision in December, officially requesting the designation "F-35" under the Mission Design Series as it was in keeping with what Aldridge had stated. The USAF Nomenclature Office forwarded the request onto the Directorate of Programs (the final authority) in April 2002, unable to resist putting in a paragraph recommending the "F-24" designation. By then however, the "F-35" designation had gained a momentum all of its own and it was officially confirmed in June 2002.[54] The designation fuelled rumours of black projects as the designations "F-19" and "F-21" had still not been assigned (the "F-19" designation had been used by the media to refer to the F-117A Nighthawk before its true designation became known) but at least the aircraft's lineage to the X-35 would be apparent.[55]

The SDD Phase officially started on 1 November 2001 with a milestone called 'Authority to Proceed' (ATP) with the design being finalised on 27 June 2002. However, two years into the SDD Phase, it was discovered that all three variants were overweight, the worst transgressor being the plane that could afford it the least, the STOVL variant (F-35B). This was around 1,000kg overweight while the other two variants were around 630kg overweight. While the CTOL (F-35A) and CV (F-35C) variants were still predicted to meet their range requirements at the greater weight, the STOVL variant was marginal in meeting its 830km mission radius requirement with a standard internal load of two 450kg bombs and two AIM-120 missiles. The extra weight would also affect its speed, manoeuvrability, endurance and bring-back load. According to Lockheed Martin, one reason for the disparity in weight from what was predicted at the start of the SDD Phase is that the aircraft's internal configuration is unique. A large proportion of the main body volume is taken up with a single, large engine tunnel and weapons bays. The weight estimates given were partly based on those from the F-22 Raptor programme which is a twin-engine design and has a central structural keel.[56]

While there were also problems with the avionics and the software codes, the weight issue was deemed so serious, that it was decided to extend the SDD Phase, thus increasing costs, which would be met by moving money from the initial procurement back into the development phase and decreasing the number of aircraft built in the early stages of Low-Rate Initial Production (LRIP).[57] Lockheed Martin, examining the problem from the perspective of the aircraft's overall power-to-weight ratio, starting looking at not only reducing weight, but reducing the drag co-efficient and uprating the engine's thrust as well. Early work identified several promising avenues, including the removal of redundant wiring, the landing gear and adjusting the structural loads carried by the aircraft's skin, the main objective being to reduce the all-up weight of the STOVL version and then carrying over as many of those reductions as possible to the other two variants.[58]

This posed immediate problems for the MoD when it revealed what was happening[59] as additional delays to the programme resulting from trying to fix the weight problem could mean that the UK's aircraft carriers, assuming they stayed relatively on schedule, could come into service without aircraft to fly on them. This was due to the Sea Harriers being retired early, in March 2006, due to problems in taking off and landing with their full load of weapons and fuel in hot climates. The only solution was to fit a better engine but the MoD decided it would be too expensive and so the Royal Navy would lose its defensive outer layer and cede the ability to mount an independent expeditionary operation against any adversary other than those with virtually no aerial capability until the JSF comes online.[60]

Lockheed Martin formed the STOVL Weight Attack Team (SWAT) and by September 2004 had shaved off some 2,700lbs. Many changes only took off a few pounds but collectively, they made a real difference. For example, down-rating the generator from 160Kw to 140Kw saved 16lbs (but would reduce the future required capability of any Directed Energy Weapon), changing the battery to a new Li-Ion one saved almost 26lbs, and reversing an earlier alteration, one which expanded the size of the weapon bays of the STOVL version so it was in line with the other variants, saved weight. The STOVL version would therefore only be able to take ordnance up to 450kg internally. Such a move was unpopular with both the US Marine Corps and the Royal Navy, as well as reducing the overall commonality between the variants but the weight saving warranted it. Indeed, the hunt for weight reduction measures was so important that Lockheed Martin paid its staff bonuses for every idea that would shave off weight, something which saw the ideas come flooding in, many of which worked.[61]

Other changes included reducing the overall drag of the aircraft, engine efficiency modifications including improved air inlet and air outlet flows, remodelled centre section couplings, realigned wing spars (for thinner skin on the wings), a revised nose landing gear, a smaller tail and an altered outer mould line to increase fuel carriage.[62] Rather than looking at this as a setback, Lockheed Martin saw it as a chance to refine the design and change the manufacturing sequence of the aircraft. These refinements were not only applied to the STOVL version but were carried over to the others, not only in terms of the weight reduction, but other changes too, that improved performance and maintainability. For example, moving from a single piece wing skin to the more traditional multi-piece skin made it easier to graduate the thickness, thus saving weight but also improved the access engineers would have to equipment on the upper side of the engine. Another example is where the wing of the CV variant was increased to lower the speed of the aircraft in the recovery phase. That, combined with other weight reductions, a reduction in drag and increased fuel capacity has meant an improvement in the F-35C's radius of action to nearly 1,300km (700nm), 100nm over that stated in the requirement.[63]

History Not Repeated

In some ways the early history of the JSF programme had mirrored that of another aircraft, the TFX from the 1960s. The programme, endorsed by the then Secretary of State for Defence Robert McNamara, began in 1961 but ended in failure. The leaders of the different services resented McNamara's demand that they work together (in this case, the Air Force and the Navy) on the programme, despite apparent misgivings. The plane that eventually came out of this collaboration, the F-111A (known as the 'Aardvark') was never fully embraced by the Air Force and completely rejected by the Navy.

The Air Force believed it to be an inferior aircraft for its requirements, in part because of the compromises it had to make with the Navy at the design stage. The Navy failed to field a single aircraft of the F-111B variant (designed to operate from aircraft carriers) and eventually developed the F-14. The programme not only engendered animosity between the services but also between the civilian and military leadership while wasting hundreds of millions of dollars. Both the TFX and JSF projects promised to achieve huge cost savings for the US DoD by consolidating the requirements of at least two services into one aircraft.

Giving these services the ability to purchase a similar system with comparable parts and maintenance requirements had the potential to achieve cost savings.

There are of course other similarities, including the development contracts for both programmes being colossal, with each being considered the largest military contract ever awarded (at the time) and in both cases, the civilian leadership played a central role in the choice of the prime contractor – at the risk of having major political backlash due to the feared impact losing the contract would have on the other suppliers. Memories of the TFX have haunted the JSF for a long time, with many critics assuming that the objective of achieving commonality between the services in terms of weapon system development was always doomed to failure. However, JSF has survived (despite a number of significant issues) with the project managing to avoid the pitfalls that befell TFX, due to:[65]

• **Firstly, the controversy over the contract award**, which in the TFX case went to the Fort Worth-based General Dynamics, led many critics to wonder whether politics had influenced the decision and there was a conflict of interest. In many ways, the award of the TFX contract to General Dynamics reflected sound, objective judgement.[66] However, many of McNamara's opponents seized on the opportunity and called his decision-making and rationality into question, causing problems for several of the civilian officials within the Pentagon. In the case of JSF, this did not happen, despite there being an even stronger appearance of favouritism and a conflict of interest, as not only did President George W Bush come from Texas and have connections to the business community in the Fort Worth area but both James Roche (Secretary of the Air Force) and Gordon England (Secretary of the Navy) both had ties to the Lockheed Martin JSF team.

Indeed, Roche had been a corporate vice president at Northrop Grumman, a key Lockheed Martin partner, before joining the administration, while England had previously worked for Lockheed Martin and had managed the plant where the fighter would be built. Boeing however declared that it had no doubts about the ability of both Roche and England to make an objective decision based on the firmly established criteria and raised no objections to the result when it was announced.

• **Second, the absence of a dominant faction** supporting Boeing within the Government greatly differentiates TFX from JSF. The Administration's political opponents refused to openly question the integrity of the decision and so even among those congressional delegations from Washington and Missouri where Boeing's operations are concentrated, criticism was muted.

• **This was because, thirdly, during the summer of 2001 evidence** was gradually building up that Lockheed Martin was the leading candidate and that the contract was awarded to them, despite both proposals being assessed as very good. According to some sources, Lockheed Martin had emerged as the frontrunner a full TWO years before the decision was officially announced, but the US DoD decided to continue the competition to stop Lockheed Martin from becoming complacent and to deliver better proposals. This approach was confirmed by Jacques Gansler, Undersecretary of Defence for Acquisition and Technology until January 2001. In contrast, there was genuine surprise when the contract for the TFX was awarded to General Dynamics, as Boeing had been the leading contender through the first three rounds. Also, military officials had been unanimous in their recommendations to select Boeing and this was the first time that civilian officials in the Pentagon had overruled their advice.

• **Fourth, is the degree of inter-service cooperation** apparent in JSF that wasn't apparent in TFX. The Navy's opposition to TFX was long-standing, entrenched and public. While it had no real choice after the Missileer was cancelled, it never liked the TFX and repeatedly argued during the design phase that the plane did not fit their requirements. McNamara however, dismissed these claims as he thought that the Navy were just trying to dismiss the project on parochial grounds. They then seized on problems with the naval variant, the F-111B, encountered during the testing phase in the hope it might kill off the project. When they began testing the F-14 which ultimately replaced the F-111B, the Navy relaxed a number of critical requirements for the aircraft they had not been willing to relax during the TFX project.

It has been argued[67] that the Navy would have obtained a workable F-111B if they had been more relaxed over its requirements but didn't do so as it wanted to scupper the project. McNamara knew of the Navy's concerns but refused to acknowledge them, leaving the Navy no choice but to sabotage the programme. Although such action could be seen as an attack on the principle of civilian control over the military, their concerns about the aircraft was not without substance. The aircraft was not designed concurrently by both services – rather the TFX was an Air Force project, managed by Air Force personnel with only scant input from the Navy, a situation which continued well after the project was forced onto them. In contrast, the management of the JSF programme has rotated between the three services. General George Muellner (Air Force) was replaced by Rear Admiral Craig Steidle (Navy), who in turn was replaced in 1996 when Major General Leslie F Kenne (Air Force) took over, being replaced by Major General Michael Hough (Marine Corps) in 1999.

A Global Programme

"The JSF program is critically important to the modernisation of the United States conventional forces ...and is also critical to the modernisation of our ally forces for coalition warfare".[68]

Many members of the US Armed Forces have underestimated the international significance of this programme. The pressure on the DoD budget throughout the first decade of the new millennium has meant that Congress has scrutinised many of the programmes being implemented to modernise the armed forces' equipment, in particular the JSF as it was designated to replace a large number of individual systems but still remain affordable. By 2002, the overall cost of providing 438 x F-22 Raptors, 1,000 x F/A-18 E/F Super Hornets and 3,000 x JSFs had been estimated at $350bn, with a peak of £18bn in 2010. Senior figures in both the DoD and the US Armed Forces recognised the need for international participation to enhance the programme's prospects of *Survival*. According to both the DoD and GAO, foreign participation in the programme, from both Government and Industry was deemed necessary because of the following reasons:[69]

* Operational – To enhance the interoperability of equipment with allies
* Political/Military – To promote foreign procurement of the aircraft
* Economic – To decrease JSF programme costs with partner contributions
* Political/Technical – To share US technical knowledge with important allies
* Technical – To gain technical knowledge and capabilities from key allies

In addition, increased foreign sales would bring indirect economic benefits as the greater numbers produced would lower unit costs by spreading the programme's fixed costs over more aircraft. By sharing technology and technical know-how, all the countries involved would maintain a lead in key niche technologies but since the terrorist attacks of 9/11, science and technology funding became more-and-more political under the Bush Administration.[70] Increasingly, US domestic politics has been at odds with its foreign policy and the USA's allies, particularly European Governments, expressing disappointment with the US Congress, for failing to ratify agreements made in principle with the Bush Administration.[71] Indeed, the 'protectionist' attitude of many US politicians was, and still is, a cause for concern, flying in the face of over thirty years of increasing globalisation[72] and according to a UK Defence Minister, threatens to see the "mutual operational, technological and industrial benefits we have enjoyed over the years evaporate, with both of us being the losers and with obvious political ramifications".[73]

As stated previously, the JSF project included significant international participation from the very beginning of the design phase – something no other US fighter project had achieved. For the CDP, there were four levels of participation, all dependent upon invitation by the United States Government:[74]

• **Level 1 – Full Collaborative Partner.** The UK is the foreign participant with the greatest involvement in the JSF project and was the only Level 1 participant during the CDP, a status that stemmed from the UK's involvement in the precursor programmes to the JSF. An initial Memorandum of Understanding (MoU) was signed in August 1994 signalling Royal Navy participation in the project (to replace the Sea Harrier) with a requirement for sixty aircraft but was modified in 1999 to include the Royal Air Force (to replace the Harrier GR7/GR9), pushing the UK's potential requirement up to 150. The UK contributed $200m to the CDP and actively participated in the process of developing JSF requirements documents. By the end of the CDP, the UK held eight country representative positions at the Program Office, as well as a National Deputy at the director level.

• **Level 2 – Associate Partner.** Three countries had Associate Partner status during the CDP – Denmark, the Netherlands and Norway. All three of these countries negotiated separate agreements that were signed during 1997, each contributed $10m with the US matching the payments for a total of $60m. Their primary purpose in joining the CDP was to influence the development of the programme's requirements as regards the CTOL version. In line with the terms of the each MOU and Memorandum of Agreement (MoA) they can influence the requirements so long as both parties perceive the outcome as mutually beneficial. Each Associate Partner was represented by one National Deputy and one technical representative during the phase.

• **Level 3 – Informed Partner.** Canada and Italy joined in January and December 1998 respectively, as Informed Partners. At this level of participation, partners do not have the ability to influence requirements but participate by highlighting what their services would be looking for in the design of the aircraft itself. In Canada's case this was mainly with regard to the CTOL variant, while Italy was looking at both the CTOL and STOVL versions. Both contributed $10m towards the programme during this phase while the US contributed $50m towards the Canadian involvement.

- **Level 4 – Foreign Military Sales (FMS)** – Major Participant. Turkey, Singapore and Israel were all in the JSF Programme's CDP as FMS Major Participants. All three countries signed Letters of Offer and Acceptance (LOA) during 1999 and are involved in the programme at a generic level, receiving large amounts of unclassified and non-proprietary data about JSF requirements and design features. Each of the three countries took part in different aspects of the phase, with Turkey contributing $6.2m, Singapore contributing $3.6m and Israel contributing $0.75 but with no equivalent US contribution.

The JSF Program Office wanted the partner countries to sign up by mid-2002 before the critical design review, scheduled for early 2003 so they could contribute to the process. Participation in the 126-month SDD Phase was to be in a similar structure as to the CDP but with only three levels of participation so that the higher level partners could have significantly more say in the strategic management of the programme than previously was the case, as well as having the opportunity to benefit from the future sales of JSF aircraft to countries other than the original partners. The levels of participation in the SDD Phase are:[75]

- **Level I – Collaborative Development Partners.** The UK is the only country is this category, signing an MoU on 17 January 2001 and has significant access to most aspects of the programme as well as the ability to influence both the requirements and design solutions. The total UK contribution makes up about ten percent of the overall budget for the SDD Phase and has ten staff members fully integrated in to the programme office. The development non-recurring recoupment charges have been waived for the UK and it will receive a share of the levies on the sale of the aircraft to interested third parties. Such levies are an attempt to redistribute some of the costs of development to countries that did not contribute towards the SDD Phase.

- **Level II – Associate Partners.** Participants at this level have limited access to the core programme and technologies and includes both the Netherlands and Italy, who signed MoUs on 14 and 24 June 2002 respectively. Their contributions are about 5 percent of the SDD Phase total each and will receive a proportional share of levies on the sale of the aircraft to third parties. Both partners can have between three and five personnel integrated in the Programme Office. The Netherlands is seeking a replacement for its F-16 AM/BM Fighting Falcons while Italy is looking at replacing the AV-8B Harrier plus aircraft in service with the Navy and possibly the Air Force's Alenia AMX light strike aircraft.

- **Level III – Informed Partners.** These countries are given enough information on the programme in order to evaluate it for their specific needs. Australia (31 October 2002), Canada (7 February 2002), Denmark (28 May 2002), Norway (20 June 2002) and Turkey (11 July 2002) fall into this category. Their funding contribution amounts to between one and two percent of the SDD Phase budget and will receive a proportional share of any levies on the sale of the aircraft to third parties but are limited to one national deputy each. The deadline for Level III participation officially closed on 15 July 2002 but Australia, after saying it was not going to join the SDD Phase, backtracked and did in fact join, effectively aborting the planned Project Air 6000 fighter competition designed to seek replacements for its aging F/A-18 A/B Hornet and F-111 C/D strike aircraft. Like Australia, Canada is looking to replace its F/A-18 Hornet fleet, Denmark and Norway are looking to replace their F-16 AM/BM Fighting Falcons while Turkey will be replacing their fleet of F-4E/RF-4E Phantom II and early TAI-built F-16 C/D Block 30 aircraft.

- **SCP – Security Cooperation Participation.** A separate category to the tiered levels, participation in this category is based on a Letter of Offer and Acceptance (LOA) with individual countries and valued at approximately $50m. Each country will be given enough information of the JSF family of aircraft to evaluate them with regard to their particular needs but will not receive shares of the levies imposed on future aircraft sales. Singapore joined in February 2003, as did Israel after failing to join as a Level III partner.

Participants have been able to contribute either in cash (for example, Turkey's contribution of $175m during the SDD Phase) or in non-monetary ways, such as Denmark, who paid $110 in cash but also donated an F-16 aircraft and its related support infrastructure for future JSF flights and the use of NATO Command and Control assets for a JSF interoperability study, which were valued at an extra $15m.[76]

The contribution and planned procurement for each partner just after the start of the SDD Phase can be summarised as:[77]

Partner Country	Partner Level	SDD Phase		Production	
		Financial Contributions	% of Total Costs	Projected Quantities	% of Total Quantity
UK	Level I	$2,056m	6.2	150	4.7
Italy	Level II	$1,028m	3.1	131	4.1
Netherlands	Level II	$800m	2.4	85	2.7
Turkey	Level III	$175m	0.5	100	3.2
Australia	Level III	$144m	0.4	100	3.2
Norway	Level III	$122m	0.4	48	1.5
Denmark	Level III	$110m	0.3	48	1.5
Canada	Level III	$100m	0.3	60	1.9
Total Partner		**$4,535m**	**13.7**	**722**	**22.8**
USA		$28,565m	86.3	2,443	77.2

(Chart values do not reflect non-financial contributions from partners)
(Percentages do not necessarily add up due to rounding)

Participation in the JSF programme does not automatically bring with it industrial benefits, as seen in more traditional projects that have utilised offset agreements, where the prime contractor agrees to place work in that country, for example, by buying components from there. One of the features of the programme is that industrial participation is offered on a best-value basis, instead of an offset-basis where a share of the work is automatically granted to participating countries. Countries do not have the ability to 'require' that their indigenous industry become part of the programme as a basis for their participation although one of the stated benefits has been that the programme should provide a means by which foreign industry is able to partner up with American business.

An example is here in the UK, where first, second and third-tier contractors have won substantial portions of JSF business. BAE Systems, a principle subcontractor to Lockheed Martin, is responsible for producing the rear fuselage and the tail of the aircraft, Rolls Royce is handling the lift system (the lift fan and roll posts are being produced over here) while a number of other companies, such as Martin Baker (who will produce the ejection seat), are contributing a variety of sub-systems and components. Around 3,500 jobs will have been created during the CDP and around 8,500 jobs during the SDD Phase.[78]

In addition, the US DoD and JSF Program Office have recognised that a major pre-requisite for many partners' participation will be ensuring not only technology transfer but also ensuring a certain level of workshare and industrial return. It was accepted therefore that there would be a degree of 'favouritism' shown to the industries of the partner countries but was subject to certain guidelines:[79]

- Participation in each phase is governed by an MoU, signed between each partner and the United States.

- Sub-contracts awarded during the SDD Phase to partners' industry will flow into Low Rate Initial Production (LRIP), Full Rate Production (FRD) and Production Sustainment & Follow-on Development (PSFD) Phases for those partners who sign the PSFD MoU.

- Signatories to the MoU will be the preferred supply source for sub-contract awards, the exception being if non-signatory industries can demonstrate a substantially more competitive solution that those from the partners.

- Second or subsequent sourcing will be restricted to the industries of partner countries on a competitive basis.

- Any major facilities that are required to be built outside the Continental United States (CONUS) will only be built in partner countries.

- Any follow-on development activity will be restricted to industries from the partner countries, on a competitive basis unless industry from a non-partner country can show much greater competitiveness.

- Industry from SCP countries will not initially have opportunities to acquire work related to the programme.

In keeping the JSF project viable, an important element has been its Strategic Management, especially as such management impacts "on the external environment, internal resources and competences, and the expectations and influences of the stakeholders". JSF's management structure can be seen as a triangle, with strategic analysis being undertaken by the US Government at the top apex (evaluating the prevailing military, defence and security environment, the resources and skills that are available, and managing the expectations and influences of key stakeholders).[80]

At the bottom left sits the programme's key stakeholders – the US Armed Forces and Partner Nations – who undertake Strategic Choice over the competing demands within the defence environment, national resources and skills as they relate these to their own requirements (single service or national) in order to generate strategic options. At the bottom right sits the winning consortium led by Lockheed Martin, who became the design and manufacturing authority for the JSF. The model's success depends upon a two-way flow between each apex of the triangle. Thus Lockheed Martin plays a critical role in informing the overall decision-making process by assessing the feasibility and potential cost of various design and manufacturing options, as well as incorporating feedback on those options into the evolving design of the aircraft.[81]

Clearly, the JSF programme is pushing the boundaries of not only technology but also those of project management and international collaboration. The diagram overleaf[82] highlights the multi-tiered relationships that exist within the project between governments and industry and how the principles of strategic management mentioned above are implemented by limiting the direct links between the parties. It also highlights the central role played by the JSF Program Office, particularly in outlining strategic choice to all the key stakeholders, as well as the importance of keeping it up-to-date by ensuring all the linkages are two-way. In addition, the practice of limiting the number of places within the Program Office according to the partnership level helps to prevent distortions in decision-making should any partners have different views.

A comparison between the CDP and SDD Phase of where each country was in terms of its level shows some interesting changes, with Italy and Turkey moving up a level (from III to II and from IV to III respectively) while both Norway and Denmark have dropped from II to III. This was partly due to the changing requirements within each country and their ultimate objectives as well as the increasing cost associated with the SDD Phase.

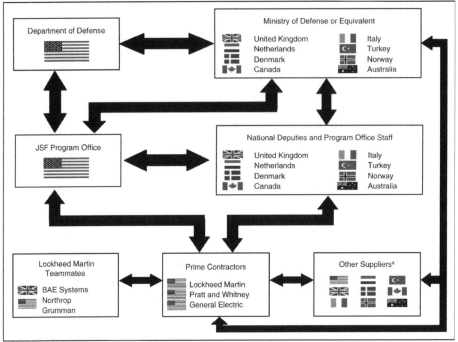

Source: GAO analysis of JSF program documents.

As discussed earlier, the primary reasons for countries to join the JSF project were economic and military but there were also undoubtedly other factors, such as operational, political and technical reasons. For example, the UK has an urgent requirement to source a replacement for the aging Harrier fleet but was also keen to partner with the USA in order to share development costs.[83]

For smaller nations, such as Denmark, Norway and the Netherlands, collaboration had already been proven worthwhile with the F-16 European Co-Production Contract, signed in on 7 June 1975, where the three countries (as well as Belgium) formed the European Participation Group and collectively ordered 348 aircraft, to be built in factories in both Belgium and the Netherlands. A third line was later established in Ankara, Turkey which has produced over 300 aircraft, for the Turkish and Egyptian Air Forces.[84] The Italians have also seen such benefits with the Panavia Tornado programme. As such, the JSF programme represents the next evolutionary stage, where partners are not only involved with the production of the aircraft but are involved with the overall design and logistics support of it as well.[85] But for many countries who have joined the programme, JSF represents an economic necessity, as much as an opportunity. For example, Norway has a long history of using offsets in order to break into formerly closed markets[86] and for a long time was trying to leverage industrial participation in both the JSF and Eurofighter programmes.[87] In the same way, both Singapore[88] and Australia[89] have ongoing national strategies to develop advanced, but niche, aerospace technology in order to become attractive to potential international partners. Italy moved up to become a Level II partner as it has an eye to becoming a Final Assembly and Check Out (FACO) site for the JSF which would maximise the return on its investment (and there are rumours that it lobbied to become a Level I partner alongside the UK, which was denied by the US), a position also coveted by Turkey, who has had experience of building a number of weapon systems under licence (including the F-16 Fighting Falcon, HK33 assault rifle, CZ75 pistol and the Land Rover Defender) and allow its "air force to take full advantage of the Joint Strike Fighter opportunities and advanced technology, logistics and training ...We anticipate that these relationships will continue into cooperative production."[90]

Partnership in the JSF project has therefore become an important opportunity for the industries of small countries to break into the US defence market, which has proven difficult to enter in the past.[91] There has also been the very real concern that the JSF Programme would push the aerospace technology boundaries to such an extent that the companies involved in it would develop the competencies, tooling and manufacturing processes that would give them a major competitive edge over the long-term.[92] In addition, those companies who have won contracts early, may well attain major advantages in terms of Revolution in Military Affairs (RMA) technology implementation by working alongside the US defence industry.[93]

Problems Emerge

"Decisions in this area will be critical because the extent of technology transfers necessary to achieve program goals will push the boundaries of U.S. disclosure policy for some of the most sensitive U.S. Military Technology."[94]

With the start of the SDD Phase in October 2001, problems quickly began to emerge. Facing budgetary pressures, the US DoD considered accelerating the programme in order to bring the aircraft into service quicker. On one hand, the Secretary of the Navy, Gordon England and the US Under Secretary of Defense for Acquisition, Pete Aldridge favoured moving things forward as quickly as possible in order to modernise the force in the fastest and most economical manner possible. On the other hand, other Pentagon officials and members of Congress suggested dropping the STOVL variant (as mentioned above). Then, in late March 2002, Aldridge suggested that the US Navy and Marine Corps purchase fewer aircraft than previously planned. The number of aircraft for the US Navy would be cut from 480 to 430 and for the Marine Corps from 609 to 350. This was taken by many as an ominous sign for the entire programme. The Administration's commitment to the project came into question and there were worries that such cuts would push the unit cost up by between five and ten percent.

There were also international repercussions – for a time, the UK began to doubt the commitment on the part of the US Armed Forces and considered delaying its continued investment in the project. On the same day that Aldridge was urging the Dutch not to delay a planned announcement to join the programme; one senior UK official told the *Defense Daily International* that there was an 'I told you so' lobby that always maintained that the USA was not a reliable collaborative partner. The Air Force were also concerned about the Navy's decision as such cuts could lead to a spiral – such cuts force the unit price up forcing other services to drop the numbers they are going to buy, increasing the unit costs, and so on. Very soon afterwards, the Air Force declared that it too would look at the total numbers of aircraft that it would buy. However, such inter-service manoeuvring failed to dissuade a number of European countries from investing in the programme.[95]

While the partner nations had all signed up to the principles of best-value resourcing, which had been a pre-condition to joining the JSF programme, it became apparent that the defence industry and acquisition structures of many partner countries were unprepared for its implementation.[96] Indeed, defence procurement in both the USA and UK has been constantly reported as having cost and time overruns,[97] while over a third of contracts awarded in the USA are considered to be uncompetitive.[98] As mentioned earlier, the whole *raison d'être* behind the JSF programme is to cut development times, reduce unit costs and fully integrate industry into the supply chain in order to ensure that the latest Commercial Off-The-Shelf (COTS) technology is incorporated into the project.

But if the JSF suffers cost and time overruns similar to previous projects, the COTS technology that is incorporated will be approaching obsolescence by the time the aircraft is fielded, given the pace of change in the consumer electronics market. Only those countries that are willing to reform their acquisition doctrine and processes will be able to take advantage of the current Revolution in Military Affairs, a key consideration being the closer integration with suppliers across the logistics supply chain.[98] However, it was clear that during the SDD Phase contract awards that the acquisition structures of many partner countries, particularly European ones, were still based on the concept of defined workshare agreements arranged at the inter-governmental level[100] and had yet to change to an entrepreneurial mentality from an entitlement mentality.[101] Many countries did not understand the best-value concept and did little to help (initially anyway) their industry to win contracts, while countries such as the UK, Canada and Australia,[102] who created specific JSF teams involving industry in order to develop an engagement strategy all captured more significant business in the early stages than could have been expected under traditional workshare agreements. However, even in these countries, the inadequacies of the procurement structures were brought to light, with the Joint Combat Aircraft (JCA) IPT Leader commenting on that, within the MoD, there are too many people empowered to say 'no' without being made accountable for their decision, which even after Smart Procurement was introduced, "slows down our processes".[103] Such a sentiment was echoed by Mike Cosentino, the JSF International Programme Director who underlined the fact that domestic politics with regard to acquisition was a problem for all the partner countries, including the US.[104]

According to the US DoD, in order to fully realise its plans to implement the RMA, the US has to find the equivalent of £30bn of cost and efficiency savings.[105] While the JSF programme is viewed by many as a 'flagship' project for the DoD to push through acquisition reform by the use of 'Cost As an Independent Variable' (CAIV) principles and best-value resourcing, it has been argued[106] that what is really needed is a Revolution in Business Affairs (RBA) with a move towards similar methodology that was introduced for Smart Procurement ...something along the lines of 'smarter, faster, cheaper'. But the problem is not limited solely to the United States and in an era of decreasing defence budgets and coalition warfare, both the USA and Europe must look to move away from traditional procurement structures and establish some form of long-term partnering or collaboration across the whole spectrum of RMA technology, if they are to modernise their respective armed forces in an affordable manner.[107] The gradual consolidation and globalisation of the defence industry in recent years shows that this has gradually been understood at the industrial level and has been implemented during the JSF programme. However, it is important that the US Congress and Administration are committed to the long-term reform of defence procurement[108] but initiatives such as the 'Buy America Act' call this into question.[109]

During the first five years of the SDD Phase, it was clear that the JSF Program Office was sticking quite rigidly to the industrial participation rules and the principles of best-value resourcing. Although contracts had been awarded to companies from non-partner nations, such as Japan, India, Israel, Poland, Russia, Spain, France and Germany, this worked out to be a very small percentage of the overall total. For example, by the end of 2003, out of the $14bn worth of subcontracts awarded, only £44m or 0.3% had been awarded outside the partners.[110] In addition, the implementation of a 'life of programme' model means that there is very little opportunity for external industries to become involved even if their country becomes a JSF purchaser prior to around 2012, as the industrial linkages already present cover over 98% of the aircraft's construction.[111]

During the early years, Canada proved to be a clear winner, being on track to receive an approximate return of around 4000% throughout the JSF programme.[112] In terms of actual orders however, the UK was the winner with 24.2% of the SDD subcontract awards compared to an outlay of 6.2% of the development costs.[113] As mentioned above, both of these countries were among the few who had conducted extensive reforms of their defence procurement structures and where the Government actively helped their industry to compete for orders. This contrasted with rest of the partner countries, who received about 1.6% of the total value of SDD subcontracts[114] as compared to an outlay totalling some 7.4% of the development costs.[115]

As could be expected, this huge disparity soon drew widespread complaints from a number of partners, including Denmark, Norway, the Netherlands and Turkey. Norway eventually backed off from a serious threat to withdraw from the programme, although it did schedule a debate and a vote in Parliament on continued participation. Lockheed Martin partially defused the situation by announcing new work offers to Kongsberg, one of four companies selected to offset a $1.2bn shortfall in composite materials production capacity. In addition, the JSF Program Office countered by highlighting the fact that US Industry was only receiving 73.9% of the orders when the US DoD had paid for 89% of the development work within the project as a whole up until that point and also reminded the partners that the US DoD had absorbed all of the $7bn rise in the costs of JSF development phase as opposed to asking the partner nations to contribute on a pro-rata basis. Despite this, the Undersecretary of the Turkish Defence Industries Secretariat complained that using best-value resourcing wasn't a fair method of allocating contracts and that as the Turks would be spending $8bn - $10bn in the purchase of JSF aircraft, they should be given a greater share of the work – clearly highlighting that not everyone had understood the implications inherent in the SDD Phase MoUs.[116]

This contrasted with the approach taken not only by the UK and Canada, but by the Netherlands and Australia as well, both of which began government-led engagement strategies in order to win additional business. Although slow to take off, the industry in both countries gradually improved its position,[117] for example, by the end of 2006, some twenty-one Australian companies had signed agreements for work worth $90m (with another $110m pending) that should lead to substantial opportunities during the PSFD Phase, while Dutch companies were in line to win some $5.5bn in contracts related to the PSFD Phase. However, Lockheed Martin have increasingly resorted to using what they term 'strategic best-value sourcing methodology' which allows for a number of contracts to be placed with companies in partner nations where contract awards have not met national expectations. This is because Lockheed Martin has the responsibility to balance partner expectations against programme goals and in some cases, has had to ensure that the requirements of international politics are met in order to secure support and so has sacrificed some of its profits in order to ensure continued partner participation and funding towards development costs. Some partners expressed concern that this was the beginning of a move away from the competitive best-value sourcing model and one towards prescribed work share.

However, Lockheed Martin is responsible for the subcontracting decisions as well as balancing partner expectations and the design, development and production of the aircraft. The use of such methods, which involved no cost premium to the programme and the arrangement of which was within the industrial participation rules already agreed, was seen as a response by Lockheed Martin to partners' concerns about the return on their investments. As things stand, most partners have a clause in their agreements that allow for their withdrawal during that phase of the project if industrial participation had not been deemed satisfactory. If a partner had decided to leave the programme, the DoD would have lost the additional development funding expected from that partner and Lockheed Martin could have eventually faced lower than expected international sales and a lower total number of units sold. At the same time however, directed work share arrangements do not always lead to the best programme outcomes, for example, the F-16 suffered additional cost premiums due to the use of foreign suppliers.[118]

So far therefore, it seems as if the JSF Programme has been working as planned, at least with regard to the participation of industry, the relative amount of workshare that different countries are getting and the return on their investment. As such, it could be argued that it provides a model for future procurement programmes, especially as the original nine-government membership has survived intact.[119] However, rumbling on since 2005 has been the almost universal criticism of the US Government's reluctance to share sensitive technologies that was promised at the start of the programme.[120] This 'technology transfer' issue is to some extent understandable given the events in 2005, vis-à-vis China and Israel,[121] but this shift in attitude has infuriated many of the partners, some of whom had already shared their technology with US businesses participating in the JSF programme.

This issue is strongly linked to the role of domestic politics, especially in the case of the United States, where even Lockheed Martin acknowledged that the biggest challenge to the programme as it entered the start of negotiations for the PSFD Phase was partner access to potentially sensitive US technology,[122] a view endorsed by the US Government Accountability Office.[123] Partner countries have highlighted that the existing MoUs all contain sections designed to safeguard US technology and that a significant element of the cost increase associated with the SDD Phase has been the installation of anti-tamper technology to the design in order to stop foreign buyers from reverse engineering sensitive technology.[124] Lockheed Martin has also argued that this reluctance to share technology ignores the reality of the global marketplace where high-technology manufacturing is a world-wide industry.[125]

Nevertheless, strong domestic political forces within the United States continued to coalesce around this issue with Duncan Hunter, Chairman of the Armed Services Committee of the House of Representatives showing concern over "the proliferation of licences on critical technologies on the JSF. Here is a stealth aircraft. This is not another F-16",[126] backed by Lisa Bronson, the Deputy Undersecretary of State for Defence – Technology, Security Policy and Counter-Proliferation.[127] This was just after an increase in tension over the EU's debate over lifting the arms embargo on China with the US DoD formally warning the EU that any such action would 'shut the door' on all transatlantic technology transfer, endangering cooperative programmes like the JSF[128] while the National Security Agency (NSA) vetoed European industrial participation in the next generation 'Link 16' communications link for NATO.[129] Such an attitude only continued to incense the partners, especially as the Western European Union (WEU), which is led by Germany and France, criticised them for "undermining achievement of the European defence and aerospace industries".[130]

The Australian Experience

"The White Paper identifies the F-35 Joint Strike Fighter as the ultimate solution."[131]

Domestic and international political issues have also surfaced in Australia. At the turn of the 21st Century, Australia was faced with the impeding decision to replace its thirty-three F-111 strike aircraft and seventy-one F/A-18 multi-role fighters, the cost of maintaining which, was steadily rising due to age, wear and tear. The Defence White Paper in 2000 stated that maintaining an air combat capability was the single most important issue in Australia's future defence requirements.[132]

The Australian Government planned to acquire its new capability of around 100 airframes via Project Air 6000 – later renamed the New Air Combat Capability (NACC) project – from 2012 onwards. The project was not only about the procurement of new combat aircraft but the acquisition of a whole raft of combat capability, including the introduction of new technologies, new avionics, sensors, weapon systems, communications and methods of processing data as well as uninhabited aerial vehicles (UAVs), long-range/stand-off missiles and other non-fast jet platforms. The project was going to consist of three phases, the scope and timing of which would be determined by the withdrawal of the F-111s in 2010 and the F/A-18s between 2012 and 2015:[133]

- **Phase One** – the first phase envisaged a decision in around 2006 and 2007, with an in-service date of between 2012-14 and a value of between A$4.5bn and A$6bn. This would procure forty new aircraft, to replace the first batch of F/A-18s being retired.
- **Phase Two** – the second phase envisaged a decision between 2010 and 2013 with an in-service date of between 2015-17 with a value of between A$4.5bn and A$6bn. This would procure thirty-five new aircraft and would replace the remaining F/A-18s still in service.
- **Phase Three** – the third phase envisaged a decision between 2014 and 2017 with an in-service date of between 2018-20 and a value of between A$2.5bn and A$3.5bn. This would procure approximately twenty-five new aircraft (with a greater degree of specialisation for the strike role).

As laid out, the project represented the most expensive procurement programme ever undertaken by the Australian Department of Defence with estimated costs of A$12 - A$16bn. Until mid-2002 the project was at an early stage that included defining the requirements and assessing generic options that might satisfy the capability requirement rather than determining if there was a specific weapon system or platform. Then suddenly, the whole focus of the programme changed. On 27 June 2002, Defence Minister Robert Hill announced Australia's intention to take part in the JSF SDD Phase at a cost of some A$300m and that the JSF would be the only likely contender to replace the F-111 and F/A-18 fleet. This move ended the options study phase of the project and the possibility of any form of marketplace competition, focusing the project on a single-platform solution. It was at this time the programme was renamed the NACC project but this did not stop the decision to effectively remove the need for a competition from attracting controversy. Firstly, it caught the potential European contenders, Dassault and Eurofighter, completely flatfooted with both sets of officials in the early stages of a four-year marketing campaign ahead of the expected competition in 2006 and both offering the Australians participation in their respective industrial programmes.[134]

Dassault for example, had just finished an extensive briefing programme with Australian industry on the opportunities associated with the Rafale project while BAE Systems Australia had put in at least two formal proposals on behalf of the Eurofighter Consortium in the first half of 2002 offering involvement in the Eurofighter Enhancement programme. Such moves had also received support from one of Australia's major defence industry associations, the Australian Industry Group, which had sent a letter on 3 June, signed by Executive Director Leigh Purnell, to the Undersecretary of State for Defence Acquisition, Michael Roche, stating that joining the JSF SDD Phase must not pre-empt the outcome of the Project Air 6000 requirement before a competition and that prematurely opting for a single solution before a complete value-for-money assessment had taken place could have serious consequences. BAE Systems Australia was still being assured by the RAAF in the weeks leading up to the announcement that the Australian Government was not considering a sole-source approach. Their CEO, Jim McDowell, in a letter attached to their 7 June proposal made it clear that they feared the worst and were not being treated on a par with Lockheed Martin – the Department of Defence had refused to engage with them on the possibilities and opportunities within the Eurofighter Enhancement programme – and appealed that the process be as fair and even-handed as possible. When the decision was announced, they were still awaiting a reply.[135]

What seemed to have hurt the prospects of exporting Eurofighter to Australia, was not industry's lack of commitment, as even then, they had well-defined and consistent views on how the aircraft should be developed from Tranche 1 (initial operational capability) to Tranche 2 (full operational capability) and finally Tranche 3 (enhanced operational capability) but the lack of commitment from the participating Governments. In stating that the main reason for joining JSF was that, in 2010, there would be no major rival to the American aircraft, Richard Hill admitted that it was Australia's view that the Eurofighter, or indeed the Dassault Rafale, would not be matured to the level that industry intends and that the market requires. Canberra obviously did not believe that the partner Governments would be looking to develop the aircraft beyond 2006 – and the decision by Austria to plum for the Eurofighter made little difference.[136]

Even without the international ramifications regarding the decision, there were domestic questions over the wisdom of committing to the JSF so early in the project's development. Such criticism prompted the release of a report through the Government-funded Australian Strategic Policy Institute (ASPI) by the Chief of the Air Staff, Air Marshall Angus Houston defending the decision,[137] citing the need for Australia to acquire a fifth-generation capability to maintain a competitive edge over the fourth-generation aircraft then coming into service across Asia, such as the Russian-built MiG-29 Fulcrum and Su-27/30 Flanker[138] and to achieve benefits for its industry.[139] Questions remained however, some of which are still relevant currently:[140]

• **Access to the Source Code** – as already mentioned, one major question has been access to the source codes, the millions of lines of computer code that enable the aircraft to fly and fight. This has been, and continues to be, a potential showstopper for international clients as under US legislation, the transfer of *Military Technology* to even close allies such as the UK and Australia, can be blocked via the US International Traffic in Arms Regulations (ITAR).

Without access to the code, Australia (indeed, any international partner) will be unable to modify or even maintain the aircraft independently – something they had done successfully with the F-111s for many years. Maintenance of the aircraft will be undertaken at a regional logistics and maintenance centre run by Lockheed Martin which is planning to massively expand the size of its Australian operation (from just under 500 employees to between 1,500 and 2,000) by 2017.[141] Without access to the source code, Australia (and everyone else for that matter) may be put in the invidious position of having to pay whatever Lockheed Martin asks during future contract negotiations for the ongoing maintenance of the JSF fleet. Not only that but joint operations between US Forces and allied forces could be hampered by the lack of technological standardisation, hindering interoperability and technology cooperation between national industries.[142]

•   **Cost Implications of Reduced Levels of Participation** – while international participation in the programme remains substantial, as alluded to above, a number of countries have wavered in their participation for a variety of reasons, including technological transfer issues and industrial workshare. Should any country pull out, especially if it was a major purchaser such as the UK, Italy or Australia, then the unit cost for the remaining partners would start to climb significantly. This could lead to a cyclical problem where the higher unit costs either force countries out of the programme or more likely, decrease the numbers of aircraft they are able to purchase – this then further increases the unit cost. The JSF programme is easily the largest defence procurement in Australian history with an approximate price tag of $16bn but questions remain as to how many airframes that will buy.

•   **How Many and How Much** – while the stated Australian requirement is for about 100 aircraft, a paper published in late 2005 by the Kokoda Foundation[143] set out the arguments for and against various force levels, from as low as seventy-five to as high as 120, while making assumptions on the number of aircraft per squadron. By this stage in the programme, the first test aircraft was still yet to fly and questions still remained over the eventual through-life cost of each aircraft. Despite the Australian procurement budget for its fleet increasing from A$12bn to A$16bn, the 'fly-away' cost of each JSF was expected to be around $100m each and this could still rise as the programme matures. Considerable development was still being undertaken, such as fully maturing seven out of eight critical technologies; completing the designs and releasing the engineering drawings for all three variants; the manufacture and delivery of fifteen flight test aircraft and seven ground test airframes; finishing and testing over 19 million lines of software code; completing a seven-year, 12,000 hour flight-test programme.

The cost of each aircraft is likely to be affected by the final overall number of airframes produced and whether access is granted to the source codes. If the overall cost of each JSF rises until it is either politically or economically unaffordable (whichever occurs first) then what alternatives are there? The demise of the competitive aspect of Project Air 6000 means that there is no 'hierarchy of merit' of alternatives to the JSF. Some have argued that the F-22 is a suitable fifth generation alternative while others have pointed towards the ongoing development of current-generation aircraft, such as the F-15, F-16 or F/A-18 as providing a reasonable capability.

• **The Available Alternatives** – with the USA in the process of shutting down F-22 production, some might have questioned why Australia didn't opt for this aircraft. Basically there are three reasons: firstly, the F-22 is decidedly an air-superiority aircraft with some strike capabilities (which it may be possible to enhance through in-service upgrades) whereas the JSF is much more of a multi-role aircraft, capable of both air-to-air and air-to-ground missions equally well; secondly, the F-22 is undoubtedly the 'crown jewel' in the USAF arsenal and there has been very little chance if any, of export, even a 'dumbed down' version, given US sensitivities over technology transfer; thirdly, the price of an F-22 is around double that of a JSF and with a relatively small budget, Australia could only afford a limited number of aircraft and given that the production process was well underway by the time this question was raised, there would have been very little opportunities available for Australian industry.

What about other aircraft? Some countries, such as Singapore and South Korea, have faced questions similar to that of Australia regarding their air defence capability needs. In both cases, they decided that the JSF was not the only alternative, with both opting for advanced variants of Boeing's Strike Eagle, the F-15K (South Korea) and F-15SG (Singapore), which includes upgraded engines and avionics, AESA radars and 'network-centric warfare' capabilities.[144]

In addition, there are the capabilities offered by Unmanned Aerial Vehicles (UAVs). The global 'War on Terror' has seen the speedy deployment of both reconnaissance and strike versions of these platforms as both have the capability to loiter for extended periods over the battlefield, a feature critical in the newly emerging network-centric warfare of the 21st Century, and their loss to enemy action will not have the same repercussions politically as the loss of a manned aircraft.

•   **Exactly How Much Stealth?** – the stealth capabilities of the JSF has been touted as a major selling point, but it is worth remembering that Australia, and indeed all partner nations, will only have 'third-tier' stealth capabilities. The 'first-tier' capability is held by the F-22 Raptor, while the 'second-tier' will be the US-version of the JSF. Technical commentary about the stealth design in the JSF seems to indicate that the aircraft is optimised against a particular set of radar types, in this case, X-band and Ku/K/Ka-band. These radars are predominantly found in other fighter aircraft and in battlefield air defence systems.

The design might however be less-optimal against what are known as S- and L-band radars, which are used predominantly in static or semi-mobile early warning radars as well as Airborne Early Warning and Control System (AEW&CS)[145] aircraft, such as the Boeing 737 also being procured by Australia, the US E3-Sentry and Russian A-50 Shmel (NATO reporting name – Mainstay) which is operated for example, by both India and China. Such capabilities are spreading across Asia and it must be wondered how much these will nullify the stealth advantage offered by the JSF, especially if bought in tandem with upgraded fourth generation aircraft (such as the Su-30 Flanker) with advanced Beyond Visual Range Air-to-Air Missiles (BVRAAM).[146]

•   **Conventional or STOVL?** – The Australian procurement programme envisages acquiring the conventional version of the JSF, primarily because this is the version with the longest range, the largest payload capability as well as being the cheapest. With the Royal Australian Navy in the process of procuring two very large amphibious warfare ships,[147] some in the Australian defence community are suggesting they acquire a small number of the STOVL variant for use aboard these ships thus providing a dedicated air warfare capability to embarked forces. While buying two versions may complicate the maintenance and support afforded the JSF fleet, it may answer another nagging question as how Australia can maintain an expeditionary strategy, even within an era of coalition warfare, if forward bases are unavailable or too vulnerable to enemy action?

Since then, the Australian Government has consistently reaffirmed its commitment to the JSF programme,[148] despite early concerns over workshare and ongoing concerns over cost[149] and technology transfer. Due to the slippage of the JSF programme and the up-and-coming retirement of the F-111s however, a capability gap was starting to appear. This forced the Australian DoD to quickly finalise the purchase of a squadron of twenty-four F/A-18F Block II Super Hornets[150] with possible additional purchases if further delays are encountered. This was after a comprehensive review of defence, the Defence White Paper 2009[151] outlining that Australia is committed to an initial purchase of no-less than seventy-two JSFs with a fourth squadron being bought later, depending on the timetable withdrawal of the Super Hornets (some of which could be converted into E/A-18G Growlers). This is in addition to planned spending on a raft of new air warfare capabilities, including another two C-130J Hercules aircraft, another ten new tactical airlifters to replace the de Haviland DHC-4 Caribous, five EADS KC-30A multi-role tanker aircraft, eight new maritime patrol aircraft and seven new high-altitude, long-endurance UAVs replacing the AP-3C Orions and six Boeing 737-based Wedgetail AEW&CS aircraft.[152]

Current UK Issues

"While we recognise the need for MoD to discuss issues relating to the transfer of information and technology with the US administration, the US Congress is where the issue needs to be addressed as individuals within the US Congress appear to be the main opposition to allowing information and technology transfer."[153]

F-136 Alternate Engine

As noted previously, there have been attempts in the FY2007, FY2008 and FY2009 budget submissions by the DoD (supported by the Administration) to terminate the F-136 alternate engine programme, a project involving General Electric, Rolls-Royce and Allison Engines. This is not due to any particular issues related to the programme but due, critics argue, to short-term budget considerations. This was highlighted in a letter by the Director, Acquisition and Sourcing Management from the Government Accountability Office dated 22 May 2006, to the Chairmen of both the Senate Committee on Armed Services and the House Subcommittee on Tactical Air and Land Forces (Committee on Armed Services).[154]

This letter argued that the DoD's decision was based on a need to identify sources of funding cuts to pay for other areas of the defence programme that have a higher priority. In making such a decision, the department did not conduct any new or comprehensive analysis, relying on data from studies done in 1998 and 2002. The focus was mainly on up-front savings in procuring the engine in the immediate future without looking at the potential long-term savings that might accrue from keeping the competition going within the overall programme, in providing support and maintenance through the entire life of the engine. Indeed, both studies had recommended keeping the competition going as the benefits of competition would outweigh the cost savings generated by cancellation. Officials also cited the favourable progress made by the primary engine and its predecessor (the engine for the F-22A Raptor) which reduced the risks involved to moving to a single source of supply. By this point in time however, the primary engine had only completed a small portion of its scheduled ground tests and the F-22A engine only about ten percent of its hours for 'system maturity' and failed to meet several reliability goals.

Such a view was reinforced the following year by testimony before the Subcommittees on Air and Land Forces, and Seapower and Expeditionary Forces of the Committee on Armed Services, US House of Representatives.[155] The estimated life-cycle cost of the JSF engine programme at this time was put at around $53.4bn, while ensuring competition within the programme by continuing with the alternate engine programme could add between $3.6 and $4.5bn. However, life-cycle savings from the competitive pressures retained within the overall programme should the alternate engine be continued were estimated to be between 10.3 and 12.3 percent, working out to between $5.5bn and $6.5bn, thereby recouping the investment and even generating additional savings.

There could be other non-financial benefits as well, such as better engine performance and reliability, improved industrial base stability and greater responsiveness from contractors. DoD experiences from other engine programmes, such as the F-16 fighter in the 1980s, have shown that competitive pressures can generate financial savings of up to twenty percent during the life-cycle of a project as well as having other benefits. A *Third Report* from the GAO[156] mentioned that the National Defence Authorization Act 2007 mandated the OSD and Comptroller General to undertake three independent cost-benefit analyses on having an alternate engine programme for the JSF, these being tasked to the OSD Cost Analysis Improvement Group, Institute for Defense Analyses and Government Accountability Office. While all three studies differed as to the break-even point (due to differences in opinion as to what costs should or should not have been included), they all agreed that there would be non-financial benefits from the engine competition, including fleet readiness, contractor responsiveness, sustainment of the industrial base, the potential impact on defence acquisition reform as well as the impact on international cooperation and the United States' relations with her allies.

From a British perspective, obviously the most important factor is the last one and any potential cancellation will have an impact on its defence industrial base. Unlike the F-22, the F-35 has been designed from the outset with a view to it being exported. Allied participation in the development of the aircraft and sales coming from such participation have been actively encouraged by the USA as a way of defraying some of the costs of development and production. Congress insisted that the JAST programme included the ongoing efforts of DARPA to develop an advanced STOVL aircraft, opening up the strong likelihood of British participation.

In fact, the programme has generated huge international interest with the eight development partners pledging some $4.5bn for the initial SDD Phase and all signing the PSFD MoU stating their intention to buy the aircraft. Israel and Singapore have both joined at the Security, Cooperation and Participation (SCP) level contributing $50m each while Spain, Greece, Japan and South Korea have all expressed interest in purchasing the JSF. Such participation could drive production to over 5,000 units.[157]

British companies have benefitted from the close UK – USA collaboration on the JSF and the UK being the only Level I participant, contributing $200m towards the 1997 – 2001 CDP and $2bn in the SDD Phase. BAE Systems is a major partner to Lockheed Martin, providing the aft fuselage, Empennage and Electronic Warfare suite. In addition, Rolls Royce is heavily involved with the programme, being under contract to Pratt & Whitney to provide the Lift-System components of the F135 STOVL propulsion system, while being partnered with General Electric on the F136 engine. The contract with Pratt & Whitney (signed in 2001) is worth $1bn to Rolls Royce over ten years and regardless of which engine is chosen, Rolls Royce's lift system will provide the vertical lift for all STOVL aircraft.[158]

However, cancelling the F136 programme would mean a large loss of revenue for Rolls Royce and while the company is involved in the F135 programme, this business appears to fall far short of the 40% partnership the company enjoys with General Electric. Rolls Royce will also be opening a new 1,000-acre facility in Prince George County, Virginia to make parts for a broad range of defence aerospace projects, including the F136 JSF engine.[159] It is unclear as to how or to what extent, termination of the F136 engine programme would harm international cooperation but each time the issue has arisen, allied reaction has not been favourable, especially in the UK. For example, the head of the CBI criticised the decision by the Bush Administration as sending out "all the wrong signals" over defence collaboration.[160]

Dr Liam Fox, the Conservative Party's Shadow Defence Minister stated that such a decision would "invariably effect future procurement decisions, with seriously negative consequences that may not be fully appreciated on this side of the Atlantic."[161] The UK's Defence Procurement Minister, Lord Peter Drayson, also expressed concern over the US Administration's unilateral decision to terminate the programme without consultation with London[162] and hinted that any such move, alongside a continued refusal to share technologies that the UK deemed necessary to operate a JSF fleet independently, would cause the UK to seriously reconsider its participation in the programme.[163] Such a view was undoubtedly mirrored across other European capitals as their defence industries have made significant inroads into both the F135 and F136 programmes.

For example, the Netherlands has seventy-four companies and research labs involved in the JSF project[164] and as European companies secure more contracts, the debate about the need for a second engine will become more complex. As alluded to earlier, friction has existed since the early stages of the programme between the DoD and the foreign partners and such manoeuvring over the alternate engine can only complicate matters. At one time or another, Italy, Denmark, the Netherlands, Norway and Turkey have all expressed dissatisfaction with the quality and quantity of the work that their companies have been awarded on the programme. These countries have all hinted at reducing their participation in the programme or even purchasing other European fighters instead of the F-35. The governments of both Italy and the UK have lobbied hard for F-35 assembly lines to be built in their countries. The level of international participation is currently around twenty percent but Lockheed Martin has said that such participation could rise to thirty percent once the programme has moved to full-rate production with an expansion of the supplier-base.[165]

The issue was raised once more in late 2009 when it emerged that Rolls Royce would have to redesign a key component of the F136 engine, fuelling criticism that this was another in a string of failings that would lead to cost increases and delays in the JSF programme adding to the determination of the DoD and Administration to cancel the F136 engine.[166] The Lexington Institute added to the controversy by issuing a brief saying that the alternate engine programme could be running as much as one year behind schedule and had managed just fifty-two hours of testing during the year, with the engine not being ready to benchmark against Pratt & Whitney's F135 until 2016.[167]

A source at Rolls Royce/GE in the US denied this was the case and that the re-design was a relatively simple task and full testing would be resumed early in 2010. He went onto say that the programme had so far kept to budget and m*et al*l the major milestones on schedule – the engine having accumulated over 550 hours of test time with over 800 hours of tests since the programme began.[168] The issue has once again became a major point of debate between Congress and the DOD/Administration,[169] with the Congress passing the FY2010 DoD Appropriations Bill on 19 December 2009 as P.L. 111-118[170] that included $6.8bn for thirty F-35 aircraft as well as $465m in Navy and Air Force R&D funding for continued development of the F136 alternate engine programme,[171] a move also backed by the Dutch.[172] The DoD/Administration decision to continue pressing for the cancellation of the programme has, once again, been criticised as 'near-sighted' by Ike Skelton, Chairman of the House Armed Services Committee, with the Pentagon's analysis failing to "consider the risk that relying on a single engine would present not only to our fighter force, but to our national security, given that the F-35 will account for 95 percent of our nation's fighter fleet".[173] Without doubt, it is a debate that is set to rumble on.

Technology Transfer

Another issue that is set to rumble on is that of the UK's access to the JSF source codes. The millions of lines of software code control everything from weapons integration, through radar performance to flight dynamics and so is a vital part of the overall systems' capability. This issue, which first arose to prominence in late-2004,[174] was supposedly settled after it had escalated to the President – Prime Minister Level during 2007, but was spectacularly re-ignited towards the end of 2009.

Back in late-2004/early-2005, relations between the USA and its European partners in the JSF programme had become strained due to the EU considering a lifting of an arms embargo on China.[175] The Chairman of BAE Systems, Dick Olver, criticised both the US technology transfer policy regime and the 'Buy America' campaign that was finding favour with many Senators at the time, during a speech given at the Woodrow Wilson International Center for Scholars in Washington DC on 12 July 2005.[176] This followed calls in June from BAE Systems' Chief Executive, Mike Turner, for the UK to withdraw from the programme if it didn't receive the information it requires.[177] The Armed Forces Minister, Adam Ingram, revealed during a debate in Parliament on 18 October that the UK and US had failed to reach an agreement on the transfer of technology but that negotiations were continuing. Unease was also growing in the UK Defence Industry generally, with Paul Everitt, Director of Communications at The Society of British Aerospace Companies (SBAC) complaining that the Government had not been open with industry over the technology transfer issue.[178]

Tensions continued to mount with a hardening of both the Government's and the MoD's position over the issue with Adam Ingram responding to a parliamentary written question on 13 February 2006 that the UK's technology transfer demands must be satisfied before the UK signed up to the next stage of the programme. This followed a speech by General Sir Timothy Grenville-Chapman, Vice-Chief of the Defence Staff at a conference in London on 7 February that it was "inconceivable that the IAB [the MoD Investment Approvals Board] will approve an aircraft where we cannot be utterly clear that we know enough about its technological make-up to be assured of all matters about safety".[179]

This was followed by a call from Mike Turner for the MoD to fund further feasibility studies into the costs of producing a naval variant of the Eurofighter Typhoon, as he gave evidence before the House of Commons Defence Committee on 28 February. Such a study had been undertaken in the late 1990s as part of an analysis into options for what was then the Future Carrier Borne Aircraft programme and which became the Joint Combat Aircraft programme, but would obviously need updating in light of the developments since then.[180] The issue then moved across the Atlantic when both Lord Drayson, the Defence Procurement Minister and Air Chief Marshall Sir Jock Stirrup, Chief of the Air Staff, testified before the Senate Armed Services Committee in mid-March, warning that a failure by the US to transfer the required technology, coupled with the cancellation of the alternate engine programme (see above) would force the UK to withdraw from the programme and adopt a 'Plan B'.[181] These sentiments were echoed by other partners, including Italy and Australia.[182] The House of Commons Defence Committee then weighed in with a flurry of reports over the remainder of that year, the first looking at the Future Carrier and Joint Combat Aircraft Programmes itself, where it recommended that:

"...the ability to maintain and upgrade the JSF independently is vital. We would consider it unacceptable for the UK to get substantially into the JSF programme and then find out that it was not going to get all the technology and information transfer it required to ensure 'sovereign capability'. This needs to be sorted out before further contracts are signed and we expect MoD to set a deadline by which the assurances need to be obtained. If the UK does not receive assurances that it will get all it requires to ensure sovereign capability, we would question whether the UK should continue to participate in the JSF programme."[183]

Next, came a report on the Defence Industrial Strategy that stated:

"We consider it vital that the UK can maintain and upgrade equipment independently. We expect the MoD to obtain all the information and technology transfer it requires to do this. We will continue to monitor the progress on technology transfer in relation to the Joint Strike Fighter."[184]

Finally, came a report on defence procurement generally, which praised the Minister of Defence Procurement, Chief of Defence Procurement and the MoD for its ongoing efforts to obtain cast iron guarantees regarding the transfer of technology but which sadly concluded that:

"...it is still uncertain whether the United States is prepared to provide the required information. If the UK does not obtain the assurances it needs from the United States, then it should not sign the Memorandum of Understanding covering production, sustainment and follow-on development. Such an impasse on a procurement programme of such strategic importance to the UK would be a serious blow to UK-US defence equipment cooperation, which has hitherto been of such positive benefit to both our nations. If the required assurances are not obtained by the end of the year, we recommend that the MoD switch the majority of its effort and funding on the programme into developing a fallback 'Plan B', so that an alternative aircraft is available in case the UK has to withdraw from the Joint Strike Fighter programme. We must not get into a situation where there are no aircraft to operate from the two new aircraft carriers when they enter service."[185]

Added to this was their concern that the MoD, although obtaining assurances on the technology transfer question from the USA, would not give details as to what those assurances were. It also recommended steps to facilitate industry-to-industry technology transfer as well as government-to-government technology transfer.[186] As it turned out, such assurances as were given, surrounded the signing of a Production Sustainment and Follow-On Development (PSFD) MoU on 12 December 2006 following a visit to Washington DC by Lord Drayson.[187] This committed the UK to the next phase of the programme, a move that had been viewed with concern by the House of Commons Defence Committee. Next came the signing of the UK-US Defence Cooperation Treaty by President George Bush and Prime Minister Tony Blair in June 2007 and published on 24 September.[188] It was hoped that this treaty, while aimed at UK-US defence trade generally, along with the assurances given before the signing of the PSFD MoU, would ease the transfer of technology in the JSF programme.

Just when everyone thought that the issue had been settled, it began to raise its head again in October 2008 when the Shadow Defence Minister, Gerald Howarth MP, spoke at a Jane's UK Defence Conference in London. The Shadow Defence Minister was delivering a speech entitled 'A 20-year Political Perspective' and in an aside to the US delegates attending the conference, he indicated that a future Conservative Government would abandon the F-35 if the UK was still waiting to secure access to technology that would ensure full 'operational sovereignty'.[189]

In August 2009, *The Daily Telegraph* reported that BAE Systems, a major contractor on the programme working in partnership with Northrop Grumman, was still battling for access to certain technologies and in negotiation over how much servicing can be done in the UK.[190] It would clearly be impractical for the RAF to fly the aircraft all the way over to the USA for repair and servicing but the two Governments have yet to reach an agreement. Once again, this issue is at the heart of what the UK considers it needs to retain 'operational sovereignty' over the aircraft,[191] with Chris Garside, BAE's Chief Engineer on the project and Director of Development, stating that "To allow the UK customer to operate the aircraft the way they would need to, we need access to various technologies".[192]

Then, in late November, the Reuters News Agency revealed[193] that the United States had decided that it was going to keep the software codes all to itself, with none of the partners getting access to them. Jim Schreiber, Head of International Affairs at the JSF Program Office, had disclosed this to the agency during an interview, admitting that the decision had been universally unpopular with the eight partner countries, saying that "nobody's happy with it" – a situation that was also reported to Congress.[194]

Schreiber claimed that all of the partners' requirements had been taken into account and they had provided ways in which they could upgrade their JSFs even without access to the codes. Such a move was also seen as a rebuff to Israel who had wanted the technology as part of a possible purchase of up to seventy-five aircraft. Instead, the USA plans to set up a maintenance and support facility, possibly at Elgin Air Force Base in Florida, to continue the development of the JSF software including upgrades, integrating them with the US aircraft at the facility and then distributing them to the other users. Representatives from the British Defence Staff in Washington DC did not return requests for comments on the decision but a statement released by the MoD on 6 December said that:

"The Joint Strike Fighter (JSF) is progressing well and the UK currently has the JSF data needed at this stage of the programme, and is confident that in future we will continue to receive the data needed to ensure that our requirements for operational sovereignty will be met. This remains the basis of the agreements reached with the US in 2006."[195]

Such a statement is obviously trying to put a positive, even an optimistic gloss on a disappointing turn of events and belies the concern that must be felt, not only in London, but in all eight partner countries. Such a situation does not bode well for the supporters of multi-national defence procurement programmes, especially those led by the USA, and comes at a critical time for the MoD as it prepares the ground for the imminent Strategic Defence Review, the last being in 1998.[196] Most of the papers[197] released so far address the issue of the defence relationship with the USA and few projects have come to epitomise the UK commitment to such a relationship quite like the JSF, so the US decision to reopen the source code issue could not have come at a worse time. For example, the 'Defence Green Paper' states that:

"...the UK has a range of close bilateral security and defence relationships. None is more important than with the United States. The relationship is based on common values and interests which will endure in the 21st Century, to our mutual benefit. The UK benefits greatly from bilateral co-operation in the nuclear, intelligence, science, technology and equipment fields."[198]

In line with this, the MoD statement in December 2009 avoided a direct confrontation over the issue, but it is not unreasonable to assume that the issue will quickly rise to renewed prominence after the election and with the undertaking of the SDR. Such a turn of events was noted by the House of Commons Defence Committee in that a

"...further issue affecting the JSF programme is technology transfer, which would provide the UK with operational sovereignty. In our Defence Equipment 2009 *report we noted that the MoD considered that the technology related to the programme was being transferred on schedule. However, recent newspaper reports have suggested that US officials have rules out the technology transfer of source codes on the JSF, even to close allies such as the UK."*[199]

This reigniting of the source code issue is likely to be more intense than the first due to three complicating factors. The first is the continued pressure by the US Administration to cancel the F136 alternate engine programme, as noted above. The second is the rise in cost of the JSF. With the UK's defence budget coming under increasing pressure, there has been speculation as to whether only one of the future aircraft carriers will be fully fitted out with F-35s in order to save money. The other options being to order the CV version of the JSF instead of the STOVL version (which would mean equipping the carriers with catapults and arrestor gear), ordering aircraft already in production (buying 'off-the-shelf'), upgrading older aircraft or having one of the carriers act as an assault carrier, equipped with helicopters, Royal Marine Commandos and UAVs.[200]

With little prospect as to savings *vis-à-vis* the carriers themselves; the only savings can come from what's placed on them. The third factor is the US-UK Defence Trade Cooperation Treaty signed in 2007,[201] but still technically on hold, something that has drawn criticism from the House of Commons Foreign Affairs Committee in that they are "disappointed that despite promises to do so, the US Senate has not yet ratified the UK-US Defence Trade Cooperation Treaty."[202]

The aim of the treaty was to cut red tape in the bilateral exchange of defence-related goods, services and information and would have been a major step forward – whether it would have had an impact on the technology transfer issue is an open question but the fact the US has still to ratify the treaty, indicates to the UK that improving bilateral relations with the UK is not a priority agenda item. This might be because the vast majority of license applications for US-UK transactions are approved (an estimated 99.8 percent, equal to some 8,500 items with a value of $14bn in 2006), indicating (at least to the US) that the bilateral relationship is operating normally,[203] especially with both US and UK firms having acquired aerospace and defence businesses on the opposite side of the Atlantic or set up established operations of their own. With US companies having bought twenty-seven UK businesses between 2001 and 2006, worth £5.1bn and UK firms having acquired fifty US businesses worth $7.3bn in the same time period, it might seem that the UK-US defence interconnectivity has increased substantially.[204]

The US Export Control Regime[205]

"Today's export control system is a relic of the Cold War and must be adapted to address current threats. The current system impedes cooperation, technology sharing, and interoperability with allies and partners."[206]

Due to the significance of the source code, technology transfer and operational sovereignty issues to the UK, it is worth looking at the US Export Control Regime and its overall impact on multinational defence procurement programmes, of which the JSF is just one. As it stands, the UK and indeed most partner countries would expect the USA to protect its cutting edge *Military Technology*, but what may not be appreciated in Washington DC is the extent to which US export control and technology transfer policies have impacted upon countries' desire to participate in a US-led defence procurement programme. For if:

"...the United States and the UK, the two closest of allies, are unable to overcome the continuing obstacles to the efficient sharing of defence-related technologies, what hope is there for broader transatlantic defense industrial and technological cooperation? Bilateral U.S.-UK cooperation in the fields of intelligence, nuclear defense, and military deployments is unprecedented in U.S. alliances. And the U.S. and UK defence industrial bases have become increasingly intertwined through investment and trade. And yet, the U.S. and UK governments have proved unable to institute a more open system for exchanging and transferring defense technologies, despite the stated intent of senior political leaders and extensive efforts by both sides."[207]

In addition, there is a perception that the USA does not value the contribution the UK is making to the JSF programme and indeed is making to the broader defence and security environment, even though:

"...the UK can bring – as it already has brought – valuable technologies to the table for the United States. The UK's track record of useful Military Technology innovation includes the contemporary examples of the vertical, short take-off and landing engine system and the anti-IED capabilities now deployed in Iraq. A common perception of the threats also means that the UK defense science and technology establishment is focused on solving problems in areas that are of value to the United States, such as counter-terrorism and net enabled warfare. And UK investment in cooperative programs such as the JSF can lessen the development cost for an increasingly-strained U.S. defense budget while decreasing the per unit costs of the system once they go into production. Furthermore, while incomparable in size to the U.S. marketplace, the UK defense market does offer opportunities for major defense contracts".[208]

Even the Pentagon itself recognises the damage which is being done to the USA's defence and security interests by the export control and technology transfer regime. An excerpt from the latest *Quadrennial Defense Review* states that:

"Today's export control system is a relic of the Cold War and must be adapted to address current threats. The current system impedes cooperation, technology sharing, and interoperability with allies and partners. It does not allow for adequate enforcement mechanisms to detect export violations, or penalties to deter such abuses. Moreover, our overtly complicated system results in significant interagency delays that hinder U.S. industrial competitiveness and cooperation with allies.

The United States has made continuous incremental improvements to its export control system, particularly in adding controls against the proliferation of weapons of mass destruction and their means of delivery. However, the current system is largely out-dated. It was designed when the U.S. economy was largely self-sufficient in developing technologies and when we controlled the manufacture of items from these technologies for national security reasons...

The global economy has changed, with many countries now possessing advanced research, development, and manufacturing capabilities. Moreover, many advanced technologies are no longer predominantly developed for military applications with eventual transition to commercial uses, but follow the exact opposite course. Yet, in the name of controlling the technologies used in the production of advanced conventional weapons, our system continues to place checks on many that are widely available and remains designed to control such items as if Cold War economic and military-to-commercial models continued to apply.

The U.S. export system itself poses a potential national security risk. Its structure is overly complicated, contains too many redundancies, and tries to protect too much. Today's export control system encourages foreign customers to seek foreign suppliers and U.S. companies to seek foreign partners not subject to U.S. export controls. Furthermore, the U.S. government is not adequately focused on protecting those key technologies and items that should be protected and ensuring that potential adversaries do not obtain technical data crucial for the production of sophisticated weapons systems.

These deficiencies can be solved only through fundamental reform. The President has therefore directed a comprehensive review tasked with identifying reforms to enhance U.S. national security, foreign policy, and economic security interests."[209]

Such a change in the Global Economy and the 'flow' of technology was highlighted recently by the purchase of 2,200 Sony Playstation 3 videogame consoles by the US Air Force to form the basis of a new supercomputer.[210] Recent research in the USA has pointed to how European firms are responding to the difficulties imposed by the US policy regime. One study reports that "virtually every interview we conducted highlighted U.S. defense trade controls as a 'barrier' significantly impeding Transatlantic cooperation...".[211] It cites specific repercussions from the US regime, which make it difficult for others to participate in US projects which involve ITAR complications. There are four key concerns. The first is limits on operational sovereignty, which has been noted previously as of particular significance to the UK, regarding the F-35. In fact, the report states that "The UK, one of our closest allies, as well as France and Italy, expressed strong concerns about this issue".[212] The other three points are: reliance on ITAR-controlled systems generating risks of schedule delays and increases in costs; re-export restrictions; and the complications the regulations generate for multi-national facilities.[213]

The report also comments that "There is clear evidence, beyond rhetoric, of a behavioral shift in Europe toward 'designing around' or designing out components or subsystems regulated by the U.S. International Traffic in Arms Regulations (ITAR), which has a particularly adverse impact on U.S. subsystem and component suppliers".[214]

Expanding on the point, the report notes:

"Over more than a decade, one study after another has highlighted the problems inherent in U.S. export controls – notably the ITAR. While the specifics of these ITAR issues are beyond the scope of this study, the impact of ITAR on the Transatlantic defense market relationship is not. Market participants, U.S. and foreign, consistently report that ITAR slows the speed of obtaining licenses needed for sales and collaboration, limits the release of U.S. technology, creates business uncertainty, and generally makes the process of Transatlantic defense industrial cooperation difficult. Fairly or not, most European governments are concerned about relying on ITAR systems and subsystems because they potentially limit their operational autonomy over major systems (especially in real-time crises), introduce program delays and risks, and curtail their export flexibility for systems with U.S. components.

Years of European talk of 'designing around' or 'designing out' ITAR have now begun to translate into action, according to market participants – with increased evidence that U.S. ITAR policies and practices, for better or worse, are limiting opportunities for U.S. firms competing in Europe (especially at the subsystem level). This is increasingly true even among our staunchest allies.

The ITAR also inhibits U.S. firms from working with foreign firms on domestic U.S. programs and creates challenges for foreign firms seeking to enter the U.S. market. By declining to release certain information on technologies, the acquisition community can effectively preclude foreign participation.

While strong and well-enforced export controls are an important tool of U.S. national security, it is clear that the U.S. failure to address these concerns will curtail the extent of Transatlantic defense technology sharing, defense cooperation and the development of an open and competitive Transatlantic defense market."[215]

The report provides specific examples of this development. On a policy level, the French White Paper "explicitly cites the need for non-ITAR-controlled electronics components to avoid limitations on French freedom of action".[216] In another instance, a country ensured operational sovereignty by "requiring that the program be staffed with domestic engineers free of ITAR restrictions".[217]

And European firms have developed policies specifically aimed at avoiding the use of ITAR items, developing "dual track" production lines of ITAR and non-ITAR items and favouring suppliers of non-ITAR components.[218] Certainly if the ITAR items are significantly superior to the non-ITAR items, the added complications may be worth the added capability. However, if the gap is not that great, the ITAR requirements can be a consideration. As noted by the report, "Where the differential is not great, European governments and firms are increasingly opting for the non-ITAR choice".[219]

In fact, complaints about US export control and technology transfer policy have been around for some time and it is not clear how heavily they will weigh on the decisions of other nations to work with the US on military projects when compared to the costs of developing their own cutting-edge *Military Technology* programmes. As another report points out, "very few national military establishments can generate sufficient orders to sustain a weapons source of efficient size in any category".[220] And with the rapid growth of military technology (and the concomitant growth of costs) the importance of US participation in any development program will clearly increase.

However, the US should consider whether it can afford to be indifferent to the willingness of other nations to participate in, and carry some of the costs of, such defence programs. Spreading the burden of large development costs would presumably be appealing to the DoD. Increasing costs also have ramifications with regard to the production phase of future programs. It is an open question whether DoD contracts alone would be sufficient to sustain US military contractors. The Congressional Research Service has noted that while the US aviation industry is positioned to compete in the growing global market for civil aircraft, "the extent to which such economic conditions may preserve an adequate US defence industrial base for the development and production of combat aircraft is debatable, however, given the significant differences between civilian and military aircraft requirements and technologies".[221] Even American firms and the DoD may need to focus more on overseas sales to sustain defence-related procurement programmes and if the US wishes to generate significant sales to other nations, then it is important that such equipment address the fundamental issues of operational requirements and sovereignty which have been critical to the UK in the JSF project.

Ultimately, there will always be a concern in the USA, as in all countries, about the leakage of technology as a result of multinational programs. However, despite that risk, there are clear benefits which also need to be considered:

"Governments seeking to strengthen the transatlantic defense relationship must weigh the potential benefits of cooperation against the risks that technology shared with allies might eventually leak to hostile states or subnational groups. Although technology transfer to allies generally involve acceptable risks, those responsible for safeguarding U.S. Military Technology and for preventing the spread of conventional arms and WMD view technology transfer, even to close allies, as a potential threat and, therefore, as something to be tightly controlled. Close transatlantic cooperation, however, may also reduce the potential risk of technology leakage as it improves each side's perception of sensitive issues and encourages adequate levels of protection."[222]

Clearly, export control and tech transfer regulations and policies are a product of political choices. The report notes that:

"Although often detrimental to transatlantic defense cooperation, the policy preferences of DOD and the agency regulatory processes involving technology transfer and export controls reflect political realities in the United States. There is widespread concern about risks that U.S. technological secrets or certain defense technological leads may be lost to competitors abroad or to hostile countries, with negative consequences for U.S. national security. Doubts about the effectiveness of European export control systems are also widespread. These concerns can translate into strong resistance, especially in the U.S. Congress, which is deeply involved in arms exports, export controls, and defense investment issues." [223]

There is a clear feeling that "there continues to be broad consensus that the U.S. export control system attempts to control too much in light of the widespread diffusion of technologies with defense-related applications". [224] Indeed, one of the key criticisms in the UK is that the US approach is far too broad and does not distinguish between high- and low-tech equipment. As already noted, it is accepted in the UK that the US wants to, and has every reason to, closely protect cutting-edge military technology which it developed at significant time and expense. The problem is that the export control and technology transfer policies extend far beyond such sensitive technology to adjacent areas which to non-American eyes do not appear to warrant such Orwellian control.

However, while the JSF has its faults, compared to past transatlantic defence projects, this has been fairly co-operative. The report notes that some JSF participating states were not pleased with the limitations on their access and participation compared to the US and UK. "Nevertheless, the overall interchange of technical data and coordination of requirements among JSF partners is unprecedented, with the latter providing leading-edge technology, state of the art manufacturing processes, and value-based products". [225] However, it is clear that transatlantic co-operation cannot work in the long-term unless the technology transfer and export control issues are addressed. As the report clearly states:

"Export control difficulties go to the core of the problems that hamper transatlantic defense cooperation. Changes in the United States and in Europe will be necessary if this critical impediment to enhanced cooperation is to be removed. The U.S. export control system is broken; its technology transfer rules increasingly self-defeating and out of step with broad trends in the global and European economies. Export control reforms in the United States

are therefore imperative, including shrinking the U.S. Munitions List to critical items, instituting greater corporate self-governance with government audits of performance and creating a stronger appeals process for disagreements."[226]

As such, if these issues are not resolved then the UK will find itself in a difficult position, either having to 'tow-the-line' and accept the aircraft as it stands and pay the price of having potential future constraints on its 'operational sovereignty' with regard to its JSF fleet, or, decide to leave the programme and look for other alternatives, with consequences for both it and the programme as a whole, not to mention the UK-US 'special' relationship. Possible alternatives are the French Dassault Rafale-M,[227] Boeing's F/A-18E or even a naval variant of the Typhoon. The relationship as it stands however, is changing, given the changing global political and economic structure with the:

"...UK's influence both globally and with the US looked set to decline. As Professor Clarke stated, "the Cold War was undoubtedly good for Britain's influence in the world [but the] present environment of disparate power and great uncertainty does not provide as relatively cheap and easy a vehicle for British diplomacy as did NATO in the Cold War". He argued that "for the United Kingdom, the long-term perspective suggests that its natural influence with the United States will be diminished". Similarly, Heather Conley and Reginald Dale believed that the combination of structural changes which will shift the US focus away from Europe with reductions defence or diplomatic capabilities will, over the longer term signal an end to the UK's 'disproportionate influence in world affairs'."[228]

Conclusion

"The JSF may be 'too big to fail' but the same was said of other programs in the past. With the budgetary shortfalls facing the U.S. in the next ten years, the F-35 may be one of the biggest casualties of the 2008 recession." [229]

The Joint Strike Fighter Programme started life as a collection of separate programmes run by each service in the US Armed Forces. The end of the Cold War meant that the West could not sustain the levels of defence spending that had occurred during that confrontation due to the clamour for a 'peace dividend' and a release of public to other areas of Government spending.[230] This was so not only in the UK but in the USA as well. So gradually, the various programmes that had been undertaken by each service gradually coalesced until it became a joint programme across the US Air force, US Navy and US Marines, looking to replace a host of legacy aircraft with a new, 5th generation capability, becoming the largest procurement programme in history.

On top of that, the programme heralded a new approach to procurement, not only in terms of being across all three US services that operate fast jet aircraft, but also in terms of international and industrial participation. Very early in the programme, participation was offered to the United States' allies, with eight countries taking up various levels of membership (with the UK being the only Level 1 participant) which afforded different rights, responsibilities and amounts to contribute. In addition, Lockheed Martin tried to stick to the principles of best-value resourcing and awarded contracts out to different companies whom they felt offered greatest value. Problems quickly emerged however, with a number of allies being unhappy at the amount of workshare their companies were receiving and so Lockheed Martin used the leeway it had with awarding contracts to award contracts to countries' industry where the return from their investments had not met expectations.

There have been other problems too. In the move towards Low-Rate Initial Production (LRIP) and Full-Rate Production (FRP) one of the most controversial aspects is that of concurrent development. The US DoD could procure up to 383 aircraft, at a cost of around $54bn before the final phases of flight testing are complete in 2014.[231] Compounding this risk is the move to speed up validation of F-35 components by replacing a large percentage of flight tests with simulations, desk studies and the use of a flying software testbed known as CATBird. Such plans were highlighted as placing "very significant financial risk on the Government."[232] Such a rush to build new aircraft before the test aircraft have flown a significant amount of test flights could lead to cost increases as additional problems are found and have to be fixed over a larger and larger number of aircraft, rather than testing and modifying a small number of aircraft and then incorporating those solutions into a final production design.

Another factor that could affect costs are changes to the planned quantities customers decide to buy. In a time of recession, pressure on the West's defence budgets has been severe, along with (for some) a rise in operational commitments overseas. This has even had an impact on the biggest spender of all, the USA, and while the US is still planning to procure some 2,443 aircraft,[233] Congress could mandate a lower figure at any time or reduce the production rate for a given year. The partner nations are looking to buy approximately 730 aircraft but that number is still subject to modification until production contracts are signed. Given the rapid pace of production, any changes to the overall number to be manufactured could have a substantial impact. Despite the programme being set up to counter the ever-present effects of defence inflation that has plagued the procurement of major weapons systems since World War II, between 2001 and 2009, changes in the numbers to be produced, as well increases in the cost of labour and certain materials, as well as programme delays due to technical problems, contributed to a rise of thirty-eight percent in the JSF's unit cost, with the CTOL variant weighing in at around $45m (in 2002 US$) while the STOVL and CV variants coming in about $60m apiece.[234]

Indeed, current estimates have increased even further to over $100m with a total programme cost approaching $400bn.[235] This is inline with the DoD having to inform Congress on 26 March 2010 that the programme had committed a Nunn-McCurdy breach, detailed in the yearly *Selected Acquisition Report*, released by the DOD on 1 April 2010.[236] This relates to what is known as the Nunn-McCurdy Provision contained within the 1982 Defense Authorization Act that requires the DoD to certify to Congress that, once a programme has passed a certain cost threshold, it is still vital to National Security. Such certification was signed by the Pentagon's Under Secretary of Defense for Acquisition, Technology, and Logistics, Ashton Carter and passed to Congress on 1st June 2010.[237]

It is this pressure on defence budgets, increased operational commitments and the rise in costs that has seen murmurings across the partners, including the UK, about the total number of aircraft that may be procured.[238] As it stands, the UK is still scheduled to procure three test aircraft (two in 2011 and one in 2012) and then start receiving small numbers of aircraft for training purposes as the programme moves through the LRIP Phase and gears up for FRP. While still looking at a final figure of 138 operational aircraft of the STOVL variant, the UK has not made any final decisions as to the number of aircraft or the variant it will procure. With the PSFD MoU (signed in 2006) not contractually committing any nation to buying a specified number of aircraft, the door is open for the UK, in light of the forthcoming Strategic Defence and Security Review (SDSR), to modify its planned procurement of JSF aircraft in some way.[239]

This could be accomplished by modifying the total numbers of aircraft or the variant(s) procured, a combination of both, or in an extreme case, to leave the programme entirely. While this last option is unlikely, the signs are becoming more and more ominous, especially with the current row between the MoD and the Treasury over who will pay for the replacement of the Trident nuclear submarines. Historical precedent has meant that that the initial procurement cost has usually come out of central funds with the MoD covering the day-to-day running costs. The statement by the Chancellor (George Osborn) that the MoD should pick up the total bill, has set the two departments on a collision course. Either the Treasury backs down and accepts historical precedent, a compromise is reached, which will mean even greater pain for the MoD, or the Treasury gets its way. This last outcome will have devastating results for the MoD budget and could very well see the UK withdraw from the JSF programme.[240]

Added to this have been the challenges the UK (and other partners) has faced in securing access to the source codes that control the aircraft, for without access, there cannot be a truly independent and sovereign capability, perhaps not in using the aircraft, but in maintaining and servicing it in the long-term and modifying it to take account of local requirements. Such a row has threatened on more than one occasion to damage the Transatlantic Relationship and was spectacularly reignited in late 2009 when Jim Schreiber, Head of International Affairs at the JSF Programme Office, confirmed in an interview for Reuters that the USA would not be allowing anyone access to the source codes. This in fact has pointed to the broader problems in the US Export Control Regime, a regime that is in desperate need of reform, where even its allies have started to design around items that are subject to the USA's ITAR controls. There has also been tension over the alternate engine project, the F136, being developed by a team led by Rolls Royce and General Electric. Congress instituted the project due to what was known as the Great Engine War in the late 1980s to ensure competition with Pratt & Whitney who were developing the main engine, the F135. Despite attempts in the last few years by both the Administration and the Pentagon to drop the alternate engine project in order to save money, Congress has repeatedly reinstated funding. Such a cut in funding would be a major blow to Rolls-Royce, even though they are involved with the F135 engine as well.

It is for all these reasons that continued allied participation in the JSF programme is, while very likely, not completely guaranteed. But for any country who does decide to leave, where does it go for a replacement capability? Its options are few, especially if they want a 5th generation aircraft.

Notes

[1.] The authors would like to thank Jonathan Davies for his help in putting this case study together.

[2.] Pete Aldridge, US Under Secretary of Defense for Acquisition as quoted in *The Daily News*, Los Angeles, CA at <<http://www.thefreelibrary.com/ LOCKHEED+WINS+FIGHTER+PROJECT.-a079570926>, article dated 27 October 2001.

[3.] *See* <http://www.globalsecurity.org/military/industry/mergers.htm> for a picture of the number of mergers that have occurred to produce the four US defence companies, and <http://en.wikipedia.org/wiki/Main_Page> for articles on Boeing, Northrop Grumman and Lockheed Martin.

[4.] Burbage, T. **'The Aerospace Industry, Today & Tomorrow'** in *Proceedings of the Shepherd Air Power Conference 2002*, Shepherd Conferences, London, 30 January – 1 February 2002.

[5.] Hayward, Keith. **'The Globalisation of Defence Industries'** in *Survival*, Volume 43, No. 2, Summer 2001, pp. 115 – 132.

[6.] Kirkpatrick, David. **'Revolutions in** *Military Technology* **and Their Consequences'** in *RUSI Journal*, Volume 146, Issue 4, August 2001, pp. 67 – 73.

[7.] Wood A. **'Challenges for the European Aerospace Industry'** in *Proceedings of the Shepherd Air Power Conference 2002*, Shepherd Conferences, London, 30 January – 1 February 2002; Foss, Christopher F *et al.* 'UK Defence Industry: Going Global', posted 27 September 2002 and located on the Jane's website <http://www.janes.com>

[8.] Vance Coffman, Vice Chairman and CEO of Lockheed Martin speaking at the Air Power Conference in Washington DC, February 1998.

[9.] International Institute of Strategic Studies. **The Military Balance 2009**, Routledge, London, 2009, p. 20.

[10.] Ibid. p. 28 and pp. 111 – 163.

[11.] <http://www.defenselink.mil/comptroller/cfs/fy2004/FY_2004_Army_Financial_ Report.pdf>, p. xiii.

[12.] Caffrey, Craig. *'Fewer, but fitter: fighter aircraft programmes'*, posted 6 April 2009 and located on the Jane's website <http://www.janes.com>

[13.] O'Rourke, Ronald. **F-35 Joint Strike Fighter (JSF) Program: Background and Issues for Congress** in *Congressional Research Service Report for Congress*, 7-5700, RL30563, 24 July 2009, p. 1.

[14.] Harkins, Hugh. **Lockheed Martin F-35 Joint Strike Fighter – The Universal Fighter**, Centurion Publishing, Glasgow, 2004, p. 4.

[15.] Dain Hancock, Tactical Aircraft Systems President, Lockheed Martin, quoted in Fricker, John. **'Joint Strike Fighter: Special Report'** in Jackson, Paul. (Ed). *Jane's All the World's Aircraft 1995-96*, Jane's Information Group, Coulsdon, Surrey, 1996, p. 3.

[16.] Keijsper, Gerard. **Joint Strike Fighter: Design and Development of the International Aircraft**, Pen & Sword Books, Barnsley, 2007, pp. 7 – 8.

[17.] Op Cit. Harkins, 2004, p. 4.

[18.] Op Cit. Keijsper, 2007, pp. 7 – 8.

[19.] Michael Michellich, JAST Programme Manager for Boeing, quoted in Jane's Information Group: **'Joint Strike Fighter Nears Downselect'** in *Jane's Defence Weekly*, 08 May 1996, p. 25.

[20.] Op Cit. Harkins, 2004, p. 4.

[21.] Official JSF Website. History Webpage, located at <http://www.jsf.mil/history/>

[22.] Op Cit. Harkins, 2004, p. 4.

[23.] Noel Longuemare, US Principle Under Secretary of Defence for Acquisition and Technology, quoted in Warwick, Graham. **'Joint Endeavour'** in *Flight International*, 3 – 9 July 1996, pp. 25 – 28.

[24.] Starr, Barbara. **'Prototype Policy to Shape New Fighter'** in *Jane's Defence Weekly*, 26 June 1993, p. 4.

[25.] Mason, Tony. *The Aerospace Revolution*, Brassey's, London, 1998, p. 139.

[26.] Op Cit. Fricker, 1996, pp. 8 – 14.

[27.] Op Cit. Harkins, 2004, pp. 24 – 26.

[28.] The versions break down as A – USAF, B – USMC/RN and C – USN. Braybrook, Ray. **'JSF ...the accountant's warplane!'** in *AIR International*, February 1997, pp. 77 – 82.

[29.] Jane's Information Group. **'The JAST Takes Joint Strike Fighter A Step Forward'** in *Jane's Defence Weekly*, 10 April 1996, p. 5; Aboulafia, Richard. **'From JAST to JSF'** in *Military Technology*, May 1996, pp. 82 – 84.

[30.] Op Cit. Braybrook, February 1997, p. 82.

[31.] Op Cit. Fricker, 1996, pp. 14 – 15; Tirpak, John. **'Strike Fighter'** in *Air Force Magazine*, October 1996, pp. 22 – 28.

[32.] Op Cit. Aboulafia, 1996, p. 83. Op Cit. Fricker, 1996, pp. 8 – 15.

[33.] Op Cit. Warwick, 1996, p. 25.

[34.] Op Cit. Harkins, 2004, p. 36.

[35.] Gerry Murff, Weapon System Lead Designer for the Lockheed Martin team, quoted in Hehs, Eric. **'Joint Strike Fighter'** in *Air Force Magazine*, July 1996, pp. 41 – 45.

[36.] Starr, Barbara. **'Pentagon Makes Innovation and Cost Priorities for JSF'** in *Jane's Defence Weekly*, 27 November 1996, p. 4; Jane's Information Group. **'JSF Downselect'** in *Jane's Navy International*, December 1996, p. 7; Jane's Information Group. **'Boeing and Lockheed Martin to Demonstrate Joint Strike'** in *Jane's International Defence Review*, January 1997, p. 5.

[37.] Fulghum, David and Morrocco, John. **'Final JSF Competition Offers No Sure Bets'** in *Aviation Week* and *Space Technology*, 25 November 1996, pp. 20 – 22.

[38.] Op Cit. Keijsper, 2007, p. 31.

[39.] Byron Callan, First Vice-President of Merrill Lynch quoted in Starr, 27 November 1996, p. 4.

[40.] Velocci Jr, Anthony L. **'Major Change Looms At McDonnell Douglas'** in *Aviation Week* and *Space Technology*, 25 November 1996, pp. 24 – 25.

[41.] Starr, Barbara. **'Boeing Opts for St Louis Base'** in *Jane's Defence Weekly*, 13 August 1997, p. 19.

[42.] Starr, Barbara. **'JSF Launches Back-Up Plan'** in *Jane's Defence Weekly*, 26 February 1997, p. 6.

[43.] Norris, Guy. **'Thunder in the Desert'** in *Joint Strike Fighter: Inside the 21st Century Warfighter, Flight International* Supplement, 5 – 11 September 2000, pp. 32 – 36.

[44.] Bolkcom, Christopher. *Proposed Termination of Joint Strike Fighter (JSF) F136 Alternate Engine*, 18 February 2009, Congressional Research Service, RL33390.

[45.] Bruno, Michael. **'House defence appropriators push back on JSF engine'** in *Aerospace Daily* and *Defense Report*, 17 February 2006.

[46.] Amick, Karl G. *The Next Great Engine War: Analysis and Recommendations for Managing the Joint Strike Fighter Engine Competition*, Naval Postgraduate School, Monterey, CA, 2005, p. 8.

[47.] Op Cit. Keijsper, 2007, pp. 35 – 82; Op Cit. Harkins, 2004, pp. 75 – 151.

[48.] Sweetman, Bill. *Ultimate Fighter: Lockheed Martin F-35 Joint Strike Fighter*, Zenith Press, St Paul, MN, 2004, p. 94.

[49.] Aldridge, as quoted in Anon. **'Lockheed Martin JSF Team Wins $200b Contract'** in *St Louis Front Page*, 26 October 2009 located at <www.slfp.com/102701BIZp.htm>

[50.] Op Cit. Sweetman, 2004, p. 92.

[51.] Sirak, Michael. **'Lockheed Martin tops Boeing to build Joint Strike Fighter'** in *Jane's Defence Weekly*, 7 November 2001.

[52.] Op Cit. Sweetman, 2004, p. 94.

[53.] Sweetman, Bill. **'JSF – how the battle was won'** in *Jane's Defence Weekly*, 07 November 2001, posted on 01 November 2001 at <http://www.janes.com>

54. Op Cit. Sweetman, 2004, p. 94.

55. Op Cit. Keijsper, 2007, p. 119.

56. Editorial. **'JSF gains weight and loses time'** in *Jane's International Defence Review*, April 2004, posted on 11 March 2004 at <http://www.janes.com>

57. Op Cit. Keijsper, 2007, p. 123.

58. Brown, Nick. **'Bloated JSF on crash diet as flight tests pushed back'** in *Jane's Navy International*, May 2004, posted on 2 April 2004 at <http://www.janes.com>

59. Evans, Michael. **'Overweight carrier fighters give MoD £10bn headache'** in *The Times*, 17 May 2004.

60. Scott, Richard. **'Retirement of the UK Sea Harrier Force – Ski Jump to History'** in *Jane's Defence Weekly*, 8 March 2006, posted on 1 March 2006 at <http://www.janes.com>

61. Op Cit. Keijsper, pp. 124 – 125.

62. Sirak, Michael. **'Answers found to F-35 weight problem'** in *Jane's Defence Weekly*, 22 September 2004, posted on 17 September 2004 at <http://www.janes.com>; Brown, Nick. **'Slimmed-down JSF back on track'** in *Jane's Navy International*, October 2004, posted on 21 September 2004 at <http://www.janes.com>

63. Warwick, Graham and Norris, Guy. **'JSF Special: Future Fighter'** in *Flight International*, 27 June 2006, available at <http://www.flightglobal.com>

64. Preble, Christopher. *'Joint Strike Fighter: Can a Multiservice Fighter Program Succeed?'*, Policy Analysis, 5 December 2002, No. 460, Cato Institute.

65. Preble, Christopher. *'Joint Strike Fighter: Can a Multiservice Fighter Program Succeed?'*, Policy Analysis, 5 December 2002, No. 460, Cato Institute.

66. *See* Art, Robert J. *The TFX Decision: McNamara and the Military*, Little & Brown, Boston MA, 1968.

67. *See* Shapley, Deborah. *Promise and Power: The Life and Times of Robert McNamara*, Little & Brown, Boston, MA, 1993; Coulam, Robert F. *Illusions of Choice: The F-111 and the Problems of Weapons Acquisition Reform*, Princeton University Press, Princeton, NJ, 1977.

68. Letter from Secretary of Defense, William S. Cohen to the Honourable Jerry Lewis, Chairman, Subcommittee on Defense, Committee on Appropriations, US House of Representatives, dated 22 June 2000, quoted in Birkler, John *et al. Assessing Competitive Strategies for the Joint Strike Fighter: Opportunities and Options*, RAND Corporation, MR-1362-OSD/JSF, 2001, p. iii.

69. US GAO. *Joint Strike Fighter Acquisition: Managing Competing Pressures Is Critical To Achieving Program Goals*, testimony from Katherine V. Schinasi, Director – Acquisition and Sourcing Management, 21 July 2003, GAO-03-1012T, p. 4.

70. Wikipedia. *'Politicization of Science'* webpage at <http://en.wikipedia.org/wiki/Politicization_of_science>

71. House of Commons Defence Committee. *Strategic Export Controls, Fifth Report*, Session 2004-05, HC145, TSO, London, 24 March 2005, pp. 49 – 50.

72. House of Commons Defence Committee. *Defence Procurement, Sixth Report*, Session 2003-04, HC572-I, The Stationery Office, London, 28 July 2004, pp. 40 – 43.

73. Chuter, Andrew. **'U.K. Threatens Limits To U.S. Firms' Access'** in *Defense News*, 28 June 2004.

74. Birkler, John *et al. Assessing Competitive Strategies for the Joint Strike Fighter: Opportunities and Options*, RAND Corporation, Arlington, VA, 2001, pp. 16 – 17.

75. Cook, Cynthia R. *et al. Assembling and Supporting the Joint Strike Fighter in the UK: Issues and Costs*, RAND Europe, Santa Monica, CA, 2003, pp. 4 – 7; Op Cit. Harkins, 2004, pp. 36 – 38.

76. Op Cit. GAO, 21 July 2003, p. 5.

77. Ibid.

78. Op Cit. Cook, 2003, p. 7.

79. Moss, Sqn Ldr D S. *JSF and the Evolution of International Defence Procurement, MDA Dissertation*, No. 11 Executive MDA Programme, July 2005, Cranfield University, UK Defence Academy, Shrivenham, p. 60.

80. Johnson, G. and Scholes, K.S. *Exploring Corporate Strategy: Text and Cases*, 6th Edition, FT/Prentice Hall, Harlow, 2002, p. 16.

81. Op Cit. Moss, July 2005, pp. 5 – 6.

82. Op Cit. GAO, 21 July 2003, p. 2.

83. Op Cit. Cook, 2003, pp. 1 – 2.

84. *See* the F-16 Fighting Falcon Wikipedia entry at <http://en.wikipedia.org/wiki/F-16_Fighting_Falcon>

85. Lok, Joris J. **'JSF best for Jointness, says Dutch Air Chief'** in *Jane's Defence Weekly*, 2 February 2005, posted on 28 January 2005 to <http://www.janes.com>

86. Hagelin, Björn. **'Nordic offset policies: changes and challenges'** in Brauer, Jurgen and Dunne, J Paul. *Arms Trade and Economic Development: Theory, Policy and Cases in Arms Trade Offsets*, Routledge, Abingdon, 2004, pp. 137 – 148.

87. Cambell, A and Wastnage, J. **'Out of the Cold: A Review of Nordic Aerospace'** in *Flight International*, 10-16 February 2004.

88. Kemp, Damien. **'Tay Kok Khiang – President of Singapore Technologies Aerospace'** in *Jane's Defence Weekly*, 10 November 2004, posted on 05 November 2004 to <http://www.janes.com>

[89.] *See* the *Joint Strike Fighter: Australian Industry Capabilities* presentation located at <http://www.industry.gov.au/Industry/AerospaceandDefence/Pages/ TeamAustraliaJointStrikeFighterJSFIndustryCapabilities.aspx>

[90.] US DefenseLink Website. <http://www.defenselink.mil/transcripts/transcript. aspx?transcriptid=3547>, news transcript dated 11 July 2002.

[91.] Hewson, Robert. **'Joint Strike Fighter – Happily Ever After?'** in *Jane's Defence Weekly*, 13 October 2004, posted on 06 October 2004 to <http://www.janes.com>

[92.] Sirak, Michael. **'JSF Partners are 'on track' for long-term boom'** in *Jane's Defence Weekly*, 05 May 2004, posted on 30 April 2004 to <http://www.janes.com>

[93.] Ibid.

[94.] Op Cit. GAO, 21 July 2003, p. 9.

[95.] Op Cit. Preble, 5 December 2002, pp. 8 – 9.

[96.] La Franchi, Peter and Trimble, Stephen. **'Share Dealing'** in *Flight International*, 13 – 19 July 2004, pp. 70 – 74.

[97.] In terms of the UK, see for example: National Audit Office. *Major Projects Report 2000*, HC970, HMSO, London, 2000, pp. 3 – 11; National Audit Office. *Major Projects Report 2005*, HC595-I, TSO, London, 25 November 2005, pp. 1 – 3; House of Commons Defence Committee. *Defence Equipment 2009*, HC107, *Third Report* (Session 2008 – 09), TSO, London, 26 February 2009, p. 81. For the USA, *See*: Jane's Information Group. **'Procurement: United States'** in *Jane's Sentinel Security Assessment* – North America, posted 21 October 2009 to <http://www.janes.com>

[98.] Mulholland, David. **'Most DoD deals not competitive'** in *Jane's Defence Weekly*, 13 October 2004, posted on 08 October 2004 to <http://www.janes.com>

[99.] Kincaid, Bill. **'Smart Procurement: Revolution or Regression?'** in Matthews, Ron and Treddenick, Jack. (Ed) *Managing the Revolution in Military Affairs*, Palgrave, Basingstoke, 2001, pp. 175 – 190; Hartley, Keith. **'Defence Reform for the 21st Century: Defence Acquisition Reform in Europe'** in *Jane's Defence Weekly*, 01 June 2001, posted on 01 June 2001 to <http://www.janes.com>

[100.] Mulholland, David. **'US Industry: Bucking the Trend?'** in *Jane's Defence Weekly*, 25 February 2004, posted on 18 February 2004 to <http://www.janes.com>

[101.] Op Cit. La Franchi and Trimble, 13 – 19 July 2004, p. 73.

[102.] *See* Op Cit. Cook, 2003; Mulholland, David *et al.* **'Not All JSF Partners are Reaping Contract Awards'** in *Jane's Defence Weekly*, 26 May 2004, posted on 21 May 2004 to <http://www.janes.com>; Op Cit. Hewson, 13 October 2004.

[103.] Henley, S. **'Joint Combat Aircraft Training'** in *Proceedings of the Shepherd Air Power Conference 2002*, Shepherd Conferences, London, 30 January – 1 February 2002.

[104.] Op Cit. La Franchi and Trimble, 13 – 19 July 2004, pp. 70 – 71; Hartley, Keith. **'Offsets and the Joint Strike Fighter in the UK and the Netherlands'** in Brauer, Jurgen and Dunne, J. Paul. *Arms Trade and Economic Development: Theory, policy and cases in arms trade offsets*, Routledge, Abingdon, 2004, pp. 117 – 136; Di Domenico, Lt Col Stephen G. *International Armament Cooperative Programs: Benefits, Liabilities and Self-Inflicted Wounds – The JSF as a Case Study*, US DoD Research Paper, February 2006, pp. 35 – 54.

[105.] Chinworth, Michael W. **'The RMA: A US Business Perspective'** in Matthews, Ron and Treddenick, Jack. (Ed). *Managing the Revolution in Military Affairs*, Palgrave, Basingstoke, 2001, p. 134.

[106.] Op Cit. Chinworth, 2001, pp. 133 – 156.

[107.] Taylor, T. **'Europe's Revolution in Defence Industrial Affairs'** in Matthews, Ron and Treddenick, Jack. (Ed) *Managing the Revolution in Military Affairs*, Palgrave, Basingstoke, 2001, pp. 208 – 225; Uttley, Matthew R H. **'Technology Transfer and the RMA: The Scope and Limitations of Licensed Production for the United Kingdom'** in Matthews, Ron and Treddenick, Jack. (Ed) *Managing the Revolution in Military Affairs*, Palgrave, Basingstoke, 2001, pp. 191 – 207; Op Cit. Wood, 30 January – 1 February 2002.

[108.] Op Cit. Chinworth, 2001, p. 154.

[109.] Sirak, Michael. **'Duncan Hunter, Chairman, Armed Services Committee of the US House of Representatives'** in *Jane's Defence Weekly*, 11 February 2004, posted on 05 February 2004 to <http://www.janes.com>

[110.] Op Cit. La Franchi and Trimble, 13 – 19 July 2004, p. 71.

[111.] Jane's Information Group. **'Asian Aerospace 2004: New Member Poised to Join JSF Project'** in *Jane's Defence Weekly*, 03 March 2004, posted on 27 February 2004 to <http://www.janes.com>

[112.] Op Cit. Lok, Joris J and Mulholland, David, 26 May 2004, p. 21.

[113.] Op Cit. La Franchi and Trimble, 13 – 19 July 2004, p. 71.

[114.] Ibid.

[115.] *See* the table illustrated on Op Cit. GAO, 21 July 2003, p. 5.

[116.] Dodd, Thomas. **'Danish companies consider quitting JSF programme'** in *Jane's Defence Weekly*, 14 January 2004, posted on 09 January 2004 to <http://www.janes.com>; Op Cit. Mulholland *et al*, 26 May 2004; Op Cit. Hewson, 13 October 2004; Lok, Joris J. **'Frustration Mounts among JSF Partners'** in *Jane's Defence Weekly*, 24 March 2004, posted on 18 March 2004 to <http://www.janes.com>; Sariibrahimoglu, Lale and Sirak, Michael. **'Turkey may withdraw from JSF programme'** in *Jane's Defence*

Weekly, 10 November 2004, posted on 05 November 2004 to <http://www.janes.com>; Trimble, Stephen and Warwick, Graham. **'Joint Journey'** in *Flight International*, 28 June – 4 July 2005, available at <http://www.flightglobal.com/articles/2005/06/28/199983/joint-journey.html>; Anon. **'Norway to back out of F-35 JSF Over Industrial Share?'** located at <http://www.defenseindustrydaily.com/norway-to-back-out-of-f35-jsf-over-industrial-share-01969/>, posted on 03 March 2006.

117. Kemp, Damien. **'Australia stands behind Joint Strike Fighter'** in *Jane's Defence Weekly*, 23 March 2005, posted on 16 March 2005 to <http://www.janes.com>; Lok, Joris J. **'Dutch confident in JSF business volume'** in *Jane's Defence Weekly*, 26 January 2005, posted on 21 January 2005 to <http://www.janes.com>; Lok, Joris J. **'Dutch bid to win more JSF work'** in *Jane's Defence Weekly*, 09 June 2004, posted on 03 June 2004 to <http://www.janes.com>; Grevatt, John. **'Australian JSF companies face best-value challenge'** in *Jane's Defence Weekly*, 01 December 2006, posted on 16 November 2006 at <http://www.janes.com>; Lok, Joris J. **'Netherlands set to win $5.5bn in JSF business'** in *Jane's Defence Weekly*, 16 June 2004, posted on 09 June 2004 at <http://www.janes.com>

118. Op Cit. GAO, 21 July 2009, p. 8.

119. Op Cit. La Franchi and Trimble, 13 – 19 July 2004, p. 70; Op Cit. Trimble and Warwick, 28 June – 4 July 2005.

120. Op Cit. HCDC, 24 March 2005, pp. 49 – 51.

121. Alon, Ben-David. **'China issue undermines US-Israeli defence ties'** in *Jane's Defence Weekly*, 27 April 2005, posted on 22 April 2005 at <http://www.janes.com>; Goldberg, Marc. **'Israel reaches understanding with US over after fall-out over arms sales to China'** in *Jane's Defence Industry*, 01 October 2005, posted on 17 August 2005 at <http://www.janes.com>

122. Op Cit. Trimble and Warwick, 28 June – 4 July 2005.

123. Op Cit. GAO, 21 July 2003; US GAO. *JOINT STRIKE FIGHTER: Management of the Technology Transfer Process*, March 2006, GAO-06-364, Washington DC.

124. Merle, Renae. **'Price Tag Jumps for Aircraft: Joint Strike Fighter to be Delayed'** in *The Washington Post*, 19 February 2004, p. E.03; Sweetman, Bill. **'JSF Security Technology costing up to $1bn'** in *Jane's International Defence Review*, May 2004, posted on 05 April 2004 at <http://www.janes.com>

125. Op Cit. Mulholland, 25 February 2004.

126. Op Cit. Sirak, 11 February 2004.

127. Warwick, Graham. **'Europe ponders consequences of US threat'** in *Flight International*, 05 April 2005, at <http://www.flightglobal.com/

articles/2005/04/05/196180/europe-ponders-consequences-of-us-threat.html>; Sharman, Alan. **'Opinion – Sealing the Special Relationship'** in *Jane's Defence Weekly*, 28 September 2005, posted on 23 September 2005 at <http://www.janes.com>

[128.] Mulholland, David. **'Dropping EU embargo may jeopardise JSF'** in *Jane's Defence Weekly*, 02 March 2005, posted on 25 February 2005 at <http://www.janes.com>

[129.] La Franchi, Peter. **'Europe barred from Link 16 Development'** in *Flight International*, 28 June – 4 July 2005, located at <http://www.flightglobal.com/articles/2005/05/03/197463/europe-barred-from-link-16-development.html>

[130.] La Franchi, Peter. **'Europe warned on JSF role'** in *Flight International*, 28 June – 4 July 2005, located at <http://www.flightglobal.com/articles/2005/06/28/199920/europe-warned-on-jsf-role.html>

[131.] Davies, Andrew. **'Australia's Defence White Paper 2009'** in *RUSI Defence Systems*, June 2009, p. 36.

[132.] Borgu, Aldo. *A Big Deal: Australia's Future Air Combat Capability*, Australian Strategic Policy Institute, February 2004, p. 9.

[133.] Ibid. p. 10.

[134.] La Franchi, Peter. **'RAAF Decision shocks Europeans'** in *Flight International*, dated 09 July 2002, available at <http://www.flightglobal.com>

[135.] Ibid.

[136.] Editorial. **'Wasted Opportunity?'** in *Flight International*, dated 09 July 2002, available at <http://www.flightglobal.com>

[137.] Houston, AM Angus. **'Is the JSF Good Enough?'**, *ASPI, Strategic Insights Paper* No. 9, August 2004.

[138.] Bostock, Ian. **'Australian air force chief defends JSF choice'** in *Jane's Defence Weekly*, 01 September 2004, posted on 26 August 2004 at <http://www.janes.com>

[139.] Kemp, Damien. **'Australia stands behind Joint Strike Fighter'** in *Jane's Defence Weekly*, 23 March 2005, posted on 16 March 2005 at <http://www.janes.com>

[140.] Tewes, Alex. *The F-35 (Joint Strike Fighter) Project: Progress and Issues for Australia,* Australian Parliamentary Library – Foreign Affairs, Defence and Trade Section, 09 June 2006 (Research Note No. 32, 2005-06).

[141.] Grevatt, Jon. **'Australia approves JSF purchase'** in *Jane's Defence Industry*, posted on 25 November 2009 at <http://www.janes.com>

[142.] Warwick, Graham. **'UK and Australia threaten to pull out of Lockheed Martin JSF over tech transfer delays'** in *Flight International*, dated 16 March 2006, available at <http://www.flightglobal.com>; Bostock, Ian. **'Australia voices concern over JSF information sharing'** in *Jane's Defence Weekly*, 5 April 2006, posted on 29 March 2006 at <http://www.janes.com>; La Franchi, Peter. **'Australia demands JSF resolution'** in *Flight International*, dated 4 July 2006, available at <http://www.flightglobal.com>

143. Nicholson, Peter and Connery, David. *Australia's Future Joint Strike Fighter Fleet: How Much is Too Little?*, October 2005, The Kokoda Foundation (Paper No. 2); *See* also Bostock, Ian. **'Australia's JSF plans inadequate, says report'** in *Jane's Defence Weekly*, 19 October 2005, posted on 14 October 2005 at <http://www.janes.com>

144. **'Singapore's RSAF decides to fly like an Eagle'** in *Defense Industry Daily* dated 10 May 2009, located at <http://www.defenseindustrydaily.com>

145. Also known as AEWACS, and often pronounced 'A-Wacks'.

146. **'The Australian Debate: Abandon F-35, Buy F-22s?'** in *Defense Industry Daily* dated 17 March 2008, located at <http://www.defenseindustrydaily.com>

147. Royal Australian Navy. *'First steel cut for LHD-02 Amphibious Ships'*, dated 2 February 2010, located at <http://www.navy.gov.au/First_Steel_Cut_for_LHD_02_ Amphibious_Ship>

148. Grevatt, Jon. **'Australian DoD reaffirms JSF commitment in the face of criticism'** in *Jane's Defence Weekly*, 1 December 2006, posted on 2 November 2006 at <http:// www.janes.com>; Kemp, Damien. **'Australia justifies JSF purchase'** in *Jane's Defence Weekly*, 28 March 2007, posted 21 March 2007 at <http://www.janes.com>; Grevatt, Jon. **'Australian defence chief adds further support to JSF buy'** in *Jane's Defence Industry*, posted on 3 December 2008 at <http://www.janes.com>; Govindasamy, Siva. **'Joint Strike Fighter purchase tops Australian defence priorities'** in *Flight International*, dated 6 May 2009, available at <http://www.flightglobal.com>; Grevatt, Jon. **'Australia approves JSF purchase'** in *Jane's Defence Industry*, posted 25 November 2009 at <http://www.janes.com>

149. Davies, Andrew. **'How much will the Joint Strike Fighter cost Australia?'**, *ASPI, Policy Analysis Paper*, No. 27, 12 May 2008.

150. Karniol, Robert. **'Australia will purchase 24 F/A-18F Block II Super Hornets'** in *Jane's Defence Weekly*, 24 March 2007, posted 6 March 2007 on <http://www. janes.com>; Govindasamy, Siva. **'Australia to stick with Super Hornets'** in *Flight International*, dated 17 March 2008, available at <www.flightglobal.com>; Govindasamy, Siva. **'Australia commits to 100 F-35s in defence white paper'** in *Flight International*, dated 2 May 2009, available at <www.flightglobal.com>

151. Australian Department of Defence. *Defending Australia in the Asia Pacific Century: Force 2030, Defence White Paper,* <http://www.defence.gov.au/whitepaper/docs/ defence_white_paper_2009.pdf>, 2009.

152. Opt Cit. Govindasamy, 2 May 2009.

[153.] House of Commons Defence Committee. *Future Carrier and Joint Combat Aircraft Programmes*, HC554, *Second Report* (Session 2005 – 06), The Stationery Office, London, 21 December 2005, p. 30.

[154.] Sullivan, Michael J. *Tactical Aircraft: DOD's Cancellation of the Joint Strike Fighter Alternate Engine Program Was Not Based on a Comprehensive Analysis*, 22 May 2006, GAO-06-717R, Letter to the Honorable John Warner (Chairman, Committee on Armed Services, United States Senate) and the Honorable Curt Weldon (Chairman, Subcommittee on Tactical Air and Land Forces, Committee on Armed Services, US House of Representatives).

[155.] Sullivan, Michael J. *Defense Acquisitions: Analysis of Costs for the Joint Strike Fighter Engine Program*, 22 March 2007, GAO-07-656T: Testimony before the Subcommittees on Air and Land Forces, and Seapower and Expeditionary Forces of the Committee on Armed Services, US House of Representatives.

[156.] Op Cit. Bolkcom, 18 February 2009.

[157.] Munoz, Carlos. **'JSF Program Leaders Expect Surge in International Participation'** in *Inside the Navy*, 27 August 2007.

[158.] Op Cit. Bolkcom, 18 February 2009.

[159.] *See* <http://www.rolls-royce.com/northamerica/na/about/us/index.jsp> (11 November 2009).

[160.] Vogel, Ben. **'US plans put future of JSF at risk, says CBI head'** in *Jane's Defence Industry*, 01 March 2006, posted on 07 February 2006 at <http://www.janes.com>

[161.] Pringle, Rodney. **'JSF Engine Rumblings'** in *Military AeroSpace Technology*, 8 October 2006.

[162.] Vogel, Ben. **'US's solo decision to cut JSF engine programme stirs UK concern'** in *Jane's Defence Industry*, 01 April 2006, posted on 15 March 2006 at <http://www.janes.com>; Trimble, Stephen. **'UK issues ultimatum on JSF technology transfer'** in *Jane's Defence Weekly*, 22 March 2006, posted on 17 march 2006 at <http://www.janes.com>

[163.] Cahlink, George. **'U.K. Procurement Chief Warns Backup Engine Dispute Threatens JSF Deal'** in *Defense Daily*, Volume 229, Issue 50 (15 March 2006); Op Cit. Warwick, 16 March 2006.

[164.] Lok, Joris J. **'Double Dutch: Pratt, Rolls involve more Dutch partners in F135, F136 Programs'** in *Aviation Week* and *Space Technology*, 11 February 2008, Volume 168 Issue 6, p. 49.

[165.] Op Cit. Bolkcom, 18 February 2009.

[166.] Hotten, Russell. **'US threat to Rolls-Royce's fighter engine'** in *The Independent*, 13 December 2009.

[167.] Thompson, Dr Loren B. *'Alternate Engine Problems Prove Critics Were Right'*, 17 November 2009, available at <http://www.lexingtoninstitute.org/alternate-engine-problems-prove-critics-were-right?a=1&c=1129>

[168.] Op Cit. Hotten, 13 December 2009.

[169.] Editorial. **'The F136 Engine: More Lives Than Disco?'**, *Defense Industry Daily*, 25 February 2010, available at <http://www.defenseindustrydaily.com/the-f136-engine-more-lives-than-disco-03070/>

[170.] Gertler, Jeremiah. *F-35 Joint Strike Fighter (JSF) Program: Background and Issues for Congress*, 22 December 2009, Congressional Research Service, RL30563.

[171.] Obey, David R. *Summary: FY2010 Defense Appropriations, Committee on Appropriations*, available at <http://appropriations.house.gov/pdf/FY2010_Defense_Appropriations_Bill_Summary.pdf>

[172.] Wall, Robert. **'Dutch Back Alternative F136 Engine For JSF'** in *Aerospace Daily & Defense Report*, Volume 233, Issue 35 (23 February 2010), ISSN: 1553-3859.

[173.] Shalal-Esa, Andrea. *'Pentagon F-35 engine analysis "near-sighted": lawmaker'*, 25 February 2010, available at <http://www.reuters.com/article/idUSTRE61O5Z020100225>

[174.] Hewson, Robert. **'JOINT STRIKE FIGHTER – Happily Ever After?'** in *Jane's Defence Weekly*, 13 October 2004, posted on 6 October 2004 at <http://www.janes.com>

[175.] Op Cit. Mulholland, 25 February 2005.

[176.] Murphy, James. **'BAE Systems says US export laws stifle US-UK technology sharing'** in *Jane's Defence Industry*, 1 August 2005, posted on 13 July 2005 at <http://www.janes.com>

[177.] Op Cit. Trimble and Warwick, 28 June – 4 July 2005.

[178.] Murphy, James. **'Issues remain in US and UK talks on JSF technology'** in *Jane's Defence Industry*, 1 November 2005, posted on 19 October 2005 at <http://www.janes.com>

[179.] Murphy, James. **'UK demands JSF technology'** in *Jane's Defence Weekly*, 22 February 2006, posted on 17 February 2006 at <http://www.janes.com>

[180.] Anderson, Guy. **'BAE calls for naval Typhoon study'** in *Jane's Defence Weekly*, 8 March 2006, posted on 3 March 2006 at <http://www.janes.com>

[181.] Harris, Francis. **'Computer codes row threatens £12bn jet order'** in *The Daily Telegraph*, 15 March 2006.

[182.] Op Cit. Trimble, 17 March 2006.

[183.] Op Cit. HCDC, 21 December 2005, p. 29.

[184.] House of Commons Defence Committee. *The Defence Industrial Strategy*, HC824, *Seventh Report* (Session 2005-06), The Stationery Office, London, 10 May 2006, p. 36.

[185.] House of Commons Defence Committee. *Defence Procurement 2006*, HC56,

First Report (Session 2006-07), The Stationery Office, London, 8 December 2006, p. 23.

[186.] Anderson, Guy. **'UK committee expresses JSF and DIS concerns'** in *Jane's Defence Weekly*, 21 February 2007, posted on 16 February 2007, available at <http://www.janes.com>

[187.] Baldwin, Tom. **'Secret codes clash may sink £140bn fighter deal'** in *The Times*, 11 December 2006, available online at <http://<www.timesonline.co.uk/tol/news/world/us_and_americas/article667124.ece>

[188.] House of Commons Defence Committee. *UK/US Defence Trade Cooperation Treaty*, HC107, *Third Report* (Session 2007-08), The Stationery Office, London, 11 December 2007.

[189.] Anderson, Guy. *'UK should insist on JSF operational sovereignty, says shadow minister'* in *Jane's Defence Industry*, posted 28 October 2008 at <http://www.janes.com>

[190.] Wilson, Amy. **'F35 jet raises tensions with US over technology sharing'** in *The Daily Telegraph*, 30 August 2009, p. B7.

[191.] Op Cit. HCDC, 21 December 2005, pp. 28 – 30; Ministry of Defence. *Defence Industrial Strategy*, Defence White Paper, Cm6697, The Stationery Office, London, December 2005.

[192.] Op Cit. Wilson, 30 August 2009.

[193.] Wolf, Jim. **'US to withhold F-35 fighter software codes'**, Reuters News Agency, <http://www.reuters.com>, posted on 24 November 2009.

[194.] Op Cit. Gertler, 22 December 2009.

[195.] Jones, Rhys. **'UK confident U.S. will hand over F-35 fighter codes'**, Reuters News Agency, <http://www.reuters.com>, posted on 7 December 2009.

[196.] Ministry of Defence. *The Strategic Defence Review*, Cm3999, The Stationery Office, London, July 1998.

[197.] *See* Ministry of Defence. *Adaptability and Partnership: Issues for the Strategic Defence Review*, Cm7794, The Stationery Office, London, February 2010 – also known as the *'Defence Green Paper'*; Ministry of Defence. *The Defence Strategy for Acquisition Reform*, The Stationery Office, London, Cm7796, February 2010; Ministry of Defence. *Strategic Trends Programme: Global Strategic Trends – Out to 2040*, 12 January 2010, accessed on 26 February 2010 and available from <http://www.mod.uk/DefenceInternet/MicroSite/DCDC/OurPublications/StrategicTrends+Programme/>; Ministry of Defence. *Strategic Trends Programme: Future Character of Conflict*. 12 February 2010, available at <http://www.mod.uk/DefenceInternet/MicroSite/DCDC/OurPublications/Concepts/>

[198.] Op Cit. MoD, Cm7794, February 2010, p. 15.

[199.] House of Commons Defence Committee. *Defence Equipment 2010*, HC99, *Sixth Report* (Session 2009-10), The Stationery Office, London, 4 March 2010, p. 24.

[200.] Norton-Taylor, Richard. **'MoD to slash jet fighter orders as it struggles to save aircraft programme'** in *The Guardian*, 12 January 2010.

[201.] Op Cit. HCDC, 11 December 2007.

[202.] House of Commons Foreign Affairs Committee. *Global Security: UK-US Relations, HC114, Sixth Report* (Session 2009-10), The Stationery Office, London, 28 March 2010, p. 30.

[203.] Franck, C., Lewis, I., & Udis, B. *Echoes Across the Pond: Understanding EU-US Defence Industrial Relationships*. Monterey, California: Naval Postgraduate School. *Report NPA-GSBPP-09-016*, 9 August 2009, available at <http://edocs.nps.edu/npspubs/scholarly/TR/2009/NPS-GSBPP-09-016.pdf>, p. 87.

[204.] Op Cit. Chao and Niblett, 26 May 2006, pp. 20 – 21.

[205.] Moore, Dr D., Young, S., Ito, P., Burgess, Dr K. and Antill, P. **'US Export Controls and Technology Transfer Requirements – A UK Perspective'** in *Excerpt from the Proceedings of the Seventh Annual Research Symposium, Thursday Sessions*, Volume II – Acquisition Research: Creating Synergy for Informed Change, 12-13 May 2010, located at <http://www.acquisitionresearch.net/cms/_files/FY2010/NPS-AM-10-068.pdf>, published on 30 April 2010.

[206.] U.S. Department of Defense. *Quadrennial Defense Review 2010*. February 2010, available at <http://www.defense.gov/QDR/images/QDR_as_of_12Feb10_1000.pdf>, pp. 83 – 84.

[207.] Op Cit. Chao and Niblett, 26 May 2006, p. 3.

[208.] Ibid. p. 6.

[209.] Op Cit. U.S. Department of Defense. Quadrennial Defense Review 2010, p. 83.

[210.] Editorial. **'The military-consumer complex'** in *The Economist*, Volume 393, Issue 8661 (12 December 2009), p. 16.

[211.] Bialos, J., Fisher, C. and Koehl, S. *Fortresses and Icebergs: The Evolution of the Transatlantic Defense Market and the Implications for U.S. National Security Policy*. December 2009, Center for Transatlantic Relations, Washington DC, available at <http://transatlantic.sais-jhu.edu/bin/c/s/us-eu_report_final.pdf>, p. 37.

[212.] Ibid. p. 113.

[213.] Ibid.

[214.] Ibid. p. 2.

[215.] Ibid. p. 20.

[216.] Ibid. p. 114.

[217.] Ibid.

[218.] Ibid.

[219.] Ibid.

[220.] Op Cit. Franck, Lewis & Udis, 9 August 2009, p. 17.

[221.] Op Cit. Bolkcom, 18 February 2009, p. 17.

[222.] Bechat, J., Rohatyn, F., Hamre, J. and Serfaty, S. *The Future of the Transatlantic Defense Community: Final Report of the CSIS Commission on Transatlantic Security and Industrial Cooperation in the Twenty-First Century.* The Center for Strategic and International Studies, Washington DC, January 2003, available from <http://www.ciaonet.org/book/bej01/>, p. 19.

[223.] Ibid. pp. 27 – 28.

[224.] Ibid. p. 30.

[225.] Ibid. p. 44.

[226.] Ibid. p. 52.

[227.] Editorial. **'Joint Strike Fighter Hits Turbulence'** in *Warships Magazine*, May 2006, available at <http://www.warshipsifr.com/navalNewsAnalysisMay06.html>

[228.] Op Cit. House of Commons Foreign Affairs Committee, 28 March 2010, p. 73.

[229.] Potter, Matthew. *'JSF Cost Increases May Imperil the Fighter program – and Lockheed'*, 25 May 2010, located at <http://industry.bnet.com as of 26 May 2010.>

[230.] Moore, David and Antill, Peter. **'Integrated project teams: The MoD's new hot potato?'** in *The RUSI* Journal, Volume 145, Issue 1 (February 2000), pp. 45 – 51.

[231.] Sullivan, Michael. *Strong Risk Management Essential as Program Enters Most Challenging Phase*, Testimony before the Subcommittee on Air and Land Forces, Committee on Armed Services, House of Representatives, Government Accountability Office, 20 May 2009, GAO-09-711T.

[232.] Sullivan, Michael. *Accelerating Procurement Before Completing Development Increases the Government's Financial Risk*, Government Accountability Office, March 2009, GAO-09-303.

[233.] Harrington, Caitlin. **'The JSF Big Bet: F-35 Joint Strike Fighter Update'** in *Jane's Defence Weekly*, posted on 05 June 2009 at <http://www.janes.com>

[234.] Ibid.

[235.] Potter, Matthew. *'JSF Cost Increases May Imperil the Fighter program – and Lockheed'*, 25 May 2010, located on <http://industry.bnet.com>; Grant, Greg. *'JSF Price Tag Now $112 Million Per Plane; Program $382 Billion'*, 01 June 2010 at <http://www.dodbuzz.com/2010/06/01/jsf-price-tag-now-112-million-per-plane/#comment-25859>

[236.] *See* US DoD. *'Department of Defense Announces Selected Acquisition Reports'*, News Release No. 248-10 at <http://www.defense.gov/releases/release.aspx?releaseid=13425>; Capaccio, Tony. *'Lockheed F-35's Projected Cost Now $382 Billion, Up 65 Percent'*, 01 June 2010, located at <http://www.businessweek.com/news/2010-06-01/lockheed-f-35-s-projected-cost-rises-to-382-billion-update1-.html>

[237.] *See* <http://pogoblog.typepad.com/pogo/2010/06/get-it-while-its-hot-joint-strike-fighter-nunnmccurdy-certification.html> for news of the breach, and <http://pogoarchives.org/m/ns/jsf/f35-nunn-mccurdy-certification-20100601.pdf> for a copy of the letter itself.

[238.] Pfeifer, Sylvia. **'F-35 cuts would hurt UK's status'** in *The Financial Times*, 19 July 2010; Op Cit. Norton-Taylor, 12 January 2010.

[239.] Email from DE&S, dated 02 August 2010; JSF Program Office. *Joint Strike Fighter Production, Sustainment & Follow-On Development Memorandum of Understanding*, signed by the UK on 12 December 2006, located at <http://www.jsf.mil/downloads/down_documentation.htm>

[240.] Rayment, Sean. **'Armed Forces Stunned By Trident Bill'** in *The Telegraph*, 31 July 2010; Sieghart, Mary A. **'What Deterrence Needs Is Ambiguity'** in *The Independent*, 02 August 2010.

Case 2.6 // JSF Alternatives

Multi-Service Procurement: Revenge of the Fighter Mafia – Alternatives Beyond the JSF Programme to Meet UK Carrier-Bourne Aviation Requirements

Jeffrey Bradford

Defence and National Security Consultant.

Introduction

The most expensive combat aircraft program in defence procurement history to-date is the US-led Joint Strike Fighter (or JSF) program.[1] In its current configuration the program has a production run of some 2,457 units across three variants (-A, -B, -C) for air force, short-take off & vertical landing (STOVL) and naval use by the US armed forces and the projects international partners.[2] The cost of the overall program is currently forecast at some $323 billion with a per-unit fly away cost (the F-35A variant) of around $60 million.[3]

The Joint Strike Fighter was conceived in the mid 1990s as an eventual replacement for the F-16 Fighting Falcon with the F-22 Raptor replacing the F-15 Eagle, creating a new "high-low" mix for the Twenty-First Century. The procurement of the F-16 was bedevilled in its early stages by politics within the US Air Force, who favoured greater purchases of F-15s. The "Eagle Drivers" faction promulgated reports and analyses internally, suggesting the F-16 did not have sufficient capability given its small size – which was eventually proven wrong with the F-16 being one of the greater export success stories of its kind.[4]

The Programme

At the time of writing, the programme is approximately 10 years in gestation since down-selection favoured the Lockheed Martin aircraft design. The most controversial element of the program, aside from the cost and time overruns, has been technical difficulties with the F-35B STOVL variant for the US Marine Corps and Royal Navy/RAF. The F-35B will serve as successor to the AV-8B Harrier "Jump Jet" and in early 2011 was put on a two-year "probationary" period by the US Department of Defense as part of a spending review.

Probation implies that should time and cost performance not improve substantially that the STOVL variant would be cancelled.

For the Royal Navy, the Future Aircraft Carrier program ("CVF") was originally designed 'for-but-not-with' steam catapults for launching more conventional naval aircraft (such as the F-35C naval variant). Such pragmatism may well forecast the demise of the F-35B variant which, whilst militarily offering a unique capability, could be seen as not surviving in a period of stringent Government finances.[5]

However, a greater challenge for the JSF program is the evolution of unmanned aerial vehicles (or UAVs). The pace of development of these aircraft in terms of payload, range and overall performance is mimicking Moore's Law from the microprocessor realm.[6] One can imagine that were the JSF be in operational service today that many missions conducted by UAVs would still remain UAV driven for cost and political risk issues.

Air Forces world-wide tend to appoint their senior leadership today from amongst their fast-jet community (over transport and bomber staffs). The UAV is a cultural threat to the institution and the JSF represents their salvation for the next generation. The incoming Chief of the Defence Staff in the United Kingdom made the point from an Army perspective that, "One can buy a lot of Unmanned Aerial Vehicles or Tucano [trainer] aircraft for the cost of a few JSF (Joint Strike Fighters)".[7]

Despite the base logic of statements of this kind, the JSF still has a role. Existing fighter fleets are ageing and require replacement of some kind due to operational as well as balance sheet depreciation. Confidence in UAV platforms to perform the air interceptor mission is not present militarily, neither (and more importantly) is the proven technology in place.

The UK Requirement

Turning to the United Kingdom requirement for the Joint Strike Fighter, the UK procurement programme for a new generation of aircraft carriers to succeed the 1970s vintage 'Through Deck Cruisers' was the key platform of the expeditionary warfare strategy defined and articulated by the then newly elected Labour Administration in the Strategic Defence Review (SDR) of 1998.[8]

Whilst the Future Aircraft Carrier ("CVF") programme is making steady industrial headway since the 2003 announcement in the House of Commons,[9] one of key questions requiring attention is increasingly the question of the principal aircraft to be flown from the two aircraft carriers entering the manufacturing phase.

The current generation of aircraft carriers use the Harrier, a unique aircraft capable of taking off and landing vertically, whose design dates back to the 1960s. Whilst plans for a supersonic version of the aircraft were a casualty of 1960s defence cuts, the Royal Navy used a naval version of the Harrier from 1980 until 2006 when retirement forced the RN to employ RAF Harrier GR7s. Whilst a highly successful design, and the beneficiary of several capability enhancements, the Harrier is fast approaching the end of its operational life and will most definitely be obsolete by the time the new aircraft carriers enter service in the 2015 – 2020 time frame.

The successor programme to the Harrier is an international programme led by the United States, known as the Joint Strike Fighter ("JSF") or by the Royal Navy as the Joint Combat Aircraft ("JCA"). This programme is currently the largest being undertaken by the US Department of Defence and currently valued at in excess of $240 billion (£160 billion or some twenty times the annual UK equipment programme).[10]

The UK has made a significant financial contribution to the programme and the Royal Navy is the only customer (alongside the US Marine Corps) for the VSTOL aircraft known as the F-35 B. The US Marine Corps however, is battling to bring into service the troubled V-22 Osprey tilt-rotor aircraft, which combines the benefits of a helicopter with an aircraft for transporting US Marines into battle with high mobility.[11] As one observer notes, "moving from a helicopter-borne assault force to a tilt-rotor-borne assault force takes priority over the move from conventional to ASTOVL fighters."[12]

The recent decision of the US Department of Defense to place the STOVL variant of the JSF on "two-year probation" highlights the budgetary and technical difficulties, not to mention engine procurement politics which collectively place the '-B' model at risk.[13]

However, the JSF programme is fighting for its own survival in the wake of extended operational commitments in the Middle East putting pressure on the budget as well as an Air Force customer who is more interested on maintaining the production line for the F-15 Eagle successor aircraft, the F-22 Raptor. Annually a decision is taken by the US Department of Defense to pursue only one engine design with Pratt and Whitney, eliminating work being undertaken by GE & Rolls Royce (the latter expert in the design of VSTOL engines for the original Harrier). In each case thus far, DoD has failed to persuade Congress, which has reinstated funding principally for industrial reasons, it could be suggested, to protect GE and keep Pratt & Whitney from exercising a monopoly position on the program. More recent reports of a $2.8 billion increase in the Pratt & Whitney programme do little to assist the situation.[14]

Whilst the bureaucratic battle within the US Air Force is reminiscent of the fight between the F-15 and F-16 aircraft lobbies, the challenges across the US Department of Defense suggest the likelihood of the programme being stretched, numbers cut and prices rising.[15] The original price per JSF was suggested as being $29-34m per aircraft (in 2001), whilst 2007 US budget estimates suggest the current price will be $33m, near the top end of the scale, before large scale production is finalised.[16]

However, these figures seem (charitably) optimistic at a minimum, up to deliberate distortion. The US General Audit Office March 2008 assessment of military programmes identified a total programme cost of $239 billion (as of December 2006), which based on a quantity of 2,458 (as at 12/2006) would give a programme unit cost of $97 million (as of December 2006). Taking in general and defence inflation, a unit price above that of $100 million today seems realistic. Where the £2 billion committed by the UK to become a level 1 partner comes in is unclear, but a UK acquisition cost of $16 billion for 150 aircraft was quoted in a RAND 2006 study. This was equivalent to the £10 billion also quoted by the JSF UK Industry Team.

Whilst drift in the JSF programme can be used to tactical advantage by UK policy makers (witness the pushing back of the CVF programme by a year under stiff budgetary pressure due to operations in Iraq and Afghanistan), sustained difficulties in the programme create a real risk of cancellation or a need for the "Plan B" mooted by former procurement minister, Lord Drayson.[17] Cancellation of international programmes unilaterally by the USA is not without precedent. The *Blue Streak* cancellation of the early 1960s forced Britain to move from an air-based to a sea-based nuclear deterrent using the Polaris missile.

One of the central questions for UK defence policy makers looking at future force structures is likely to be that of what platform will fly from UK aircraft carriers in support of expeditionary operations to provide close air support and air superiority for UK forces.

Options

This paper considers the following five options, which policy makers will likely have to consider should the JSF programme outright, or the VSTOL variant specifically, are cancelled:

1. Extending the service life of the Harrier.
2. Stay the course with the Joint Strike Fighter.
3. Purchasing an 'off-the-shelf" alternative.
4. Converting the Eurofighter for maritime use.
5. Pursuing unmanned aircraft (UAVs).

1. Extending the Service Life of the Harrier

The Harrier VSTOL "jump jet" was procured in the 1970s and a navalised version, the Sea Harrier entered service in 1979. On the 28th February 2002, the Ministry of Defence announced that the Sea Harrier would be withdrawn from service between 2004 and 2006 and the Harrier by 2012, with the intention of replacing both aircraft with the Joint Strike Fighter ("JSF").

Up until recently, the Fleet Air Arm of the Royal Navy was currently flying Royal Air Force Harrier GR7 aircraft from its two remaining in-service aircraft carriers HMS *Ark Royal* and HMS *Illustrious*. HMS *Invincible* is currently inactive through 2010, moored at HM Naval Base Portsmouth.[18]

The Harrier will be approaching fifty years since its design concept and whilst both unique and substantially enhanced from its original capabilities, will be increasingly harder and expensive to keep airworthy as spare parts become harder to obtain and airframes reach the end of their design life.

In addition, the Harrier aircraft being used by the Fleet Air Arm are not navalised as were the original Sea Harrier and their utility in classic carrier aviation roles in anything other than benign environmental conditions could be suggested as being questionable. The principal use of RN Harriers since the Sea Harrier was withdrawn from service has been in support of land operations in Afghanistan, where the aircraft have been based on land.

In summary, the Harrier is not going to fare well in the likely scenario where JSF introduction dates lengthen or the programme is cancelled. It is both an elderly aircraft, expensive to maintain, low on capability and not prepared for operations in the maritime environment.

2. Stay the course with the Joint Strike Fighter.

Three variants of the JSF are currently being procured, the F-35A, B and C. The F-35A is a land-based, conventional aircraft. The F-35B incorporates VSTOL engine technology which the Royal Navy and US Marine Corps are keen to acquire. The F-35C is a conventional aircraft, navalised with the intention of replacing US Navy aircraft currently in service.

Whilst the F-35B VTOL aircraft has not been cancelled, a very good option for the UK is to convert the CVF carriers for conventional flight deck operations and purchase the F-35C which has a lower through-life management cost plus greater range and payload characteristics. The benefits of procuring F-35B or C would be to obtain modern aircraft, interoperable with the US Marines and/or US Navy, in which the UK also has a defence industrial stake. Whilst the US Marine Corps may lose heart in developing the F-35B, the US Navy will be absolutely determined to acquire the C variant, operable from larger carriers of the kind which the UK is now engaged in procuring.

A Government Accounting Office survey of major programs in 2008 identified (as at December 2006) a total program cost of $239 billion and order quantity of some 2,458. This would crudely equate to an effective unit cost of $97 million per aircraft (including obviously the entire costs of R&D and other investment). In summary, this should be considered the default option, although with more flexibility than immediately apparent.

3. Pursuing an 'off-the-shelf' alternative

In the 1960s, following cancellation of a replacement programme for its Sea Vixen aircraft, the Royal Navy purchased F-4 Phantom aircraft from the United States. In deference to political considerations, the aircraft were re-engined with Rolls Royce power plants, which led to continued in-service difficulties with the aircraft.

Today there are several off-the-shelf options of which only two are practical, and politically viable – the French and Americans.

Russia developed aircraft during the cold war for its emergent blue water aircraft carrier strategy based on the MIG-29 Fulcrum and SU-27 Flanker. More recently, Russia has sold some of the MIG-29 aircraft to India in order for the Indian Navy to replace their Sea Harriers, purchased from the UK. The age of the aircraft, spares access and political considerations make neither of these a practical solution.

France developed the Rafale M for use on its nuclear powered aircraft carrier, *Charles de Gaulle* built in the 1990s. With a second aircraft carrier planned for construction, sharing the design of CVF creates a political scenario for shared procurement and subsequent creation of a joint carrier force combining similar aircraft carriers and combat aircraft. Britain and France have collaborated on aircraft in the past, most notably the Jaguar which was retired as Eurofighter entered service.

Politically, this would be seen as a very strong signal to Washington regarding British priorities, especially given the French stance over Iraq and Afghanistan and the symbolism in American power projection of the aircraft carrier. A possible compromise lies in the recent Anglo-French defence accord which aims to enable access to French naval aviation assets to practise operations with a view to potentially establishing a joint aircraft carrier capability.[19]

The US Navy has several aircraft that are combat proven and have gone through substantial capability upgrades, which could be of interest to the UK, if available. The

British purchased the F-4 Phantom in the 1960s and declined the chance to acquire F-14 Tomcat aircraft in the 1970s. However, in the latter instance the Royal Air Force chose instead to modify the Tornado aircraft creating the F.3 interceptor.

Given the current exchange rates an early decision for cancelling the JSF and purchasing the F-18 Hornet could be perceived as an astute move.

With both the French and American options there would be the requirement to modify the CVF to be capable of conventional flight deck operations, through the addition of arrestor wires and steam catapults for which the CVF is designed, though as currently conceived, does not include.

4. Converting the Eurofighter for Maritime Use

In 2000, the UK Government admitted that it had requested an analysis on converting or 'Navalising' the Eurofighter aircraft for use on the future aircraft carrier.[20] Potentially, the navalisation of Eurofighter is a pragmatic method for utilising some of the 232 aircraft to which the United Kingdom is contracted to purchase. Given the Royal Air Force propounded 'future fast jet front line' of some 70 combat aircraft, additional use of Eurofighter would assist in reducing the cost of spares and support for the aircraft fleet.

However, adapting a land-based aircraft for use at sea is potentially a substantial engineering challenge:

- The undercarriage has to be strengthened to withstand punishing carrier deck landings, a tail hook has to be added to be used with arrestor wires to bring the aircraft to a stop rapidly. Weapon pylons and wings require strengthening to handle the stresses of carrier life. All of these deduct from the payload potential of the aircraft.

- The materials of which the aircraft is made need to be checked for their ability to withstand sea corrosion and changed if necessary.

- Carrier landings are made at low speeds and High Angles of Attack for which land-based aircraft may not be optimal (or indeed designed for) requiring special pilot aids or control surfaces to be added. However, systems to automate carrier landings are in-service to take the pressure from F-18 Hornet pilots. These are being developed – in the USA, the first GPS landing occurred in 2001 and in the UK by QinetiQ who mounted a successful demonstration of a VTOL carrier landing in July 2005. The American Automated Carrier Landing System (ACLS) is in the process of being of succeeded by JPALS which further enhances the system by integrating GPS and other navigation aids.

• The future aircraft carriers are currently designed for, although are at the moment without, steam catapults and arrestor wires as the operational requirement was based on the use of VSTOL aircraft (Harrier, succeeded by JSF). Bringing a conventional take off and landing aircraft to the CVF will be possible but will require rectification of these two design omissions which the recent Strategic Defence and Security Review (SDSR) has indicated will occur.

An additional factor for the Fleet Air Arm and the RN is that they have lost their operational expertise at conducting conventional flight operations on deck. The F-4 Phantom and Buccaneer stopped flying at sea with the retirement of HMS *Ark Royal*, the last catapult launch being in November 1978, over thirty years ago.[21] Whilst it is entirely possible that the Eurofighter could be modified for use on aircraft carriers, the potential cost and trade-offs in capability are the key unknowns for decision makers.

5. Pursuing Unmanned Aircraft ("UAVs")

Whilst it can be easily dismissed today, the rate of development of the technology underpinning unmanned aerial vehicles is fast. A decade ago, UAVs were unarmed, large and on the whole, unwieldy. Today UAVs come in a variety of shapes and sizes with the US Predator and Reaper vehicles having been used in combat operations. Taking one metric, that of payload – Predator A designed in 1997 has a payload of 440lbs. Reaper (or Predator B) from 2007 has a payload of 3,500lbs. Extending this development path suggests by 2017 (midway between the two CVF completing construction) that a UAV could have a payload of around 6,600lbs which would be in the order of magnitude of the original Sea Harrier.

Given the danger inherent in carrier operations, an aircraft carrier would potentially be an ideal environment for the operation and deployment of UAVs and worthy of very serious consideration.

Summary

The Joint Strike Fighter programme is a traditional, long-running, big budget procurement program which is meeting all the normal technical, industrial and political challenges associated with the endeavour. The challenge for the program is the impact of delays due to economic strain and also technological innovation. The JSF is in a footrace with unmanned vehicles and could well not see serious operational use for sometime to come.

For the United Kingdom, a major international participant in the project, fortunes have waxed and waned. Whilst participation at high level has yielded research benefits, much of the support was predicated on the unique needs of the UK and US Marine Corps for the F-35 STOVL variant. Economic challenges and resource prioritisation have removed this concern for the United Kingdom.

It could be suggested that perhaps the UK decision to in effect withdraw from the F-35B program was precipitous. On the one hand, the ability to influence the battle over engine supply to the program is massively diminished. Alternatively the length of the British CVF aircraft carrier program makes for little need to make a final decision at present as the platform was designed from the outset for, although not with, steam catapults. Signalling withdrawal now leaves no option to re-engage should the B variant improve its own prospects. Additionally, Royal Navy aviators have not used steam catapult and arrestor systems consistently since the early 1970s.

Currently, there are several strategic options open to decision makers which, at first glance, suggests a lack of need to make a choice. Option 1, development of JSF, continues though may be delayed, with a conventional naval version being developed should the VSTOL version ultimately fail to be procured or indeed produced.

Cancellation of the entire JSF programme, or the F-35B and C variants opens the option to buy off-the-shelf, or convert the Eurofighter Typhoon. Both of these options are contingent on the prevailing political winds and the economic opportunity of the moment.

UAV development continues and probably deserves some injection of funds into ensuring that UAVs can be used from an aircraft carrier.

In terms of one of the key design metrics, that of the thrust-to-weight ratio of in-service and aircraft-under-development, next generation aircraft offer better performance although existing aircraft, whilst of similar ilk, offer greater range and payload as conventional versus VTOL aircraft design.

Appendix: Comparison of key carrier-borne aircraft in service and under development [22]

	Sea Harrier[23]	Harrier GR7[24]	JSF STOVL (F-35B)[25]	"Naval" JSF (F-35C)[26]	F-18 Hornet E/F
Empty weight	13,070lbs	14,800lbs	22,450lbs	23,950lbs	24,650lbs
Maximum weight	26,145lbs	30,800lbs	49,940lbs	49,940lbs	51,480lbs
Engines/ Thrust	1 x 21,800lbs	1x 21,750lbs	1 x 40,000lbs	1 x 40,000lbs	2 x 11,000lbs
Range	620m	600m	600m	600m	330m
Thrust: Weight ratio[27]	0.83	0.71	0.80	0.80	0.42
Payload	7,980lbs	9,000lbs	12,950lbs	16,900lbs	13,600lbs
Wing span	7.70m	9.20m	10.66m	13.10m	12.30m
Length	14.17m	14.10m	15.40m	15.50m	17.1m
Height	3.71m	3.45m	4.7m	4.7m	4.7m
In-service	1980	1990	2012 +	2012 +	1983
Unit price (US$ m)	$32.0m (2006)	$21.6m	$28-$35m +	$28-$35m +	$41.0m (2008)

	Dassault Rafale M	MIG-K 'Fulcrum'	SU-33K 'Flanker'	Eurofighter Typhoon
Empty weight	21,275lbs	24,200lbs	40,480lbs	24,200lbs
Maximum weight	47,300lbs	46,200lbs	72,600lbs	34,100lbs
Engines/ Thrust	2 x 11,250lbs	2 x 11,110lbs	2 x 14,110lbs	2 x 13,500lbs
Range	1,100m	430m	3,000km	864m
Thrust: Weight ratio[28]	0.47	0.48	0.38	0.79
Payload	13,180lbs	7,700lbs	17600lbs	14,300lbs
Wing span	10.80m	11.40m	14.70m	10.95m
Length	15.27m	17.37m	21.18m	15.96m
Height	5.34m	4.73m	5.72m	5.28m
In-service	2000	1983	1995	2003
Unit price (US$ m)	$62.1m (2006)	$20.0m +	$30.0m +	$122.5m (2007)

Notes

[1.] The aircraft in-service designation with the US Armed Forces is the F-35 Lightning II.

[2.] Current units on order by US armed service/country are: United States 2,443 (USAF 1,763, USN & USMC 680), United Kingdom (RAF & RN) 138, Italy 131, Netherlands 85, Turkey 100, Australia 100, Norway 48, Denmark 30 and Canada 65. (Source: Lockheed Martin).

[3.] General Accounting Office, *Joint Strike Fighter: Significant Challenges and Decisions Ahead* (US GAO GAO-10-478T: March 2010).

[4.] The politics of the "fighter mafia" and internal dispute over the capabilities of the F-16 Falcon are well documented in two books concerning the development of the F-18 Hornet, O. Kelly, *Hornet: The inside story of the F/A–18,*1991, Airlife Publishing, and J. P. Stevenson, *The Pentagon Paradox*, 1993, Naval Institute Press.

[5.] Cm7948, *Securing Britain in an Age of Uncertainty: The Strategic Defence and Security Review*, Her Majesty's Stationery Office, 2010. Part 2, pp. 23. Para. 3.

[6.] Moore's Law, named for its inventor Gordon E. Moore, states that the number of transistors on a microchip (and therefore processing speed) will double about every two years.

[7.] D. Barrie citing British Army Chief of the General Staff David Richards, **"Careful with that axe"** in *Aviation Week*, 19 January 2010.

[8.] Cm3999, *Strategic Defence Review,* Her Majesty's Stationery Office, 8 July 1998. Para 115.

[9.] Hansard, House of Commons (Rt. Hon. Geoff Hoon MP), 30 January 2003.

[10.] GAO. *Defense acquisitions: Assessments of selected weapons programs* (GAO-08-467SP March 2008). p.105.

[11.] For more details concerning the program please see <http://www.navair.navy.mil/v22>

[12.] Aboulafia, R. '**Rethinking U.S. airpower**' in *Aerospace America*, March 2001.

[13.] Department of Defense, *Gates Reveals Budget Efficiencies, Reinvestment Possibilities,* US Department of Defense, 6 January 2011.

[14.] Flight International (15 – 21 April 2008). *Huge cost overrun hits JSF power plant.*

[15.] For a detailed discussion of the bureaucratic politics surrounding the emergence of the F-16, see James P. Stevenson, *The Pentagon paradox: The development of the F-18 Hornet,* US Naval Institute Press 1993.

[16.] United States Air Force, Committee Staff Procurement backup book, *Aircraft Procurement Air Force* volume 1 (February 2007). Exhibit P-40.

[17.] *The Daily Telegraph* newspaper. *Computer codes row threatens £12 billion jet order.* 15 March 2006.

[18.] See <http://www.royal-navy.mod.uk> under "Harrier GR7" for more details.

[19.] *The Guardian* newspaper. *Anglo-French defence agreement hailed by leaders,* London, 2 November 2010. For original text see <http://www.number10.gov.uk>, UK–France Summit 2010 Declaration on Defence and Security Co-operation (London, 2 November 2010).

[20.] Hansard, House of Commons (Paul Flynn MP to the Rt. Hon. Geoff Hoon MP), 18 January 2000.

[21.] Source: Fleet Air Arm Officers Association, notable dates.

[22.] The data has been derived from a wide variety of conflicting sources, and should be therefore treated as indicative.

23. The Sea Harrier was withdrawn from Royal Navy service on 29 March 2006.

[24.] On 24 November 2010, the Harrier made its last ever flight from a carrier. The Harrier fleet's farewell operational flights occurred on 15 December 2010, with fly pasts over a number of air and naval stations.

[25.] JSF is currently in development and not currently in service. Performance criteria are potentially subject to change.

[26.] JSF is currently in development and not currently in service. Performance criteria are potentially subject to change.

[27.] A key metric in terms of understanding performance of combat aircraft, and from a carrier-borne perspective the ability of a fully laden aircraft to take-off from an aircraft carrier. Ratio refers to the engine thrust divided by maximum weight of the aircraft. A conservative approach to thrust:weight has been taken here, using maximum weight rather than some form of mission-related weight, which would be very difficult to meaningfully compare across aircraft types.

[28.] A key metric in terms of understanding performance of combat aircraft, and from a carrier-borne perspective the ability of a fully laden aircraft to take-off from an aircraft carrier. Ratio refers to the engine thrust divided by maximum weight of the aircraft. A conservative approach to thrust:weight has been taken here, using maximum weight rather than some form of mission-related weight, which would be very difficult to meaningfully compare across aircraft types.

Part Three // Defence Procurement & Logistics Challenges

Case 3.1 // Small Arms Design

The .256" British: A Lost Opportunity

Anthony G Williams

Co-Editor, *Jane's Ammunition Handbook* and Military Technology Consultant.

Introduction

The history of British military rifle cartridges in the twentieth century has not been a happy one. At the beginning of the century, the .303", one of the first of the small-bore military cartridges, had been in service for a dozen years and was already being criticised as obsolescent and inadequate. Attempts to introduce a satisfactory replacement were initially stymied by the onset of World War 1, then in the interwar period by a lack of resources, and finally – twice – by demands for NATO standardisation. As a result, British troops throughout this century have been equipped with small arms using cartridges which have been less than ideal. Ironically, one cartridge which did see limited British service early in the century could have made a much better basis for military small arms than anything developed since; the .256" British.

The .303" cartridges in service early in the century were of various Marks (some with a hollow point to improve stopping power - withdrawn soon afterwards for fear of contravening the Hague Convention) but all had heavy round-nosed bullets fired at a low muzzle velocity. The performance of the .303" Lee Enfields in the Boer War of 1899-1902 was regarded as greatly inferior to the relatively high velocity 7mm Mauser weapons of the Boers, as the open country of the South African plains favoured a flat-shooting, long-range rifle.

As a result, work began in 1910 on a suitable replacement which emerged in 1912 as the .276" Enfield cartridge, chambered in the new 'Rifle, Magazine, Enfield .276" Pattern 1913' bolt-action rifle - a modification of the well-known Mauser rifle. The cartridge was impressive and had performance to match; not far short of present-day 7mm Remington Magnum ballistics. Not surprisingly, troop trials in 1913 revealed problems with excessive recoil, muzzle flash and barrel wear and overheating. Attempts were made to find a cooler-burning propellant, but further trials were halted by the events of 1914. As a result, the .303" was retained in service, albeit in the much improved Mark VII loading with a lighter, pointed bullet and a higher muzzle velocity to provide a better long-rage performance. This remained the standard rifle ammunition until the .303" was replaced by the 7.62x51 NATO in the mid-1950s.

At this point of the story, the .256" British makes its appearance. Early in World War I a considerable shortage of Lee Enfield rifles occurred. As a result, the UK purchased no less than 130,000 Japanese Arisaka service rifles of assorted types, all in 6.5x50SR calibre, a designation Anglicised into .256". These were mainly used for training and home defence but were also issued to some naval units. From 1916 they were all shipped to Russia, and it appears there were none left by the end of the war. The rifle was a conventional, if rather long, bolt action weapon, but the cartridge had an unrecognised potential which we will return to later.

It was fortunate that the .276" Enfield was not adopted, as even the .303" proved unnecessarily powerful for the predominantly short-range trench warfare of 1914-1918. By this time, a further problem with the .303" was becoming obvious; the rimmed case design was not well suited to most fully-automatic weapons. Much experimentation with semi-rimless and rimless cartridges took place between 1917 and the late 1920s to no avail. Despite interest being shown in the (subsequently cancelled) American Pedersen rifle and its medium-velocity .276" ammunition, the combination of severe economic problems and vast quantities of standard .303" weapons and ammunition in store after the First World War militated against the adoption of any new calibre. The 1930s rearmament programme therefore saw such weapons as the American Browning aircraft machine gun and the Czech ZB30 (better known in its British incarnation as the Bren Gun) expensively redesigned to fire the rimmed cartridge.

Experience in the Second World War reinforced the point that Britain's standard rifle ammunition, in common with that of other nations, was unnecessarily powerful for most engagements. It was capable of shooting accurately out to 1,000 yards (and ranging out to over 4,000 in the streamlined Mark VIII version for the Vickers Machine Gun), yet almost all fire-fights took place at less than 300 yards. This might not have mattered, except that the advantages of fully-automatic fire in shoulder-fired weapons had also become obvious, and this was not feasible with full-power military rifle cartridges because of their heavy recoil.

The 1939-45 conflict saw both the acceptance of the sub-machine gun as an important military weapon and the introduction of its eventual replacement; the assault rifle. The need for a weapon with the automatic fire of a submachine gun but the range to cope with most infantry engagements was first realised in Germany. This led to the introduction of the MP44 Sturmgewehr (assault rifle), firing a new 7.92x33 Kurz (short) cartridge with a performance intermediate between the standard 7.92x57 rifle/MG round and the 9mm sub-machine gun cartridge. The Soviet Union soon followed suit with the 7.62x39 M1943 round, due for international fame in the Kalashnikov AK and AKM assault rifles but also used in other weapons including light machine guns.

After the Second World War, both the USA and the UK gave much thought to the introduction of new small arms cartridges. The British developed a 7x43 round (also known as the .280" and, in slightly modified form, the .280/30") and the associated EM-2 "bullpup" assault rifle, similar in configuration to the current 5.56mm L85A2 Army weapon. The calibre, bullet weight and muzzle velocity were carefully calculated to combine the low recoil necessary for automatic fire with a sufficiently long range to replace the .303", by all accounts successfully. The compact bullpup layout was chosen so that the EM-2 could replace the 9mm Sten SMG as well as the Lee Enfield rifle.

The US Army also produced a smaller round to replace their .30'06 and as a competitor to the British 7mm for the standard NATO small arms cartridge, but appeared to suffer from considerable confusion over the purpose of the new round. They decided both to retain the .30" calibre and to provide sufficient power to replace the .30'06 in machine guns, thereby ensuring that the new cartridge was too powerful for an assault rifle. Selective fire was provided in some rifles, but the recoil made automatic fire uncontrollable. The British protested but American military and economic power prevailed, saddling NATO with the 7.62x51 NATO cartridge which was no real advance over the 50-year old 30'06.

Having committed NATO to a cartridge too powerful to permit the development of assault rifles to match the Kalashnikov, the US Army decided as a result of experience in Vietnam that it had made a mistake and went to the other extreme in adopting the .223" (5.56x45) round for the M16 assault rifle. This acquired a controversial reputation in Vietnam, with tales of the devastating effect of the high-velocity bullets conflicting with other accounts criticising the lack of range and penetration. As a result, the 7.62mm NATO remained in US service as a machine gun cartridge.

Other NATO countries began to consider replacing the 7.62mm (at least in the standard infantry rifle) in the 1970s, and a competition for a new cartridge was held. The British proposed a 4.85mm round but ensured that it was based on the 5.56mm cartridge so that the weapons designed for it could easily be converted in the event of the US cartridge winning the NATO competition; which (to no-one's surprise) it did. Since then, much further research into future military rifles involving such exotic concepts as caseless ammunition and flechette (steel dart) rounds has taken place but it seems that the 5.56mm NATO will be with us for the foreseeable future.

So where does the .256" British fit into this? The answer lies in a series of experiments conducted by British, American and (possibly) Russian agencies to discover the ideal military small arms calibre. In the late 1960s, the Royal Small Arms Factory at Enfield began a detailed theoretical analysis of the striking energy needed to disable soldiers with various levels of protection, and the ballistics required to deliver that energy at battle ranges for a number of different calibres. The conclusion was that the optimum calibre would lie between 6mm and 6.5mm, and an experimental 6.25x43 cartridge (based on the abortive 7mm round) was developed which was claimed to have significant advantages over both the 5.56mm and 7.62mm calibres. Performance proved to be virtually equal to the 7.62mm at up to 600 metres, with recoil and ammunition weight much closer to those of the 5.56mm.

At the same time, the US Army realised the need for a light machine gun with a longer effective range (out to 800 metres) than the 5.56mm cartridge could provide but appreciably less weight than the 7.62mm M60 MG. Their research led to the development of the 6mm SAW (Squad Automatic Weapon) cartridge. A relatively heavy bullet combined with a moderate velocity were selected for the optimum long-range performance. In the event, weapons firing improved 5.56mm ammunition were selected instead, largely to avoid the supply problems created by the use of three small-arms calibres. More recently, it has emerged that Russian armament firms, who had earlier copied NATO in producing a small-calibre (5.45mm) cartridge, are now offering weapons in a new 6mm calibre.

Returning to the .256" British, it is now clear that the 6.5mm calibre is a much better compromise for a general-purpose military rifle and machine-gun cartridge than either the 5.56mm or the 7.62mm. So why did the Japanese (and most other nations with 6.5mm weapons) develop larger and more powerful calibres in the 1930s? The main reason is certainly that the need for automatic fire from shoulder weapons was not generally appreciated; cartridges were optimised for machine guns and bolt-action rifles. Another reason probably lies in the relatively ineffective loadings used in most 6.5mm cartridges at the time, usually featuring heavy, round-nosed bullets and low velocities. By 1914, the standard Japanese loading was better than most, with a pointed bullet and reasonable ballistics (interestingly, closely matching those of the 7mm EM-2). However, the RSAF Enfield experiments suggested a lighter bullet at a higher velocity as the optimum loading for the 6.5mm calibre, a specification well within the capacity of the Arisaka cartridge.

The .256" cartridge, although very compact for its power, is not entirely ideal for modern military purposes as it is semi-rimmed rather than rimless. Nonetheless, its potential was recognised in (of all places) Russia, which used the Japanese cartridge in the Federov Avtomat, a selective-fire rifle first produced in 1916. If it had remained in British service as a supplement to the .303" (preferably in a modified, rimless, form and with a lighter bullet at a higher velocity - 8g at 800 m/s - 123 grains at 2,620 fps - should have been possible) the .256" would have provided an excellent basis for a family of light, fully-automatic weapons. It's problem was that it entered British service about 30 years ahead of its time.

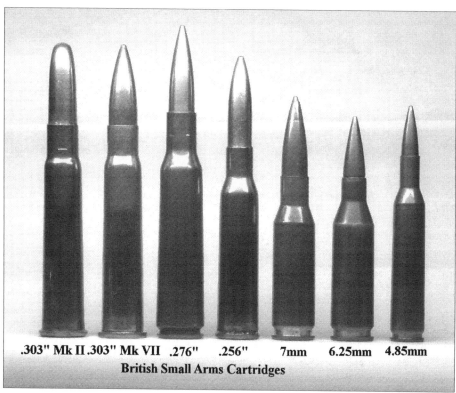

.303" Mk II   .303" Mk VII   .276"   .256"   7mm   6.25mm   4.85mm

British Small Arms Cartridges

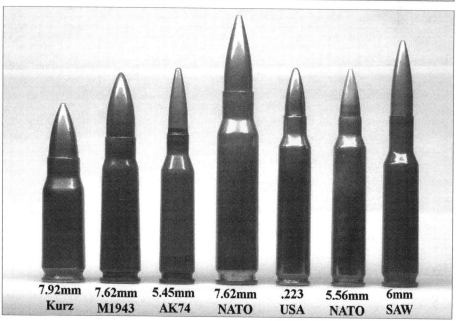

| 7.92mm Kurz | 7.62mm M1943 | 5.45mm AK74 | 7.62mm NATO | .223 USA | 5.56mm NATO | 6mm SAW |

Cartridge	Metric Size mm	Bullet Weight in Grains	Velocity fps	Energy ft/lbs
.303" Mk II	7.7x56R	215	2,000	1,920
.303" Mk VII	7.7x56R	174	2,440	2,312
.276" Enfield	7x60	165	2,800	2,888
.256" British	6.5x50SR	139	2,500	1,930
.280" British	7x43	139	2,530	1,986
6.25mm British	6.25x43	100	2,680	1,603
4.85mm British	4.85x49	56	2,950	1,088
7.92mm Kurz	7.92x33	125	2,250	1,413
7.62mm M1943	7.62x39	122	2,330	1,478
5.45mm Soviet	5.45x39	54	2,950	1,049
7.62mm NATO	7.62x51	150	2,750	2,532
5.56mm USA	5.56x45	55	3,200	1,257
5.56mm NATO	5.56x45	61	3,050	1,267
6mm SAW	6x45	105	2,520	1,488

Cartridges and Ballistics (at the Muzzle)

Bibliography

Barnes, F.C. (2009). *Cartridges of the World: A Complete and Illustrated Reference for Over 1500 Cartridges.* Gun Digest Books; 12 edition (October 20, 2009).

Dugelby, T.B. (1984). *Modern Military Bullpup Rifles: The EM-2 Concept Comes of Age.* Arms and Armour Publishing.

Ezell, E.C. (1977). *Small Arms of the World.* Stackpole Publishing, U.S.A.

Hogg, I.V. (1985). *Jane's Directory of Military Small Arms Ammunition.* Jane's Information Group.

Hogg, I.V. and Weeks, J. (1991). *Small Arms of the 20th Century.* Guild Publishing; 6th edition, revised edition.

Labbett, P. & Mead, J. F. (1988). *303 Inch: A History of the .303 Cartridge in the British Service.* Publisher unknown.

Labbett, P. (1993). *British Small Arms Ammunition 1864-1938.* Publisher unknown.

Woodend, H. (1981). *British Rifles: A Catalogue of the Enfield Pattern Room.* Her Majesty's Stationery Office Books.

Case 3.2 // Army Foundation College

An Exercise in Procurement and Marketing Management

David Jenkins and Dr David M. Moore

Centre for Defence Acquisition, Cranfield University, Defence Academy of the UK.

Introduction

This case study is intended for use as a syndicate-based learning activity and whilst it is intended for practical use in a classroom environment, it can also be utilised as to gain an insight into the issues and challenges that must be faced in undertaking an activity that requires differing perspectives of the same problem.

Since the early 1990s, the UK Government has emphasized the use of private sector funding for public sector activities. These includes the Private Finance Initiative (PFI) and the Public-Private Partnership (PPP) schemes which were seen as an opportunity to enable a commercial approach to funding, providing longer-term profit for the private organization whilst alleviating the need for immediate funding year-on-year by private sector organizations. As a fundamental part of the UK Public Sector, the Ministry of Defence (MoD) was encouraged to move away from the management and immediate funding of many of its activities to those that were undertaken by private sector organisations. This has, over the years, become a tenet that has been followed by successive governments and until very recently (2009) was a preferred approach by governmental organizations such as the MoD. This was initially undertaken with respect to support activities, such as cleaning and catering. However, it has over time been applied to activities that are increasingly close to frontline operational tasks. This can include logistics support, maintenance and security through Private Security Companies. To illustrate some of the considerations that have to be taken into account, it is useful to consider as an example, the activity of training and education which has become a major PFI activity at the Defence Academy where third parties undertake training, education and the management of facilities. This case study was prepared to provide a real-life baseline example set in the late 1990s against which to consider procurement, management and marketing issues. It should not in any way, be taken to indicate the MoD's definitive plans for education and training.

Situation

At the time, the Army was experiencing great difficulties in recruiting and retaining soldiers for the Teeth Arms (the infantry, artillery, tank and cavalry). The number of recruits was about 10,000 per annum against a target of 15,000. At the same time retention and reengagement targets were not being met. One of a number of measures being implemented to encourage young people to join the Army was the establishment of the Army Foundation College. The aim was to produce a group of committed, high quality, young soldiers. The first entrants joined the new Junior Entrant scheme courses in September 1998.

Location

The Army initially earmarked a college site at Harrogate, Yorkshire, the site of the old Royal Signal's Apprentice College. The establishment was purpose built for a capacity of 1,000 trainees. Unfortunately, the maintenance of the barracks had been minimal for the preceding five years. Significant sums of money were required to bring the facility up to an acceptable standard.

The Challenge

In the climate of that time, the Army faced a number of problems, two of which were to find the funding for the college and to staff it to an appropriate level. It was considered that the possibility of a PFI (Private Finance Initiative) approach would provide sufficient funding to develop a college (either at Harrogate or on a green field site) that met the expectations of the potential student, facilities management and the non-military staffing of the establishment.

Students

Students would be potential soldiers for the Royal Armoured Corps, Royal Artillery and the Infantry. Intake sizes were expected to grow from 684 in the first year, 960 in year two and thereafter a steady state of 1,344 Army students.

Selection of the students would be based on potential and using the British Army Recruitment Battery (BARB – an intelligence/trainability test) and evidence from the individual's National Record of Achievement; no academic qualifications are required.

Course Approach

There would be two yearly intakes in September and January (2/3-1/3 proportional intake).

Training would take place in three fourteen-week terms (forty-two weeks in total). There would be a one-week half-term break at Week 7 in each term. The main programme would be broken down as follows:

- 22 weeks military training

- 15 weeks vocational training

- 3 weeks external leadership

- 2 weeks leadership training

Students would work a five-and-a-half day week 08.00-17.00 Monday to Friday and 08.00-13.00 on Saturdays. Students would also participate in the Duke of Edinburgh's skills and sports programme two evenings a week.

Qualifications Gained

Army students would achieve a National traineeship in IT (Level 2 S/NVQ). This would include SNVQ Key Skills qualification to level 2/3 commensurate with ability. The students would also be able to commence modules to Level 2 S/NVQ in Management. These will be completed within their units in the Field Army.

The Task

"A" syndicates should identify both the policy and practical issues to be addressed by the Army sponsor in considering a PFI initiative and outline the main factors to be incorporated into a draft Statement of Requirement.

"B" syndicates are to present their preliminary thoughts on a possible consortium approach to the senior management team of a prime contractor well-versed in Facilities Management contracts for the Ministry of Defence.

Managing the Activity

It is suggested that having presented this challenge to the two syndicates, they are allowed sufficient time (typically about 2-3 hours) to discuss fundamental issues and how they can best be overcome. They should then come together and one syndicate presents to the other. This is not intended as a competition, nor is it intended that one syndicate 'score points' over the other, rather it is for both syndicates to see that the issues that are to be considered by one side in such a PFI scenario have a matching set of issues that merely reflect two perspectives of the same challenge. These 'two sides of the same coin' can enable the optimal solution to be obtained and the fundamental principles of PFI to be achieved. What can become a problem is that one side will only focus upon their issues without seeing the whole picture.

Summary

Under such a PFI, the Government seeks to gain better value for money by transferring risks relating to the construction and operation of a facility to the private sector and benefitting from the innovation of those tendering. In order for the private sector to deliver savings and efficiencies, the Ministry of Defence recognises that as well as taking on the management of services, the private sector must also be given the flexibility to design improvements to the underlying assets which are required to deliver the service.

The Ministry of Defence believes that the above is fundamental to the delivery of a successful project under PFI and considers that the need to purchase services in relation to training represents an ideal opportunity to exploit this flexibility.

Issues That Should be Identified

ELEMENTS OF CONTRACT

Provision of Training Places
Provision of Residential Places
Estate management
Catering
Cleaning
Academic support to military training
Administration and ancillary support

Sports and recreational services

Security

Teaching

RISK AREAS

Planning

Design

Construction

Commissioning

Transition

Operational performance

Volume of usage

Residual value

Regulations and Public Policy

Obsolescence

Annex: 'The Defence Helicopter Flying School', *Aerospace,* May 1997, pp. 8 – 11 (to be used as an indicative example).

The DHFS contract, worth £500m over 15 years is the largest training contract to have been awarded by the UK MoD, both in terms of value and time scale. FBS, which was formed with the sole aim of winning the DHFS contract, was one of three principal contenders. The other competitors were a Bond and Hunting Aviation team and, as a single company bid, Shorts. it is interesting to note that, as in so many other sectors of the aviation industry, companies which have formed partnerships are generally more successful in securing work.

The bid process for DHFS began in March 1995. The task of selecting suitable aircraft for the school was put entirely in the bidders' hands. It was up to the potential supplier to offer the aircraft it thought most capable of fulfilling the training role, after which the MoD would narrow down the choice of aircraft.

In FBS's case, the aircraft evaluations were carried out by Captain Geoff Connolly of Bristow Helicopters, a test pilot and former RAF helicopter instructor, and Martin Richards, FBS project manager, the longest serving FBS team members. A common flight test-schedule, based on the Invitation to Tender (ITT), was drawn up. Among the test criteria, the operation of the aircraft and its suitability as a training helicopter was fundamental, but meeting the specification (remembering that the training needs of all three services needed to be met) and the target cost was equally significant.

The challenge of coming up with a viable proposal for the DHFS was undoubtedly compounded by the MoD refining its requirement during the bid process. One such instance is the Ministry of Defence's shift in preference from a multi engine helicopter with wheels to one with skids, a preference, which had in fact been strongly advocated by FBS.

Geoff Connolly and Martin Richards carried out one sortie of 1 hour 30 minutes per aircraft, Captain Connolly conducting a pilot's assessment of each helicopter while Mr Richards assessed it from an engineer's viewpoint, as a technical response had to be provided to each requirement set down in the IFF They evaluated all aircraft which, on paper, met, or nearly met, the specification.

Several helicopters were considered including the Bell 407 and 430 were also considered, but the time frame to which FBS was working meant that neither aircraft would be available (the 407's order book has been full since before the aircraft's first flight and the 430 would not have been certificated in time). Similarly, the Eurocopter EC 120 might have been a contender as the single engine trainer had it been available in the time scale.

At the beginning of 1996 the MoD short-listed four aircraft for DHFS: the Bell 2068 or Eurocopter ASS50 as the single engine basic trainer, and the Bell 41 2EP or the Eurocopter BK117C for the advanced training role. On the basis of the flight tests FBS offered all four aircraft, in contrast to the other final round bidders. Bond-Hunting offered only Eurocopter types, while Shorts opted for the Bell types. Steve Bartlett, FBS project director, believes that the consortium had a significant edge over its competitors by offering all four helicopter types in any combination, thereby giving the MoD maximum choice. In the event, the MoD opted for the AS350 and the Bell 412EP a combination which only FIBS could provide.

Although the school opened in April, the repeated delays in announcing the decision (it was supposed to have been announced in April 1996 to allow for a one year phase in, but was put back to June and then October, when the preferred tenderer was named) meant that the lead times for procuring equipment, principally the airframes, were very short. Nevertheless, both Bell and Eurocopter have, according to Allan Brown, director commercial and technical services, Bristow Helicopters, been pro-active and responded positively to the resultant re-bidding processes.

FBS will have its full compliment of 38 single engine AS350BB by November. (The consortium is the first customer of the Arriel 1D1 powered Squirrel, hence the change in manufacturers designation, from AS3508A at the time of the bid.) During the middle of the year the delivery rate of the AS350BB is expected to reach a peak of seven aircraft a month. Four of the nine Bell 412EPs are with FBS and the remaining five aircraft will be delivered by June.

The time scale for both helicopter types to achieve the Military Aircraft Release (MAR), which is necessary to meet low flying requirements has also been very tight. All the aircraft are owned by FBS but will have military registrations (although in order to retain the airframe value, the aircraft will be maintained to civil standards). The Defence Helicopter Support Agency (DHSA), as the regulatory authority, has accepted the aircraft with a civil certificate for the basic military release. However, the specific modifications required for the aircraft to meet their military training roles - a night vision goggle (NVG) compatible cockpit for both types and instrument meteorological conditions (IMC) equipment on the AS350BB - will necessitate the MAR being amended. The AS350BB with IMC and NVG should have completed its trials at DTEO Boscombe Down early this month, while the Bell 412EP with NVG is expected to go to DTEO later in the year.

Bristow has recently gained both CAA and Directorate of Flying (MoD Procurement Executive) approval to carry out experimental flight tests, which in effect gives Bristow the same capability as that of a manufacturer's flight test centre. This makes FBS's position regarding the MAR more flexible, should problems arise, because it enables Bristow, as the design agent, to do exploratory modifications and to conduct a flight test programme at its Redhill base.

The amount of original design and trial installation work associated with both the 412EP and the AS350BB makes the term 'off-the-shelf' aircraft a misnomer Bristow is the design agent for both aircraft and is carrying out all the modifications to the 412s and to the first two Squirrels. The remaining 36 Squirrels are being converted by FR Aviation at its Hum base.

The DHFS Bell 412, the military designation of which is Griffin HT1, has two variants. Four of the aircraft will be standard fit with IMC and NVG equipment and will be based at RAF Shawbury in Shropshire. Five helicopters are being modified for search and rescue (SAR) training, two of which will operate from Shawbury side three will be permanently deployed to RAF Valley in north Wales. The winch installation and flotation gear for the SAR helicopters are new installations, which have been designed, modified and prototyped by Bristow. The MoD had fixed ideas about the winch and no off-the-shell product met the requirement. Fitted to the starboard side of the 412, the position of the winch is compatible with existing winches used in the Sea King, adding to the relevance of SAR training by making it consistent with the operational situation.

As well as the winch and flotation gear (the latter was trialled on the Isle of Wight), other more minor, but nevertheless important, modifications have been required. The seats on the 412 have had to be adapted to facilitate the use of a five-point harness, which has led to the seat height adjuster also needing a lesser modification. Furthermore, during the modification process, Bristow strip the aircraft right down to give the airframe an anti-corrosion treatment. By treating parts not normally included by the manufacturer in the anti-corrosion process, FBS is thinking ahead and considering the long term maintenance of the aircraft.

The AS350BB has undergone considerable changes too. The aircraft are delivered in their basic state from Marignane to McAlpine (Eurocopter's UK distributor) where the auto-pilot and instrument panel, made to Bristow's design, is installed. The original seat, deemed not to be robust enough, has also been replaced. The avionics kit, again all designed by Bristow and the numbers of civilian QHIs conducting SEBRW, SEARW and MEARW will reflect the percentage split as a whole. So, 40% of the SEBRW QHIs will be civilian and on each of the three SEARW flights, 40% of the instructors will be civilian. Similarly, 40% of the staff on MEARW will be civilian, but none of the civil instructors will fly OFT students at Middle Wallop.

All the instructors underwent training with the manufacturers – at Marignane with Eurocopter and Fort Worth with Bell. Instructional training then continued in the UK and will be completed by September with the phasing in of the 412s at RAF Valley. A civil-registered 412 was used for OH training at Shawbury, prior to its modification to the DHFS standard. This was carried out under the terms of a Department of Transport ruling, which allows military pilots to fly a civil registered aircraft for military duties, without holding a civil licence.

The DHFS is located at two main centres, RAF Shawbury, where FRA Aviation and Serco previously provided support services to the then RAF training school, and Middle Wallop, where Bristow had a similar role at the MC school. This geographical division of the work will remain. Not only was it a fundamental aspect of the original bid, but it is now a contractually binding arrangement.

As mentioned, a third location, RAF Valley, will be used for SAR training which, because of the phased start to the training programme, does not begin until September. Maintenance support of the three Griffin HT1s permanently deployed there will be undertaken at RAF Valley by FRA Serco. However; major maintenance will be conducted at RAF Shawbury.

At both the main locations FBS is using the existing facilities, the only new building required being the simulator centre at RAF Shawbury, which is owned by FBS (the provision and maintenance of simulators is included in the contract). The building will house the six-axis, full motion simulator supplied by Flight Safety International. This will be equipped with an Evans and Sutherland 200° horizontal and 60° vertical field of view, day-night, and Night Vision Goggle capable visual system. Two cockpit procedures trainers for the Squirrels and one for the Griffin have been manufactured by Frasca and will be located elsewhere at Shawbury.

Within FBS seven internal contracts were set up to enable it to function as a single unit. The CEO of FBS, Peter Ashleigh-Thomas (Originally with FRA Serco at Shawbury) took over the running of the contract from the FESS bid-implementation team in April. He is supported by a small management team, based at Shawbury, which comprises a chief engineer, chief pilot and various contracts and finance staff. Employees will come under the FBS banner and, since FR Aviation, Bristow Helicopters and Serco have worked together successfully during the two year bid process, it seems reasonable to assume that this will continue, says Allan Brown. Moreover, Steve Bartlett believes that the strength of the alliance between the three companies was instrumental in FBS winning the bid in the first place. a crucial importance to the integrity of FBS as a whole is its own certification as a JAR-145 approved maintenance organisation.

Currently Bristow's JAR-145 activities have been extended to Shawbury and Middle Wallop, but in the long term it is envisaged that FBS will have its own set-up. There is more to the DHFS contract than solely the support of the school, and other "duties" have been written into the main DHF-FBS contract. These include being responsible for the running of multi-activity contracts (both Shawbury and Middle Wallop fall into this category), the support of Lynx aircraft at Middle Wallop, as well as the MC historic flight, and the operation of the aircraft store at Shawbury and REME workshops at Middle Wallop.

FBS clearly believes in its capability to support the DHFS successfully, at the right price. From the outset cost drove the contract and the Ministry of Defence will save money for three principal reasons;

1. FBS is a contractor; so the Ministry of Defence will not have the burden of costs associated with full time employees.
2. Higher utilisation of aircraft.
3. Maintenance is transferred from the military to a civil organisation.

As the previous incumbent of the main sites in its constituent parts (FRA Serco at Shawbury and Bristows at Middle Wallop), it is tempting to suggest that it was inevitable that FBS should have won the contract. However, as both Allan Brown and Geoff Connolly stress, this could potentially have been a negative factor, as it is easy for an incumbent to just accept existing practices, assuming that the same set up will work equally well when applied to a new situation. FBS's approach has been to ensure that the task was looked at with fresh eyes, so that an objective assessment could be made as to what aspects of the organisation (at both sites) should be changed and what should be retained.

FBS has already proved itself to the Ministry of Defence, with the credibility of their aircraft design and installation plans and by meeting the tight into service schedule. FBS certainly faced a tough challenge during the bid, as the MoD was keen to secure a water-tight contract. Now, all being well, FBS has a further 15 years to go on proving to be an efficient provider of high quality training and support to the DHFS. The consortium's main aim will be to renew the contract after the initial 15 years, but could FBS's horizons stretch to operating other helicopter training schools in the UK and abroad?

Case 3.3 // Support Chain Management in the RAF

'An Air Bridge Too Far': Support Chain Management in the Royal Air Force

Alan Farnsworth

Royal Air Force Supply Officer, MSc DLM Course Graduate 2000.

Introduction

This real-life case is about the development and implementation of a support chain management strategy that was launched during a period of significant transformational change for the RAF, and in particular, one of severe costs reduction in light of the peace dividend following the end of the Cold War.

The case describes the background and environment that the RAF faced in the early 1990s following the 'Options for Change' force reductions and other Government initiatives. While the RAF was in the process of restructuring and co-locating its support activities into Multi-Disciplinary Groups, further organisational changes, including reductions, were required by the Government, especially amongst the MOD headquarters staff. The RAF recognised that to meet the new cost-conscious environment and growing need to further reduce support costs it would need to streamline its' processes to achieve lasting cost reductions and improved support efficiencies.

Consequently, in 1993 the RAF launched its Support Chain Management (SCM) strategy that sought to radically transform its' support activities the traditional support stovepipe management structure and view support processes holistically and so by improve the total process as opposed to discrete elements. The strategy was managed by a steering group, working group and implementation teams at a strategic, tactical and operational level whose task, between them, was to implement the concept.

Situation

The Royal Air Force is one of three military arms of the Ministry of Defence (MOD), responsible for the delivery of Air Power in support of MOD's defence roles. In December 1993, the Air Force Board, sanctioned the commencement of a strategy that was aimed at adopting a radical approach to providing support to front-line forces that was entitled: Support Chain Management. The strategy was designed to reduce costs and improve service but without impairing any additional operational risk. It also aimed to reform the support business processes in preparation for the design and implementation of a major new logistics information system, named LITS that was planned to become operational, in tranches, from 1994 onwards. LITS was costing over £400 million that had been funded from the RAF's support budget based upon the projected savings of improved visibility and transparency of information across the support environment. The strategy was simple: to view processes from a holistic perspective as opposed to the traditional functional perspective and thereby make decisions based upon the whole support chain as opposed to single elements. The strategy was in line with current academic thinking and was seen by senior RAF management as an opportunity to bridge the growing gap between Government funding and planned RAF in-service programme costs.

RAF Support Chain

The RAF's support environment is a series of complex support chains, which in the early 1990s controlled over 1 million line items. This figure has been reduced, in 1999, to 835,000 line items supplied by over 2,000 companies. End-to-end support chains are not managed by any individual or single agency, but by an organisational structure, controlled by several budget holders situated across command boundaries. The result is that information flows are extended, decision-making remains slow, and the support chains are mainly sub-optimised. Also, elements within the support chain, notably at unit and depot, are not routinely involved in support chain decisions, principally as a result of the current organisational support structure.

The generic RAF support chain in 1993 consisted of five main elements: upstream suppliers and their suppliers, 4th Line commercial repair and overhaul companies, some of whom may also be suppliers, 3rd Line Service repair and overhaul facilities, a storage depot (16 Maintenance Unit) at RAF Stafford, and downstream: Main Operating Bases and Deployed Operating Bases.

The 3rd Line facilities, depot, and transportation to link the nodes were under the command and control of the recently formed Headquarters Logistics Command (HQ LC). Main and Deployed Operating Bases were under the command of Headquarters Strike Command (HQ STC) the operational and fighting arm of the RAF. The 2-way flow of information across the support chain was extended due to the hierarchical command structure and the lack of an integrated information system. The management of in-service equipment within the RAF is the responsibility of delegated Support Authorities (SAs) that cover: airworthiness, maintenance policy, repair programmes, new spares buys, contracting and post design services. The SAs form part of a Multi-Disciplinary Group (MDG) with similar responsibilities, for example, aero-engines or avionics. A number of MDGs, usually 3 or 4, form a directorate. The Support Management Group (SMG) consists of four functional equipment directorates and a central directorate responsible for policy, plans and budgets. The SMG is an autonomous Strategic Business Unit within HQ LC.

Within HQ STC, aircraft role offices are the interface between units and the relevant equipment SA. The role offices are primarily to provide operations staff with on-hand logistic support. They have no equipment responsibilities or budgetary powers.

Environment the RAF was operating in between the late 1980s and mid-1990s

The RAF had been, and was continuing to be, subjected to a raft of primarily led Government initiatives to reduce Defence costs, seek 'value for money' and hence maximise the 'peace dividend' following the fall of the Warsaw Pact. Defence costs as a share of GDP had peaked in 1985/86 at 5.1 % and was planned to fall to 2.7% before the millennium. In 1990, under the 'Options for Change' review, the RAF were planned to lose over 16 squadrons, 14 stations and over 14,000 Service personnel (15% of the RAF); these were taken and further cuts were announced. In 1991, the PROSPECT headquarters study report required a further reduction of at least 20% of MOD headquarters staff, with particular focus in the support, training and personnel areas. To achieve these savings the RAF's headquarter structure was redesigned that resulted in extensive down-sizing, relocation of Service and civil servants and the building of major infrastructure projects.

These changes were further compounded by the Defence Costs Study that was a Treasury led cost-cutting study that solely examined the support areas of the Armed Forces. The outcome for the RAF was a further loss of £286 million over 10 years from its support budget and 7,500 Service posts, rationalisation of its sites was accelerated along with civilianisation programmes. The changes already mentioned were set against a background of earlier initiatives, namely 'Competing for Quality' and 'Next Steps' that were launched during the 1980s and were still ongoing. The former was an initiative to improve the quality of public sector services by identifying those that were suitable to be market tested; the RAF's target in 1992 was £217 million with further targets in the future. The latter was the creation of Government agencies to carry out the executive functions of

Government within a policy and resource framework. In the support area, the RAF by 1991 had launched the Maintenance Group Defence Support Agency. It comprised major repair and overhaul facilities of aircraft, aero-engines and components, a major avionics repair centre, depot storage and distribution facilities, and 5 signals units that formed a major part of the UK's Defence Communications network. The pressure on the agency from the start to reduce costs was severe that resulted in the shedding of 1400 jobs within the first 2 years of which 1250 was Service posts.

Since the end of the Cold War, the RAF's concept of operations changed. No longer did it support a Cold War citadel concept of support but moved to the current, highly mobile, expeditionary approach. It was no longer able to afford to support deployed aircraft operations with traditional 30-day Fly-Away Packs, but moved to leaner Primary Equipment Packs (PEPs) that were, on average, designed to support operations for 10 days, with re-supply being provided at regular intervals. In conjunction with PEPs, the RAF also developed an Express Chain Management strategy that was the template to identify critical spares and for the support manager to establish accelerated support chains in times of crisis to guarantee continuity of supply.

In short, the RAF support areas had been subjected to virtually continuous change since the late 1980s and throughout the 1990s that had placed severe pressure on its' resources and management whilst having to remain focussed on its' primary function of providing effective support to the front-line. By 1994, the RAF had effectively been imposed with several large mortgages as a result of LITS funding, the Defence Costs Study and the annual efficiency savings measures – by any measure this was a severe challenge.

The RAF Response

In December 1992, during this period of ongoing 'change', Chief of Logistic Support (RAF), following earlier conceptual work by a supply branch specialist, instigated a Support Chain Management (SCM) Study. The aim of the study was to identify ways of exploiting in-house and commercial 'best-practise' so as to improve the efficiency and effectiveness of RAF logistics. The Air Force Board endorsed the recommendations of the Study in 1993. Its objectives were:

- To introduce the concept of tailored, specific support chains that reflect the lead-time, impact of support chain failure and total cost of selected options.

- To develop support chains, which are affordable, flexible and robust enough to support the RAF's peacetime and training task, but which have the capability to produce additional resources in times of operational necessity.

- To encourage horizontal cost awareness and the adoption of business best practises where possible, across the organisation.

Management of Change

A Steering Group (SG) whose aim was to provide the strategic direction headed the management of change structure to deliver SCM across the RAF. It met several times after Board approval, but not after late 1994. At a tactical level, a SCM Working Group (WG) was convened of middle ranking officers representing the key stakeholders across the RAF and other Services. Its main function was to conduct research into any area of the support chain where application of SCM principles might realise benefits and decide, based upon their results, on the optimum policy for logistic support.

An action plan was produced following their work that relied upon other areas of the support chain to implement. At an operational level, SCM Implementation Teams (IT) were formed in 1995 principally within the Support Management Group that was responsible for facilitating the change. A SCM IT was embedded within each of the four directorates and they reported directly to their director who was accountable for delivering support to the front-line for specific ranges of equipment, including weapon platforms. The SG and WG did not have executive control over the SCM ITs, their work and recommendations were advisory and could be over-ruled by each director.

The problem the RAF soon realised was the lack of executive status that these committees had, as the support responsibility and budgets for in-service equipment support had been delegated, within the SMG. The RAF's support chain also crossed many functional, budgetary and even command boundaries which further served to add complexity to the management of change process, all working to potentially differing priorities.

SCM Implementation Plan

The implementation plan was based upon having mapped the existing RAF logistics system and by identifying areas for improvement. In the first year, the emphasis was on simplifying existing processes, and correcting by 'fast track' first aid measures, the more significant shortcomings of the present system, which would deliver initial improvements and reduce costs. The longer-term exploitation of SCM relied upon LITS. The implementation plan drawn-up by the working group was comprehensive and was to be realised in measurable steps by firstly conducting trials, within the Support Management Group, in four 'Showcase' SAs: Harrier, Pegasus Aero-Engine, Avionics, and a major consumable Support Authority.

These trials were designed to test the theory, secure visible short-term wins, and gauge the scope for roll-out across the SMG. The emphasis was on reviewing existing contracts and orders and cancelling those that were no longer deemed necessary. It concentrated on equipment repair loop processes and sought to apply actual data as opposed to default values, and also focussed on reducing repair times, or what was preventing times being reduced. The thrust of this work was to reduce the RAF's costly repairable inventory. The showcase trials proved successful and following Air Force Board approval in 1994, work templates were constructed for 'rollout' across the SMG, that controlled the majority of the support chain budget.

The templates were issued to the newly formed SCM ITs throughout 1995. The templates concentrated entirely on provisioning related activities. The main focus was on data management, timely contract renewal, reduction of production lead times from Industry and the improved management of equipment repaired by Industry. Other areas included the accuracy of consumption data, accuracy of available assets and accuracy of system parameters used in the system generated re-provisioning calculation. As stated, the teams were not responsible for actual implementation, but for facilitating the changes within their SMG directorates. No mechanism was established for monitoring or measuring the SCM work within the SMG.

In addition to the work templates, the RAF embarked upon other RAF-wide initiatives that were:

• **Single Point Holding.** The concept was by changing the disposition of stocks from depots to units (stations), inventory levels would be reduced and potential reductions in transport and depot storage costs were possible. Research was conducted to model the effects of single point holding. For the Harrier aircraft alone, eliminating depot holdings identified £2.6 million savings. However, to pursue such a policy was difficult because it involved a variety of stakeholders across functional, budgetary and command boundaries who all the ability to delay implementation.

• **Reduction of Consumable Spares Inventory.** Between 1988 and 1993 the proportion of consumables in the RAF inventory, by value, had risen. In 1993, they accounted for 40% of the inventory by value as against 29% in 1988. The reasons for the increase were not understood, although it was suspected that newer technology items were using a greater proportion of 'disposable' spares, and so classed as consumable. Further analysis revealed that only 13% of the consumable inventory had a demand rate of more than one a month and that 74% were not experiencing any demand activity. The consumable inventory was growing at the same time that the RAFs fleet was decreasing. In 1994, the RAF had on average, approximately 5 years of stock.

•   **Electronic Purchasing by Units or EPU.** Since 1994, the RAF had identified the scope for introducing EPU at units based upon the earlier experience of the Royal Australian Air Force. In 1995, agreement in principle was reached that work should commence on introducing EPU, but mainly due to organisational and budgetary support boundaries and limited resources, the trials did not commence until earlier this year at RAF Leeming. The objectives of the trial were to gain experience in electronic commerce, reduce supply system costs, reduce unit and depot stock holdings, devolve responsibility for budgets, accounting and ordering and provide more responsive service.

In addition, the PROSPECT study required the SMG responsible for in-service support of air stores, platforms and weapon systems, to relocate at RAF Wyton as part of the newly created HQ LC. The relocation resulted in the loss of over 800 support staff from the former Harrogate site, primarily employed on procurement activities, referred to as RMs, who declined to move. At the time, senior management viewed the loss as an opportunity to introduce new working practices, but these notions were soon dispelled as the scale of the skill and knowledge loss became apparent. Recruitment and training only partially addressed the problem and it was recognised that urgent action was needed to provide decision support tools to the RMs. Consequently, a suite of support tools was developed in-house that combined what is now known colloquially as Expert Provisioner or EP. In essence, EP provides a 'what if' tool, smarter logic for forecasting, step by step guides and repair calculation templates.

The majority of the SCM implementation plan was centred within the SMG as they controlled the majority of support chain costs. The establishment of the SMG SCM ITs cost £4 million over 4 years between 1995 and 1999. Team vacancies existed well into the first year of funding due in part to the relocation plans. In some teams, the majority of staff had not worked within the SMG or had experience on in-service support work, therefore initially, their contribution was limited. Team members did not receive any training on the concept of SCM or on change programmes, they were largely left undirected to pursue action plans that were devised by their Team Leaders, who in the main were not selected with specific skills. The result of the team's work was advisory and always subject to agreement by the respective budget holders. The RMs, who were expected to implement the changes had lost over 600 staff, leaving their experience and knowledge levels extremely low. Therefore to expect them to maintain existing support procedures, let alone embarking upon analysing and restructuring processes, proved problematic. The result was that the SCM ITs were forced to expend a considerable amount of their time in restoring basic RM knowledge levels before more proactive support chain measures could be taken.

Each IT worked independently and most embedded their staffs within the MDGs that formed each directorate. Initially, they were viewed sceptically by the MDG staffs as 'outsiders', but over time as MDG staff began to realise the true benefits the team was able to offer to the organisation, their standing and credibility improved. The work plan of each of the four teams was diverse and extensive, and all primarily concentrated on qualitative improvement measures that relied upon others to implement. The teams early work in assisting in restoring the foundation skills of the RMs by conducting seminars, workshops and on-the-job training sessions began to bear fruit as the corporate knowledge of the SMG began to improve. In the early years, the thrust of three of the four teams was mainly geared to improving data management and the purification of the data in readiness for LITS, in line with the work templates produced following the 'showcase' trials.

They also recognised that reducing the workload of the Range Managers by suppressing system generated outputs deemed as non-essential, was essential to make headroom for more proactive work and to counter the staff reductions inflicted by PROSPECT. Economic analysis of the Teams quantitative work revealed that over £25 million was saved during their 4-year life. This was mainly generated by reviewing and cancellation of outstanding orders, application of actual as opposed to default provisioning parameters, and by reviewing and amending other key data fields on the RAFs supply inventory system.

Also, substantial but less tangible savings were achieved by their qualitative improvements across a range of diverse activities. The fourth of the teams adopted a different tact that mainly concentrated, from the start, on improving the directorate processes and in particular with a stronger focus towards Industry. The differing approach was largely due to their team leader, who had received post-graduate training on the theoretical aspects of supply chain management and was therefore more inclined to adopt a strategic approach to the teams work plan. The ITs were not solely established for the SCM initiative but were also required to facilitate the other major initiatives, namely LITS development, Mobility and Deployed Support, and Resource, Accounting and Budgeting to name but some across their respective directorates. The ITs played an important and pivotal directorate role in the development and implementation of these other initiatives.

The Problem

In late 1994 the SCM SG met for the last time. By the end of 1996, the SCM SG had folded and the IT activities were beginning to stray away from the original work templates issued in 1995 due to the differing priorities of their director to respond to other centrally driven initiatives. Also, following the PROSPECT headquarter cuts, policy staffs were fragmented that had caused the SCM strategy to become unfocussed, lacking in direction and with no single change agent, champion, in charge of the SCM strategy. By 1995 senior management had become disinterested and frustrated in the strategy because of the mounting problems and their inability to implement change across organisational boundaries; some had always been sceptical of the initiative.

Senior management had failed to measure and monitor the progress of the strategy and had allowed the strategy to 'drift' without any clear direction. The RMs who were expected to implement the work plans remained relatively inexperienced and their skill levels were generally low. The SMG middle managers, by 1995 mainly engineers caused by the PROSPECT staff reductions and delaying of supply specialists from the management chain, were without the provisioning experience, knowledge, skills or time, to provide much support to their RMs. In addition, to meet the mortgages imposed by the DCS and annual efficiency targets, further programme cuts were taken that effectively increased the support risk and resulted in delayed support decisions (buys). Furthermore, LITS was proving extremely problematic that resulted in delays of over 4 years and redefinition of the specification was needed for the project to remain affordable.

Main Issues & Challenges

The main question arising from this case was: what does the RAF need to do to bring the strategy 'back on track' to meet the aim and objectives of the original strategy? In order to optimise learning opportunities, those reading this case should consider the task of developing a recovery strategy, devising operational policies and implementation plans. Specific issues could be given out for students to consider under guidance. The following separate issues should be addressed through the case:

Strategic Issues

- Consideration of the management of change structure needed.
- The availability of resource and skill levels.

- How do senior management monitor the progress of the initiative and what performance measurements should they use.

Development of Operating Policies and Systems

The main issues in this case are: maintaining operational levels but with a reduced budget and how should the operational aspects of the support chain be managed. An integrated information system was unlikely to be available in the short-term therefore any plans need to assume that existing systems will be utilised.

Case 3.4 // Exercise Saser Kangri

Mountaineering in the Karakoram Range

Ivar Hellberg

Centre for Defence Acquisition, Cranfield University, Defence Academy of the UK.

Introduction

This case study aims to test students as to their ability to consider the logistic considerations of a mountaineering expedition in Northern India. Even today the limiting factor is the ability of the human ability to carry certain loads and require certain comforts to survive in harsh conditions.

The case study is based on the experience that the author has had over a number of years in leading logistic challenges that whilst set in a specific context can be applied in the much broader strategic logistic environment. It is intended to stimulate creative thinking to maximise the ability of the team to conduct the operation whilst protecting their safety.

Task

You have been selected as a member of a Royal Geographical Society and Indian Himalayan Institute expedition to climb Saser Kangri 1 in the remote Eastern Karakoram range of Northern India. After years of climbing in both Scotland and the Alps, it has been your dream to further your mountaineering skills by climbing in the Himalayas.

You were aware that each member of a mountaineering team must play their part, not only as a climber, but also to take on at least one other responsibility. You had rather foolishly mentioned in your resume that you possessed a degree in logistics management and, inevitably the team leader has selected you to organise the logistics for this major expedition.

The team leader has asked you to have the outline logistic plan ready for the second week in January, ten weeks away. You immediately started to refresh your mind about the details of the expedition and the logistic planning constraints.

Background

Saser Kangri literally means the 'yellow mountain' and it derives its name from the bright yellow granite rock which can clearly be seen on the summit ridges. The survey of India lists the mountain as K22 at 7,672 metres. The mountain is situated in the Eastern Karakoram range in the district of Ladakh of Jammu and Kashmir State. It is the highest peak amongst a group of 7,000 metre peaks which constitute the Saser range. This range is bounded by the Nubra River to the West and the Shyok River to the East and South.

The mountain itself is surrounded by seven main glaciers and defended by particularly rugged area which has developed a culture of its own as a result of being cut off from the rest of the world for most of the year by high passes and heavy snowfalls. Early reconnaissance expeditions considered the peak to be impregnable.

The first person to have recorded the mountain was Arthur Neve who, in 1899, climbed a peak near Panamik and photographed the Saser range. Dr. Tom Longstaff (the first man to climb Trisul in the Garwhal Himalaya), led an expedition to the area in 1908 and reached a height of 5,486 metres on the Phukpoche glacier to the West of Saser 1. He recorded the difficulties of a western approach in an article for the Geographical Journal published in June 1910.

Other visitors to the area were the Vissers who mapped the area in 1929 and Lieutenant Colonel Jimmy Roberts who made the first attempt on the mountain in 1946. Although unsuccessful, Jimmy Roberts' thorough reconnaissance of the western approaches has added immeasurably to our knowledge of the mountain. Another attempt from the west was made by an Indian team led by Major N. D. Jayal. This team counted no less than twenty avalanches hurtling down the west face of Saser 1 in one day. Again they could find no way up.

Another Indian expedition in 1970 attempted the west side of Saser 1. It was led by the late Major Bahunguna (who died on the ill fated 1971 international expedition to Everest), who, with Captain Alok Chandola directed this main effort towards an approach via the North Phukpoche glacier and up the North West ridge of Saser IV (7,310 metres). Again the sustained difficulty of the route turned them back after reaching a high point of approximately 6,000 metres.

No other serious attempt was made from the west and attention was drawn to an easterly approach (advocated by Jimmy Roberts). In 1973, the mountain was at last scaled from the east by a team led by Commander Jogindar Singh after an epic approach march via the east Shyok River and the north Kunchang glacier. The western side of the mountain remained un-vanquished. You very quickly realise that the successful assault of Saser Kangri 1 from the west is going to depend on excellent logistic planning as well as mountaineering prowess.

Exercise composition

The expedition is to be a joint British-Indian expedition sponsored by the Royal Geographical Society and the Indian Himalayan Institute. The aim is to climb the both the unscaled west ridge of Saser I (7,672 m) and Saser Kangri IV (7,310 m) between April 4th and July 4th.

At the start it should be noted that Saser Kangri is in a disputed area which belongs to India but is claimed by both China and Pakistan. In consequence the area is subject to constant border skirmishes which have, in the past, prevented foreign expeditions from visiting the region. Your expedition is therefore very privileged to be allowed by the Indian government to climb in the area on a joint Anglo-Indian basis.

The expedition is to consist of nineteen British climbers and a similar number of Indian climbers and one doctor from each country. There will also be a support team from the Indian Army (because of the sensitivities of the area) which will include a base camp manager, cooks, mess and cooking tents and 4 ton Tartar trucks for the journey from Leh to Panamik. The leadership of the team is to be on a joint basis. In addition to climbing the mountain there are to be two important supplementary aims, firstly to help select a British team for an attempt on K2 in eighteen month's time and secondly to record the ecology and environment of this seldom-visited part of the world.

Timings and route planning

The expedition is planning to arrive in Delhi, India on 4th April so you, as logistics member will have to arrive a few days before to sort things out with Indian Customs and make contact with the Indian team's logistics and transportation member. After four days in Delhi the team is due to fly to Leh on board a giant Soviet built Ilushin 76 transport plane. The British team will carry out acclimatisation training based in the ancient of Leh at a height of 3,350 metres. This will also allow the two teams to get to know each other and climb together.

The workout climbing training will take place in the mountains of Zanskar to the South of Leh and to the south of the Indus River. This is very wild and remote part of the world and the home of the elusive snow leopard. One of the aims of the expedition is to try and see a snow leopard and to photograph it in the wild. You will note that photographic equipment will take up a lot of precious space. Additionally the area is known for huge wolves, Himalayan blue sheep (Burrel) and large Himalayan vultures (Lamagier). There is also an abundance of Himalayan golden eagles. The ornithological team member is expected to be quite preoccupied.

The plan is to spend three weeks in Leh before moving north across the infamous Khardung La (at 5,600 metres the highest motor pass in the world) into the Nubra Valley. The Khardung La is a masterpiece of Indian engineering which zigzags for miles up the steep snow sides of the pass and at one point actually cuts through a glacier to allow the trucks to pass. It is the supply line for the Indian Army stationed on the Siachen glacier to the north of Nubra. This is a narrow road with very few passing points which have to be cleared at the start of each day by a single snow plough. By consequence when trucks break down on occasion it is simpler to push them over the edge rather than hold up a convoy of hundreds of vehicles. The journey over the pass and into the Nubra Valley takes about twenty hours over some of the roughest roads in the world. You note from your guide book to the area:

"The scenery of the Nubra is on a vast scale and almost completely desolate of vegetation as the mountains of Zanskar to the south block-off the passage of the summer monsoons in this remote region of the eastern Karakoram. The rainfall in the region measures only 50mm per annum."

It is planned to establish the valley base at the small village of Panamik in the Nubra Valley (approximately 3,200 metres). Panamik lies a further one days' journey north, up the Nubra from the Khardung La. From Panamik, the route to base camp follows in an easterly line up the precipitous gorge of the turbulent Phukpoche Nullah. You have been told that it is likely to take two days to make the journey by foot to base camp from Panamik.

It was originally planned to use a large number of locally recruited porters to help shift the 40 plus tons of expedition supplies and equipment over the two day march from Panamik to base camp. However you have heard that it is planting time in the valley, so porters will be difficult to find. You therefore plan a worst case scenario with all the expedition members carrying the stores themselves (an unpopular idea). You will therefore need to make an estimate as to how long this will take using all thirty eight expedition members on a 'one days rest in three' basis and carrying not more than twenty kilograms each.

Further it will be necessary to calculate how many trucks you will need to make the journey from Leh to Panamik. Your calculations should include all the expedition members and forty tons of stores. Normally one truck in three will break down on the route for various reasons.

After the establishment of base camp, it will be necessary to confirm which of two routes should be taken to make the main assault on Saser Kangri 1. In effect two reconnaissance expeditions will have to be mounted from base camp, one to the north up the northern Phukpoche glacier which might eventually lead onto the north-west ridge of Saser 1. The other, more likely route, is to follow up the southern glacier which might lead onto the south-west Col of Saser 1. You should allow ten days for each of these mini-expeditions to run concurrently.

General planning figures for the advanced base camp and beyond

Whichever route is to be taken it is planned to establish an advanced base camp at around 4,800 metres on either the north or south Phukpoche glacier. The advanced base camp should have enough accommodation for at least twelve members with a mess and cooking tent and two more tents to be left empty for returning climbers from the face. Above the advanced base camp two further camps need to be established. Camp III at about 5,700 metres and then a high assault Camp IV at 6,800 metres. Camp II should have accommodation for at least eight climbers and Camp IV for four climbers. Both of these camps should also have one spare two-man tent for emergencies.

The expedition plans not to use oxygen for the climbers, so this will slow the climbers down but at least it will lighten the logistic burden. Some oxygen will obviously still have to be available for medical purposes such as mountain sickness etc. From a planning point of view above the advanced base camp you will have to assume that one day in three will be written off because of bad weather (especially winds and storms) and one day in three will be written off for medical reasons or fatigue. Also the amount that climbers can carry will be reduced considerably to around 15kg.

You will also note that fixed rope will have to be put in place on the steep ice wall between Camp III and Camp IV. It is estimated by your team leader that this will require at least five good days to achieve, involving four climbers working on the face at any one time and returning to Camp III each evening. Fixed rope and ice screws/dead men (used to fasten the fixed ropes to the ice face) are very heavy and you will need about 750 metres of rope. This come in lengths of 100 metres each weighing 12 kg.

Further requirements include hardware to go with the ropes which is estimated at around 4kg. In good conditions it is possible to put up around 300m of rope per day. Your leader has asked you to check out his five day calculation. Beyond Camp IV it is estimated that a very fit pair of climbers could make the summit of Saser 1 and return to Camp IV in about eleven hours. You will also have to consider that climbers should only operate for a maximum of eight days above Camp III before returning to base camp for a three-day rest. Other fresh climbers would then come up to take their place.

The initial logistic problem

At this early stage of planning you are uncertain as to the exact breakdown of the camp stores, the climbing equipment, the rations, the medical stores, the fuel and the general stores that you need for such a complex expedition. However you note that you will have to determine the exact quantities fairly soon, as only from that can the expedition treasurer work out exactly what needs to be purchased and how much money will need to be raised. However the more immediate problem is that the expedition leader has asked you to calculate whether it is possible to achieve your objective and return to Delhi by 1st July.

The task

To achieve this you will have to work out a plan in both spatial as well as chronological terms using all the constraints which you have already been given. Some things you do not know at this stage, so you will have to make assumptions as and when you need to. Your plan should show the loads that need to be carried to establish the various camps and the dates these camps can be established as well as the occupancy and climber turn-around rotas.

You will have to calculate the weight of supplies/rations that need to be carried to keep the momentum going on the mountain, otherwise you could waste good climbing days through poor logistics. This can be done by a critical path analysis or a flow chart – either represented in a graph or spreadsheet.

Conclusions

Climbing expeditions are complex endeavours. However what this case study has sought to do for the logistics student is to provide a real abstract problem for they and their colleagues to think through and analyse. Creativity is at a premium in terms of working out the art form of enabling this type of enterprise to succeed.

Case 3.5 // Strategic Mobility

Long Range Logistics and Joint Integration

David M. Moore

Centre for Defence Acquisition, Cranfield University, Defence Academy of the UK.

Introduction

Whilst there have been many instances of inter-service co-operation to deliver adequate and timely logistics support to frontline British forces, one of the outcomes of the 1998 policy review known as the Strategic Defence Review was been to emphasise the concept of 'Jointery' to characterise future operations. This has been subsequently reinforced through the changing emphasis in defence policy, which is turning more towards developing and enhancing the capability to intervene in crisis areas far from the European mainland.

This case is intended to allow discussion of the nature of long range logistics drawing, where appropriate, upon commercial sector concepts. Furthermore the case seeks to identify the debating points and clarify the nature of the challenges ahead for supporting this concept. As ever if the right questions can be asked the intellectual assets available to the services can solve the territorial problems to deliver logistics support in the right place at the right time, within resource constraints of the government of the day.

Some Definitions

Joint integration can be characterised in terms of:

- Firstly, that of the supply or support chain. Joint integration facilitates complete integration from manufacturer to service to theatre of operations and the reverse. By identifying the supplier – customer relationships and bargaining over territorial and political issues successfully, a clear chain can be identified rendering savings in time, and delivery of a higher quality process to the armed services.

- Secondly, as being tri-service. Formerly, the British Army, Royal Navy and Royal Air Force had their own support arms. One aspect of the 1998 Strategic Defence

Review was to appoint a Chief of Defence Logistics with a remit for integrating the three logistics arms in the most appropriate manner. Subsequently, the Defence Logistics Organisation (DLO) has been integrated with what was the Defence Procurement Agency (formerly the Procurement Executive) to form Defence Equipment and Support (DE&S) under a Chief of Defence Material (CDM). At that time, efforts towards integration were being made where there were obvious links. For example, in the late 1990s and early 2000s, the Fleet Air Arm and Royal Air Force were jointly practising the use of Harrier VSTOL aircraft from aircraft carriers.

• Thirdly, as stretching across narrow boundaries in the context of alliances such as NATO, the Western European Union (WEU) and with Non-Governmental Actors possibly in the context of humanitarian operations. Coalition or multi-national operations such as those mounted against Iraq in 1991 and 2003 can, to an extent, be dictated by the inter-operability of equipment and the complimentary nature of logistics. NATO efforts to codify the lessons learned in logistics support of peacekeeping operations suggest strongly that inter-operability is an important issue and point to the need to, 'in general, plan for the worst-case scenario'.

• Fourthly, as military activities are increasingly supported by commercial organisations, or where traditional military logistic support has been outsourced to commercial firms, the concept of being joint is not merely about coordination between the three services. It has to include civil servants within the Ministry of Defence as well as those commercial organisations who have been contracted to provide the management of certain logistic activities that range from transport and storage to repair and maintenance.

Long range logistics itself has three broad challenge areas:

• Firstly, there are geographical challenges. Rapid deployment operations of the kind seen in Bosnia, Kosovo and much further afield create problems linked to the sheer distances involved and the strain created with the physical supply and distribution of goods. The Red Ball Express run after D-Day during the campaign in Northwest Europe and the air bridges to West Berlin during the crisis of 1949 and from the Continental USA to Saudi Arabia in 1990 – 1991 (Desert Shield/ Storm) bear testament to this issue.

- The second issue in long range logistics is that of communications. Distance strains the ability of command and control networks. The downgrading of this capability puts fresh challenges on those responsible for maintaining the coherence of the force structure in its theatre of operations.

- The third issue is that of time and the future. Long range logistics refers also to the planning horizon for new equipments, processes and doctrine. In the long term the means can change and the process by which transactions occur along the supply chain can be adapted as well as the doctrine of the service generally. These all have implications for the logistician.

The word logistics itself has many different definitions, depending upon the arena in which it is used. This case proposes the use of the following definition:

"Logistics can be defined as the strategic management of the flow of resources through the supply chain. It includes the management of the procurement of materials, production planning, delivery to customers and all the information flows necessary to the management of physical material flow."

In summary, joint integration and long range logistics are terms which can be viewed in several dimensions. The case has also suggested a definition of logistics which has the following key attributes:

- It is strategic in focus.
- It is a total process from acquisition to delivery to consumption or disposal.
- Finally, logistics is about management and the skills required in managing the interfaces between suppliers and customers.

The Challenges of the Logistics Environment

Although an obvious statement, it is important to remember that the environment for conducting logistical operations is fluid and always changing. Processes are being combined as defence resources shrink. Leaders are forced to think about and plan for change rather than waiting for events to demand new ideas. Processes themselves are constantly changing. In short, the central challenge of logistics is managing change. The challenges for the logistician are many. However the main themes can be characterised as follows:

- **Increasing Integration** – as processes integrate, the inevitable spectre of organisational politics provides challenges which can interfere with reaching an optimum result. Some barriers are cultural and have to be overcome. Others need to be recognised and harnessed within a new vision in order to achieve success.

- **Improved Service** – no matter what efficiencies can be made, the theme of providing better service to customers, whether it be within a business or to a company commander in Asia still holds as being critical. Service leads to trust and greater bonds between the supplier and customer. In the military context, it could be suggested that this glue is crucial to maintaining operational effectiveness.

- **Effective Use of Information Technology** – logistics support tends to fall to the bottom of the list for receiving new technology. It is essential that the benefits which commercial distribution is reaping from the rapid integration of IT are transferred to the defence establishment. Campaigns such as the Falklands and the Gulf showed the UK Armed Forces the value of knowing what is where and when it is arriving. However, this goal has to be tempered by the need for effective IT, which will not be jammed in a high electronic warfare environment.

- **Time Compression** – as in so many activities, time is becoming increasingly crucial. Communication times are falling, the tempo of operations is becoming ever rapid with the advent of good quality night vision capabilities. This means for the logistician, the need to deliver more, on an 'around-the-clock' basis in an appropriate, timely manner.

The Central Elements of Integration and Long Range Logistics

The seven elements of integration which are central to a consideration of the concept could be suggested as comprising:

- **Specification.** The integration of specifications are central to achieving long term effectiveness in any supply chain between parts of a commercial organisation or the armed services. The NATO standardisation agency move to the WEU is an important example of international recognition of the problem. In the domestic context, differences between such apparently simple items as boots can lead to gross inefficiency in the varying types beyond those necessary supplied to the three services.

- **Procurement.** The Smart Procurement Initiative contained within the 1998 Strategic Defence Review seemed to aim for a lowering of the time and cost of procuring equipment. Whilst there are many high profile examples of international integration such as the Eurofighter Typhoon project, those in the domestic sphere are harder to track down. For example, the Royal Air Force and Fleet Air Arm operated Harrier VSTOL aircraft of different types in order to compensate for the harsher maritime environment faced by the Royal Navy. However, an integrated procurement of these aircraft between the services could have led to a larger available force of aircraft able to operate on land or sea. One imagines that this will be rectified in the Joint Strike Fighter programme currently underway with the US Armed Forces.

- **Supplier Management.** Integrated supplier management offers benefits in terms of being able to focus efforts on managing relationships with fewer suppliers in depth. In a Public Sector context however, this has to be balanced with the requirement to provide competitive provision of services as well as accountability. It remains to be seen what the ramifications of the current activity in the UK Defence Industrial Base will create for supplier management practice within the MoD.

- **Inventory Management & Transportation.** The 1998 Strategic Defence Review appointment of a CDL at the head of the DLO, followed by its integration with the DPA to form DE&S seems to be a signpost to the integration of service inventory management and transportation practices. It remains to be seen whether the integration of support will extend on a defence-wide basis to the frontline or else whether there shall still be some form of single service support at the cutting edge. Similarly, transportation mechanisms are amenable to consolidation and more importantly standardisation for the benefit of all three forces. This is especially true in the context of the maritime strategy currently emerging.

- **Communications.** The UK Armed Forces demonstrated considerable foresight in their historical post-war efforts to develop a standard communications system throughout the services. The introduction of Bowman has continued the practice of defence-wide communications. The trick however is to ensure that Information Technology provision will not fall into the trap of being tailored to single service standards creating blockages in the peacetime flow of information along the supply chain that can present problems in an operational context.

- **Managing Cultural Change.** It could be suggested that designing an optimal supply chain is relatively easy compared to making the cultural changes necessary on an iterative basis to ensure delivery of the objectives. Like any organisation, the armed services possess distinct identities and a heritage which resists active attempts to try to dismember aspects of it. The challenge for an integrated supply chain is to inculcate values, which are complimentary with the cultural paradigm of the services. This enables change to occur which is harder to resist constructively and ultimately complimentary to the single service ethos whilst delivering defence-wide logistics support.

Turning to long range logistics, what can be identified as being its key elements?

- **Time Response.** Political crises requiring a military response often have a short timescale to which the defence establishment must react. Pre-positioning of certain equipment, long range heavy lift aircraft and garrisons around the world present ways of meeting this requirement at varying costs. These costs can be the sheer economic opportunity cost for other programmes, political cost in terms of garrisoning arrangements or even environmental costs for using areas to station forces which present a threat to delicate ecosystems.

- **Flexibility.** Enabling the supply chain to react to changing operational needs is an emerging trend within the defence logistics debate. For the UK Armed Forces, the period between the Second World War and the end of the Cold War enabled clear planning for supporting their effort in Central Europe. However this led to stagnation in the thinking about the artistry of meeting defence needs in a flexible manner.

- **Agility.** An agile strategy requires agile logistics. Manoeuvre warfare and the employment of increasing numbers of helicopter-borne forces require logistics to be agile tactically to evade such efforts by the opposition but also strategically in terms of meeting needs in far-flung operational theatres.

- **Technology.** The general rate of technological progress offers new opportunities for long range logistics. The capacity of aircraft and their ability to reach isolated areas, improvements in vehicle technology to enable better delivery of complex goods enable innovative methods of meeting the logistician's challenge.

- **Information Technology.** IT deserves a separate category in terms of long range logistics. Computer aided inventory management and distribution are virtually dependent on good IT systems which are reliable, accurate, effective but most important user-friendly and intuitive.

- **Management Processes, Planning and Budgeting.** A timeless element of logistics delivery at range is management skills. Being able to manage the distribution challenge and the relationships involved is critical. Being able to plan for requirements or adverse scenarios requires creativity coupled with credibility to articulate these issues. Finally, budgeting skills are crucial. There is little point having the most effective supply delivery system if no supplies can subsequently be afforded.

- **Supplier Relationships, Coping with Demand Surge.** Some would suggest that defence logistics is different. Whereas the commercial distribution sector is at war constantly, demand in defence has a stepped level of difficulty from peace to wartime. Demand surge in operations is a problem to simulate regularly enough not to de-motivate logisticians and wear out their capability to enact the requirement for real. Long periods between testing leads to lessons being forgotten and then having to be re-learned when new operations start. Clearly personnel management and training techniques play a central role in this area.

The following four figures seek to offer the reader a debating point for considering integration and long range logistics using four types of product in a pictorial form. Figure 1 seeks to show the types of defence product in terms of the risk associated with procuring and supporting them balanced by their effectiveness. For example, a hammer is an example of a low risk, low effectiveness product which the supply chain has to handle. A multi-role combat aircraft on the other hand, represents a high risk (in technological terms), highly effective item to handle and maintain.

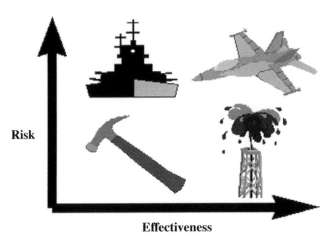

Effectiveness

Figure 1: Product types which the supply chain needs to handle.

Figure 2 builds upon this limited typology offering a broad characterisation of the problems associated with the type of good being handled. A battleship for example would take time to procure and thus could be characterised as creating a bottleneck in the system in terms of its effect upon other procurement funds and also its physically taking up a slipway as it is constructed upon which no other vessel could be built. The hammer alternatively (low risk, low effectiveness) represents a routine item being handled in the supply chain.

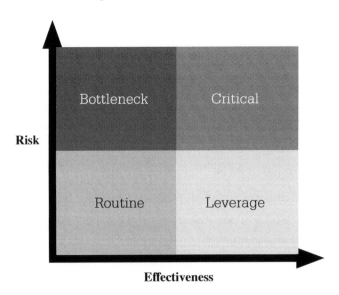

Effectiveness

Figure 2: Characterisation of key product types in the supply chain.

Figure 3 suggests the key challenges associated with each type of good. To ensure supply chain effectiveness of critical items such as Eurofighter for example, the formation of strategic partnerships offer a good way to guarantee its supply.

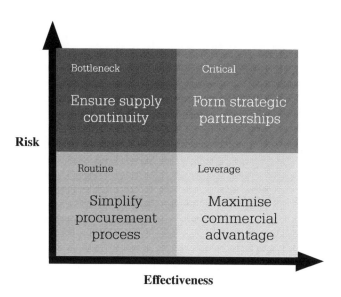

Figure 3: Strategies for coping with the challenges.

Figure 4 (overleaf) suggests tactics for implementing the goals suggested in Figure 3 to achieve effective management of key types of good within the defence supply chain.

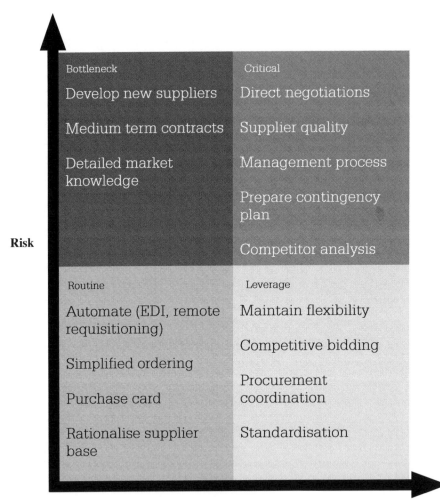

Figure 4: Tactics for operationalising supply chain strategy goals.

Conclusion

Broadly speaking there are five main conclusions to come from this analysis as to the central requirements for long range logistics and joint integration. These require consideration of the processes, maintenance and generation of effective relationships, continuous improvement efforts to consolidate and build upon the present excellence in supply chain management, the ability to model and speculate as to the nature of the future environment, and finally to change the mind set in a manner coherent with the nature of the services, but which facilitates effective logistics support on a defence-wide basis.

The author would like to thank Jonathan Davies for his help in putting this paper together.

Case 3.6 // Humanitarian Logistics

Swords, Ploughshares and Supply Chains: NGO and Military Integration in Disaster Relief Operations

David M. Moore and Peter D. Antill
Centre for Defence Acquisition, Cranfield University, Defence Academy of the UK.

Introduction

This paper will examine the background to and context of, humanitarian aid operations and looks at the interaction of the various organisations, particularly the aid agencies (Non-Governmental Organisations – NGOs) and UK military forces, as well as the supply chains used by both. It considers the use of military forces in deployed operations generally then examines humanitarian aid operations and the potential for co-ordination and co-operation between the armed forces and the other international players that take part in these scenarios. Finally it considers the challenges that will need to be faced and overcome in order to improve the provision of humanitarian aid in the most timely and effective manner.

Background and Context

In 'winning' the Cold War the Western Alliance (including the United Kingdom) has faced pressures to divert scarce resources away from defence spending back into other areas of public expenditure, such as education and health. In the case of the UK, this has meant a cut, in real terms, in the defence budget and downsizing of the armed forces. All this has occurred at a time of renewed interest in 'Out-of-Area' operations. The concentration on their role within Europe in deterring the Warsaw Pact has meant that the UK Armed Forces have had to rapidly relearn what it is to have 'strategic mobility', particularly in relation to moving ground forces, as was shown in the Gulf crisis of 1990 – 91.

What was apparent was that after two decades of concentrating on the role of supporting in-place forces to meet a possible Warsaw Pact onslaught, this aspect of modern warfare had become alien to British defence policy and planning. A small capacity for acting out-of-area had been retained (as shown by the Falklands War) but the capacity for moving

315

ground forces was strictly limited.[1] It was also shown that the logistic support afforded to the UK forces in Germany had not been particularly good in that resources had tended to be concentrated on the 'shop-window' of combat capability in order to provide maximum deterrent value. Such a system was very quickly shown to be a false economy when it came to actually using those forces in the Gulf.[2]

The *Strategic Defence Review* recognised that crises affecting British security are likely to happen anywhere in the world, given the wider definition of what British security entails (political and economic stability for example), being one of the permanent five members of the United Nations Security Council and responsibilities under international treaties.[3] Being involved with such 'wider' security interests is not a new phenomenon to the UK Armed Forces as up until the early 1970s, British troops were constantly being involved in missions that are now termed 'Operations Other Than War' or 'Peace Support Operations' such as disaster relief, handling refugees, riot control, counter terrorism and conflict prevention, as well as low and medium-intensity operations such as coalition warfare, counter insurgency and amphibious warfare.

Almost all of these operations were conducted outside Europe in Africa, the Middle East, the Gulf and Asia and thus needed an element of strategic mobility and a capacity to act out-of-area with the dispatch of British forces from the Home Base or a major regional base such as Aden, Hong Kong, Cyprus or Singapore. The end of the Cold War has meant a return to a more traditional role for the UK Armed Forces.[4] In fact, the two world wars fought in the first half of the Twentieth Century proved to be an exception for the British Armed Forces.

Up until then, they had been concerned with worldwide commitments and served outside Europe as much as they had served inside it, had been fighting low intensity conflicts against poorly armed, organised and trained opposition, had been an all-regular force and had been comparatively small. In the years after the end of the Second World War, the armed forces of the United Kingdom at first gradually reverted to the old role of policing the Empire, but were increasingly drawn back into Europe as the withdrawal from Empire proceeded.

The United Nations, Conflict and Humanitarian Aid

The first forty years or so of the United Nation's history saw the organisation sponsor some thirteen Peace Support Operations. From 1988 to 1992 it sponsored another thirteen[5] and up until 2000 sponsored a total of forty, many of which are still ongoing.[6] As the UK Armed Forces were gradually concentrating on their role within NATO as they withdrew from the Empire, these sorts of operations gradually received less and less attention. It was not until 1988 that the Army produced guidance on peacekeeping duties.[7]

The UK Armed Forces have a long history of conducting these sorts of operations in both the colonial and post-colonial eras. The new post-Cold War environment emphasises Peace Support Operations and places new pressures upon commanders (such as strict rules of engagement). These operations are often multi-national in character, involve a wide range of civilian agencies and have necessitated a rapid development of doctrine. There are still no 'standing force' arrangements within the United Nations, which to some is still the most important body for dealing with complex humanitarian emergencies.[8] The proliferation of conflict since the end of the Second World War and particularly the end of the Cold War in many parts of the developing world[9] has led to a new thinking as regards the provision of assistance in aid operations. The end of the Cold War has led to the end of the ideological conflict between Russia and the United States and therefore the withdrawal of much of the political support as well as military and economic aid that was given out to a number of 'favoured' countries, organisations or groups. There has been a massive increase in the movement of peoples due to the ever-widening gulf in national wealth between the developed and developing nations, which is likely to engender conflict.[10] The emergencies that now develop are usually characterised by a large media presence (the 'CNN' factor), which increases pressure on national decision makers to respond, a large international aid effort and in many cases, a large military presence to keep the peace.[11] This sort of effort is usually targeted at the immediate saving of lives, requiring massive material support and the employment of considerable logistic assets.

The Supply Chain in Humanitarian Aid Operations

Humanitarian relief supply chains can be seen as a systems exercise, involving the integration and co-ordination of widely scattered groups of specialists.[12] There are many different types of logistic programmes, and activities that have to be planned and implemented around the specific catastrophe that has occurred, but forward planning can be initiated through the use of an accepted template for disaster planning (see overleaf):[13]

Disaster Impact

React to Disaster Assessors' requirements and prioritise equipment movement.

Orchestrate equipment control measures.

Arrange shipment and distribution contracts.

Preparedness

Setting strategic stock items and holding required levels. Continual review of logistics strategy.

Response

Equipment receipt, packaging, labelling, issue and distribution.

Recovery

Equipment support

Mitigation

Developments in equipment specification and procurement.

Review stockholdings and locations in view of risk assessments.

Prevention

Shipment in support of preventative measures programme.

Development

Assistance with disaster area training programmes.

Disaster Management Cycle

Most agencies have specific structures to react to new emergencies, after which specifically constructed teams are assembled to take over the work. Permanent staff within each aid organisation maintain the strategic management of the aid programmes underway, actively raise funds, assess potential black spots, and manage the stocks and supply chain to meet programme requirements. Most organisations keep strategic stock at locations throughout the globe, ensuring that aid is available to an area whose infrastructure may be unable to meet the demand. This however, requires a certain level of co-operation between donors as in many cases, aid organisation logistical structures are still in their infancy.

In order to be effective, a supply chain in a humanitarian aid operation should be 'owned', that is, responsibility taken by one of the players in the scenario.[14] Such a concept would be in line with the commercial practice in, and academic theory of, supply chain management. The lack of ownership of the supply chain stems from the complexities and difficulties inherent in such operations, such as relationship issues, that impair the smooth operation of the cycle.

The aims and objectives of individual agencies are not always conducive to an integrated and co-ordinated effort. Objectives that have become highly politicised at the strategic level can impair the benefits that can be gained from a concerted and co-operative effort among the various players at the operational level. Logistic activities have, up until recently, been undertaken in a fragmented and sub-optimised manner and based upon outdated logistics philosophies.

It is possible to identify a generic supply chain that applies in many of the humanitarian aid scenarios. Such a supply chain is usually designed to allow a one-way flow of goods and equipment into the theatre of operations to where it is needed the most:[15]

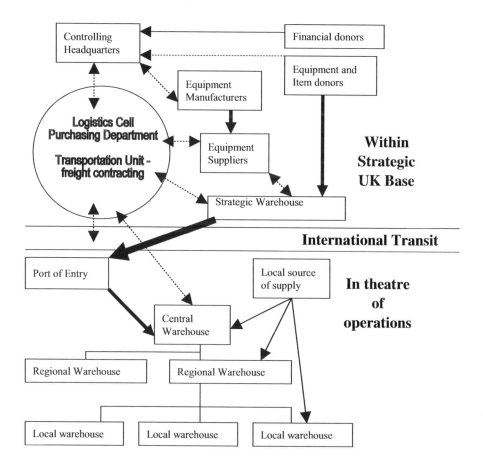

The 'players' in such humanitarian aid scenarios all have differing management styles and administrative structures and whilst the supply chain appears straightforward, the complexities in the relationships that occur as well as the impact of having different structures and procedures may conspire against the establishment of effective supply chain strategies. This is exacerbated by aid programmes often being seen as development opportunities with the contribution that supply chain management can make being overlooked or underrated. Commercial best practice and academic theory have not been widely accepted leading to an emphasis within the supply chain that is transactional rather than relational.

There are a number of additional reasons as to why the aid agencies have been slow to adopt supply chain management strategies. Firstly, relief supply chains are designed to deliver (in the main) a one-way flow of goods, equipment and material into the theatre of operations. The basis of aid operations is donation, with the majority of equipment remaining in, and becoming the property of the country in which the disaster occurred.[16]

This simple one-way chain is complicated by the internal relationships that exist. The procurement and logistics functions seem to be undervalued by the organisations concerned and the system as a whole, although there are of course exceptions. Even here, many involved in the logistics side of operations are dissatisfied at not being involved in the original assessment phases of an emergency (although this has started to change in some quarters). Reasons for this vary, possibilities being the lack of understanding in the remainder of the organisation (purchasing being seen as an everyday transactional activity), the fact that ordinary administrative staff used to carry out logistics and procurement activities (giving the impression that anyone could do it), logistics is the last link in the internal supply chain and therefore receive the blame if things are delayed as they are mainly judged solely on delivery times and that logistics is not considered a core function in many agencies.[17] This traditional and reactive approach to procurement and logistics hampers enhancement in performance.

This is exacerbated by humanitarian aid organisations suffering from regionalisation and parochialism, and while many are attempting to integrate their procurement and logistics functions, they have met with varying degrees of success.[18]

This could be overcome through planned training and education programmes. However, opportunities seem to be very limited for procurement and logistics personnel in aid organisations.[19] Training courses seem to be aimed at ensuring personnel learn the internal procedures so that they can fit into the organisation's operations as quickly as possible.[20] There is also a noticeable impact from the traditional and reactive approach to supply chain management where the logistics and procurement functions were separate. This resulted in different people (with different backgrounds) being recruited for the different streams.[21]

In this respect, integrated logistic systems are essential for performance improvement as the existing administrative systems are not designed for the 'holistic' nature of supply chain management. Professionalism will bring a demand for better management and control systems. In many of the aid agencies, even the UN, where such systems exist, their use is fairly arbitrary and suffers from poor monitoring, data input discipline and understanding.[22]

It has also been identified that there is a general concern about the way donors distribute their aid, the reactive nature of funding, the unwillingness to fund managerial overheads, the threat to an organisation's freedom of action and the increased pressure for accountability.[23]

This is exacerbated by difficulties within the area of UN/NGO co-ordination which are well documented,[24] with issues such as co-ordination verses control, competition, publicity, conflicts of interest and funding predominant. There have been however, instances of successful co-ordination through logistics and the supply chain. For example, recent operations in Ethiopia have had NGOs and UN agencies using World Food Programme (WFP) aircraft to move goods and equipment around. Nevertheless, Agencies are often reluctant to be co-ordinated but there is potential for UNHCR to track and co-ordinate the shipment of goods.[25]

NGOs constitute the main *interface* between the *relief* system and the *beneficiaries* and are almost at the end of the supply chain. The position they hold as the implementers of aid relief can conflict with their stand as advocates for such relief.[26]

Finally, in general terms, the attitude of relief organisations to the commercial sector is one of distrust and suspicion of the profit motive, with commercial transactions undertaken only reluctantly. Commercial firms may see humanitarian aid relief as a chance to make maximum returns due to the lack of commercial awareness and experience on the part of the relief agencies. Attitudes are however changing slowly with the realisation that the commercial sector has a role to play in future operations and can be brought into aid operations in innovative ways. Moreover, the movement of personnel between the relief organisations as contracts expire may be a barrier to the entry of commercial contractors. This could be eroded if professional logistics and procurement specialists enter the field.[27]

These issues highlight the complexities that must be overcome in order that the supply chain for humanitarian aid may be optimised. However, to this must be added the involvement of the military.

The Military Involvement in Humanitarian Aid Operations

The provision of aid by military forces is not a new phenomenon. Such instances have occurred since before the time of Alexander the Great and have continued through the

Napoleonic Wars, the World Wars of the Twentieth Century (particularly the Second World War and the devising of the Marshall Plan) up until the present day, including the Berlin Airlift (1948-9), the Congo, Bangladesh, Ethiopia, Sudan, Iraq, the former Yugoslavia, Rwanda and Mozambique.[28] It seems that when disasters, either natural or man-made occur, governments often turn to the military for help as the military have certain resources immediately to hand, such as food, medicine and fuel as well as transport and human assets with which to distribute them.[29]

With the end of the Cold War, the UK Government has become more involved in the provision of humanitarian aid, not only through the funding of UN Agencies and Non-Governmental Organisations (NGOs) but also with the tasking of the armed forces with a greater responsibility for undertaking such missions outside what was the traditional area for possible NATO operations. This was formally adopted under the Strategic Defence Review[30] where the Military Tasks assigned to UK armed forces were altered to reflect the changing defence environment, the armed forces' roles within that and the new foreign policy focus.

With respect to such operations, it was stated that "... in a less stable world, we have seen more operations of this type ... Britain will play its full part in such international efforts. At one end of the spectrum this might involve logistic or medical support to a disaster relief operation. At the other, it might involve major combat operations ... our forces have developed particular experience and expertise in operations of this kind".[31] Such an outlook has been continued in more recent Government documents in that the UK has "a responsibility to act as a force for good in the world"[32] and that "we can nevertheless expect continuing pressures to contribute to peace support and humanitarian operations.[33] The involvement of the UK Armed Forces is however constrained by the contraction of the defence budget, the downsizing of the armed forces, Britain's economic position and the expansion of UK commitments that has taken place since the end of the Cold War[34] and resources that are used to fund the use of military forces in humanitarian aid operations cannot be used in other areas.

In operations such as these, there needs to be a distinction between humanitarian aid (man-made) and disaster relief (natural) operations. The difference usually lies in the degree of preparedness and response time involved. Humanitarian crises (such as those in Kosovo, the former Yugoslavia, East Timor and Rwanda) rarely happen at a moments notice or overnight, and are usually monitored by the aid agencies in an attempt to give themselves time to prepare and alert the remainder of the international community if a catastrophe is about to happen. Natural disasters (such as in Mozambique, Ethiopia, Bangladesh and Turkey), while slowly becoming more predictable, can still strike with little warning and rely more on the training, education and preparedness of those in the actual disaster zone to hang on until the relevant concerned organisations and agencies can mobilise their resources and come to the rescue. Whilst this inevitably takes time, military forces are seen as a pool of prepared, disciplined and available source of assistance while the international aid community gears itself for action.[35]

However, the involvement of the military in such operations is not without challenges. A balance must be sought in allowing the civilian aid agencies a free hand in utilising the available military resources, whilst being aware that military manpower is trained to fight and engage in combat operations. There is quite a wide cultural difference between the civilian aid worker and the soldier.[36]

The Armed Forces obviously need to be involved in conflict prevention, outreach programmes, training missions, defence diplomacy, humanitarian emergencies and disaster relief operations as these all contribute to a more stable security environment. They also show that the Armed Forces can be employed in worthwhile tasks while not training for, or involved in, major combat operations and that the defence budget is being spent as a tangible force for good. There is however a challenge to prepare and train a soldier to participate in high intensity conventional operations on the one hand, while preparing and training them to participate in peace support, humanitarian aid and disaster relief operations, where the measure of success will be the numbers of lives saved as opposed to taking objectives and eliminating the enemy as a fighting force.[37]

Further, a large number of players often operate within these scenarios either from the civilian side (NGOs or UN Agencies) or the military (particularly if there has been a multi-national force put together, as in East Timor, the former Yugoslavia and Rwanda). The various players have different perspectives, agendas and Standard Operating Procedures (SOPs) and it is difficult to get them all to operate smoothly to ensure the effective management of the operation, as the Americans have found out "their successful conduct requires the US military to work with a wide variety of institutions and organisations"[38] to reduce what is called *friction*[39] in military circles but has also been called *tempo drag*.[40] This is the factor that is most readily seen in the natural time lag that occurs between a commander giving an order and the order actually being carried out. In some extreme cases it can obscure the original intention of the order and there are many factors that conspire to increase this delay.[41]

The military aid provided by the UN, international organisations or individual states can in certain circumstances, be seen as a problem by the aid agencies. The military can be seen as a challenge to the aid agencies' independence with the resulting increase in threat from the belligerents, as for example, in Operation Provide Comfort in Somalia. This is because an important factor in the delivery of aid by the agencies is their perceived impartiality and the consequent reduction in the importance of the aid in terms of conducting military operations. Association with military forces deployed under UN auspices increases the risk that the agencies will be become identified with one of the parties, even after the conflict has stabilised.[42]

The military forces that are sent are not always equal to the task of providing both logistic support and protecting themselves and the aid workers, being lightly armed and politically constrained, as was seen in the former Yugoslavia.[43] Also, not all military forces are trained to the same standard, employ the same technology, use the same procedures, speak the same language and, in addition, have different cultural backgrounds and therefore will have a different approach to and be more or less effective in, these sorts of operations.[44] Hence co-ordination becomes an increasingly difficult proposition.

Added to this, there are arguments that there is a conflict between the very nature of the respective missions and deployments of the aid agencies and the military, in that the provision of humanitarian aid, as practised by the aid agencies is strictly apolitical, while the deployment of military forces is usually in the pursuit of a political goal and if not, will fail to address the underlying political conditions that prompted the need for aid, in respect of man-made disasters.[45]

Thus, it is important that rules of engagement be drawn up before the start of an operation that allow the military to do what they have been sent to the theatre of operations to do – protect and assist the civilian aid agencies and the relief effort. They will reflect legal and political constraints however, but will always authorise the use of force in self-defence and should never inhibit a commander from taking any and all measures to protect his force.[46] Unfortunately, Operations Other Than War and Peace Support Operations (especially humanitarian aid operations) are often not taken very seriously by the military, which are reluctant to view it as proper soldiering.[47] This can be evidenced by the fact that many armed forces place minimal importance on it with a "difference between stated doctrine and an insufficient capability to match it".[48]

All of these issues are underpinned by the fact that the military and civilian aid organisations compete for similar funds (for example, the UK Armed Forces tend to be reimbursed by the Foreign Office or the Department for International Development) it should be incumbent on the military to ensure value for money in the services they offer. For example, aircraft are often used to fly in non-emergency aid when it could be undertaken much more cheaply by sealift, rail or road transport.[49]

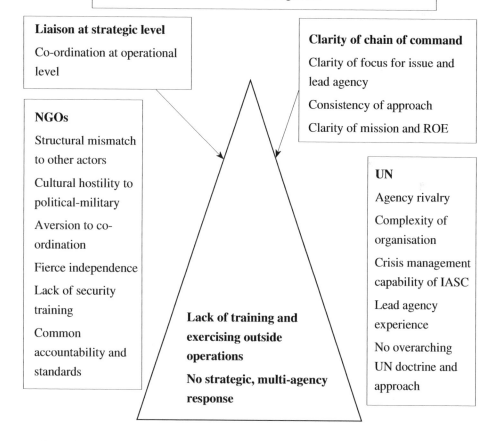

Military

Small Civil-Military Co-operation (CIMIC)/G5 capability

Lack of CIMIC/G5 focus in non-operational HQs

Lack of interagency doctrine

No CIMIC Standard Operating Procedures

Lack of Organisational Awareness/decision-making tools for Purchasing and Supply Officers

Cultural acceptance of CIMIC

Cultural dissonance with civilian agencies

Liaison at strategic level

Co-ordination at operational level

Clarity of chain of command

Clarity of focus for issue and lead agency

Consistency of approach

Clarity of mission and ROE

NGOs

Structural mismatch to other actors

Cultural hostility to political-military

Aversion to co-ordination

Fierce independence

Lack of security training

Common accountability and standards

UN

Agency rivalry

Complexity of organisation

Crisis management capability of IASC

Lead agency experience

No overarching UN doctrine and approach

Lack of training and exercising outside operations

No strategic, multi-agency response

From this it is possible to develop areas of 'mismatch' when seeking to integrate the various players in a holistic manner. These can be identified in the diagram above.[50] This is known as the 'integrating mismatch' and identifies areas of weakness and deficiencies within the system of operation, which are very relevant to the humanitarian supply chain. All these elements could be viewed as obstacles to the co-ordination of the relief supply chain. "There is insufficient direction at the strategic level and little co-operation in terms of training and exercise."[51] At its core lies the lack of a strategic multi-agency response as each organisation has its own strategic headquarters to activate the aid process. The initial focus for implementing a co-ordinated multi-agency response could be a combined base logistics centre. Solutions to logistic issues demand long-term relationships based on trust and mutual understanding enhanced by joint training and exercises. There is also room for the expansion of education, which could cover a wide range of topics including culture, law, the characteristics and organisation of NGOs and UN Agencies and the processes of in-theatre operational logistics.

Notwithstanding these challenges there are considerable potential benefits and opportunities that are presented by using the armed forces to provide, or assist in the provision of, logistic support to humanitarian aid and disaster relief operations. The armed forces can bring considerable expertise and professionalism in undertaking logistically challenging roles. They can bring a task focus that is inculcated in their modus operandi, which is rarely matched by civilian aid agencies. The essential logistic equipment, expensive for aid agencies is often readily available for the military; indeed it provides their reason d'être. Above all there is within the armed forces a ready pool of experienced logistics personnel who can act swiftly and effectively when tasked.

The opportunities can be highlighted further by a comparison of the similarities in the supply chain management processes established for military operational deployments with those for civilian humanitarian aid operations.

The Military Supply Chain for Expeditionary Operations

For much of the Cold War period, it has been generally accepted that the UK Armed Forces would have a role to play outside of the NATO area due to the special responsibilities that the UK had as the head of the Commonwealth, lingering great power sensibilities, residual imperial obligations, a member of the UN Security Council and the concern over the potential spread of communism. The UK's willingness and ability (although limited) to use the armed forces in operations far from the Home Base was shown on many occasions, such as Korea, Suez, Cyprus, Oman, Kuwait, Jordan, Malaya, Kenya, Aden, Belize, the Falklands and the Gulf. There are in fact similarities between the situation now and that which the UK faced in the early part of the Cold War[52] in that Britain has accepted that it has international obligations that may require the use of military forces overseas; the

British role overseas is not merely reactive, but should be proactive as well, featuring crisis prevention and deterrence; The capacity for independent action remains constrained (for both financial and force structure reasons); Moving military forces to a distant theatre of operations and sustaining them is a costly and demanding task; Overstretch is an ever-present problem and the cost of overseas operations remains high.

Whereas the problem of reacting to worldwide events was previously solved by a combination of garrisons (both local and regional) and a mobile strategic reserve,[53] that method is no longer available to the UK. The UK now faces the task of projecting its armed forces directly from the Home Base to almost anywhere in the world.[54] It is however vital for the MoD to recognise that there are constraints on the resources available to it and ensure that operations are supported adequately and utilise the best practises from the business sector to act as drivers in the way the forces are organised and operated.[55]

Future operational deployments will therefore be very different from the ones planned for under NATO, in that the armed forces will have to go to the operation, instead of the operation come to them, and operate in areas will have infrastructures that are nowhere near as well developed as Western Europe. In some ways, the generic model for the support of an expeditionary operation is similar to the conceptual framework that is used by the aid agencies and NGOs for the supply of humanitarian aid:

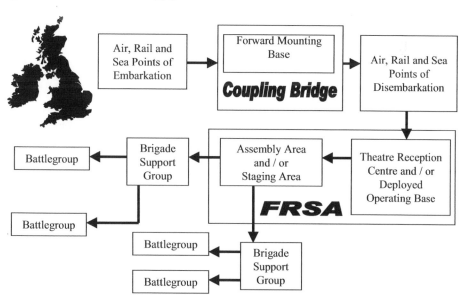

The operation starts when the force, its equipment and the matériel to support it are outloaded from the bases and depots in the Home Base, moved to the Air and Sea Ports of Embarkation (APOE or SPOE), loaded onto the transport that is awaiting them and moved through the Coupling Bridge to the Air and Sea Ports of Debarkation (APOD or SPOD). The area that will house the logistic support elements for the deployed force in the theatre of operations is known as the Force Rear Support Area (FRSA), the size of which can vary quite considerably depending where in the world the operation is, what the type of operation is and what forces have been sent to conduct the operation and will therefore contain a number of Assembly Areas, Staging Areas, Deployed Operating Bases and possibly a Theatre Reception Centre.[56]

While many aspects of the logistic support afforded to the UK Armed Forces retains elements that have been carried over from the Cold War,[57] it is starting to change in a number of important ways to accommodate the operations that are increasingly being undertaken in the post-Cold War world. One of the main changes is the greater acceptance for the role of contractors in supporting the UK Armed Forces on deployed operations far from the Home Base.[58] In itself, this can bring challenges in respect of legality, ethics, medical support and the skills and competencies of those involved, within an environment where tensions arise because of the differing operating basis of each of the 'players' (military forces – commercial organisations – aid agencies).

It can be seen from the outlines given above, the tasks and activities that are undertaken in the military supply chain are similar too, and could integrate with, the supply chain for the humanitarian aid agencies.

Conclusion

Many of the complexities that have been highlighted can be categorised as either internal or external to the relief effort. *Internal complexities* occur due to the differences that exist, not only between the aid agencies and the military forces, but between the agencies themselves and the military forces themselves. All the 'players' that become involved in these types of operations operate with their own agendas, standard operating procedures, have their own organisational cultures and history. It will be difficult to reconcile all the organisations' approaches with one another to produce and implement an effective supply chain management strategy.

Internal conditions have limited the aid agencies in their move to adopting supply chain management, a direction which the military are adopting and adapting due to the conditions of the post-Cold War defence environment. In-theatre, aid agencies may be concerned about working with the military, as despite the benefits that they can bring to an operation (a disciplined, well-equipped force that can protect itself and the aid workers and has its own logistic assets), there is always a danger that there may be a hidden political agenda at work that may become a threat to one of the aid agencies major strengths – their perceived neutrality. Such benefits come at a price though as the agencies and the military often compete for the same funds, which makes it a necessity that the military shows value for money in its involvement in aid operations. For the military, time and resources spent training, equipping and deploying on Peace Support Operations is time and resources that cannot be used for other aspects of their role in the defence of national security.

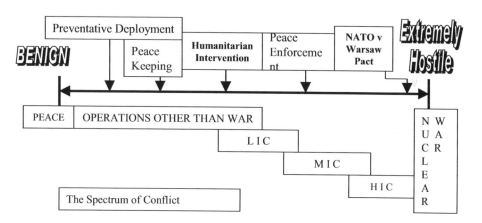

External complexities that can have an effect include the changing nature of warfare and the international environment which has accelerated since the end of the Cold War and the end of the ideological rivalry between the United States and Russia. The conflicts that have erupted in recent years are increasingly based on religion, ethnicity or race and are less amenable to Superpower control given the withdrawal of economic and military support from many areas of the world.

Such conflicts can also endanger aid operations in that unless a positive and binding political settlement has been instituted and a suitably equipped peacekeeping force established, there is a possibility that the fighting could erupt (the conflict effectively moves along the conflict scale from left to right as shown above) again endangering not only the peacekeeping force but the aid workers too.

There is also the impact of the media and the 'CNN' factor, which can increase the pressure on Governments to act in these sorts of situations, probably using the military as well as factors such as the weather and terrain that can affect operations. Arguably, some issues of the issues identified here can be overcome *relatively* easily through enhanced processes, improved systems, co-ordinated data sharing and communication. However, other issues will require fundamental reappraisal of working cultures and ethos. Although neither can be achieved without the goodwill of all concerned, the potential prize for the human race generally, is too great to ignore.

While there are obviously complications, challenges and definite problems in having civilian aid agencies and military forces working in tandem in an emergency, it is likely that "no matter what the depth of the paradoxes inherent in military assistance in humanitarian aid operations, the moral and political imperatives at work will be sufficiently strong as to ensure that military forces and NGOs engaged in humanitarian relief will need to operate, if not altogether, then in the same theatre of operations."[59]

Notes

[1] McInnes, Colin. **'Strategic Mobility: An Historical Perspective'** in *Royal United Services Institute Journal*, August 1998, pp. 30 – 34.

[2] House of Commons Defence Committee. *Implementation of the Lessons Learned from Operation Granby, HC43/928, Fifth Report*, Session 1993 – 4, HMSO, London, 25 May 1994.

[3] Op Cit. McInnes, 1998; Ministry of Defence. *The Strategic Defence Review and Supporting Essays*, Cm3999, The Stationary Office, London, July 1998; Ministry of Defence. *The Defence White Paper 1999*, Cm4446, The Stationary Office, London, December 1999.

[4] Op Cit. McInnes, 1998.

[5] Connaughton, R. M. *Peacekeeping and Military Intervention*, SCSI Occasional Paper No. 3, HMSO, London, 1992.

[6] Guthrie, Gen Sir Charles. **'British Defence – Chief of the Defence Staff's Lecture 2000'** in *Royal United Services Institue Journal*, Volume 146, Issue 1, February 2001, pp. 1 – 7.

[7] Army Field Manual. *Peacekeeping Operations*, Volume V, Pamphlet 1, 1988.

[8] Ramsbottom, O. & Woodhouse, T. *Humanitarian Intervention in Contemporary Conflict*, Polity Press, Cambridge, 1996.

[9] Croft, S. & Treacher, T. **'Aspects of Intervention in the South'** in Dorman, A. M. & Otte, T. G. *Military Intervention: From Gunboat Diplomacy to Humanitarian Intervention*, Dartmouth Publishing, Dartmouth, 1995.

[10] Snow, D. M. *Distant Thunder: Third World Conflict and the New International Order*, St Martin's Press, New York, 1993.

[11] James, A. **'Humanitarian Aid Operations and Peacekeeping'** in Belgrad, E. A. & Nachmias, N. *The Politics of International Humanitarian Aid Operations*, Praeger, Westport, USA, 1997.

[12.] Stevenson, R. S. *Disaster Management Training Programme*, DHA Logistics, DHA/94/2, GE.94-00020, 1994.

[13.] Carter, N. W. *Disaster Management: A Disaster Manager's Handbook*, Asian Development Bank, 1991, p. 52.

[14.] Moore, D. and Antill, P. **'Humanitarian Logistics: An Examination of and Military Involvement In, The Supply Chain for Disaster Relief Operations'** in *Global Logistics for the New Millennium*, Proceedings of the ISL 2000 Conference, July 2000, Iwate, Japan, pp. 51 – 57.

[15.] Hoff, A. P. *An Analysis of Disaster Relief and Humanitarian Supply Chains*, MSc Defence Logistics Management No. 1 Dissertation, Cranfield University, RMCS, Shrivenham, 1999.

[16.] Molinaro, P. *An Examination of the Changing Face of Emergency Humanitarian Relief and the Role of the Strategic Supply Chain*, MSc Defence Logistics Management No. 2 Dissertation, Cranfield University, RMCS, Shrivenham, 2000.

[17.] Ibid.

[18.] Ibid.

[19.] Craft, D. A. *An Analysis of the Current Military Training Received by the British Army to Assess its Suitability for Preparing Them to Undertake Peace Support and Humanitarian Operations*, Master of Defence Administration No. 13 Dissertation, Cranfield University, RMCS, Shrivenham, 1999.

[20.] Burke, M. C. *An Evaluation of the Potential for the British Army to Provide Skills Training to Humanitarian and Aid Agency Staff*, Master of Defence Administration No. 14 Dissertation, Cranfield University, RMCS, Shrivenham, 2000.

[21.] Op Cit. Molinaro, 2000.

[22.] Op Cit. Molinaro, 2000; Martin, J. D. V. *Developing an Integrated Approach to Strategic Evaluation and reporting: Benchmarking Best Practice from Military and Humanitarian Approaches*, MSc Defence Logistics Management No. 3 Dissertation, Cranfield University, RMCS, Shrivenham, 2001.

[23.] Op Cit. Molinaro, 2000; Thornton, M. E. *How Might the Interaction Between the Military and NGOs be Further Developed and Improved in Order to More Efficiently Deliver Humanitarian Aid*, DTC (MA) Military Studies No. 4 Dissertation, Cranfield University, RMCS, Shrivenham, 2001; Barton, N. I. M. *Logistic Support to Humanitarian Aid Operations: Logistic Solutions to Food Security*, MSc Defence Logistics Management No. 2 Dissertation, Cranfield University, RMCS, Shrivenham, 2000.

[24.] Op Cit. Molinaro, 2000; De Mello, S.V. **'The Evolution of UN Humanitarian Operations'** in Gordon, D. S. and Toase, F. H. (eds) *Aspects of Peacekeeping*, Frank Cass,

London, 2001; Op Cit. Moore and Antill, July 2000; Tomlinson, R. *Reversing the Downward Spiral: Exploring Cultural Dissonance Between the Military and NGOs on Humanitarian Operations*, MSc Defence Logistics Management No. 2 Dissertation, Cranfield University, RMCS, Shrivenham, 2000; Hawley, S. A. *Military NGO Co-ordination on Humanitarian Aid Operations*, MA Military Studies No. 6 Dissertation, Cranfield University, RMCS, Shrivenham, 2000.

[25.] Op Cit. Molinaro, 2000.

[26.] Ibid.

[27.] Ibid.

[28.] Doel, M. T. **'Military Assistance in Humanitarian Aid Operations: Impossible Paradox or Inevitable Development?'** in *Royal United Services Institute Journal*, October 1995, pp. 26 – 32.

[29.] Weiss, T. G. & Campbell, K. M. *'Military Humanitarianism'* in **Survival**, Sept/Oct 1991, Vol. XXXIII, No. 5, pp. 451 – 465.

[30.] Op Cit. Ministry of Defence, 1998.

[31.] Ibid.

[32.] Op Cit. Ministry of Defence, 1999.

[33.] Ministry of Defence. (2000) *Defence Policy 2001*, Ministry of Defence Website, URL: <http://www.mod.uk/index.php3?page=2445>

[34.] Dorman, A. M. **'Western Europe and Military Intervention'** in Dorman, A. M. & Otte, T.G. *Military Intervention: From Gunboat Diplomacy to Humanitarian Intervention*, Dartmouth Publishing, Dartmouth, 1995.

[35.] Op Cit. Hoff, 1999.

[36.] Op Cit. Hoff, 1999; Whitman, J. **'Those Who Have the Power to Hurt but Would Do None: The Military and the Humanitarian'** in Gordon, D. S. and Toase, F. H. (eds) *Aspects of Peacekeeping*, Frank Cass, London, 2001.

[37.] Op Cit. Guthrie, 2001.

[38.] Alberts, D. S. & Hayes, R. *Command Arrangements for Peace Operations*, 2nd Edition, NDU Press, Washington, 1996.

[39.] Van Creveld, M. *Supplying War: Logistics from Wallenstein to Patton*, Cambridge University Press, Cambridge, 1977 (1995 reprint); Clausewitz, Carl von. *On War*, Penguin Classics Edition, reprinted 1982 (originally 1832), Penguin Books, London.

[40.] Kiszely, Major General J. **'Achieving High Tempo – New Challenges'** in *Royal United Services Institute Journal*, December 1999, pp. 47 – 53.

[41.] Patrick, Stephen B. *NATO DIVISION COMMANDER: Command and Control in the Modern Battlefield Environment*, Simulations Publications Inc, New York, 1979.

[42.] Op Cit. Doel, 1995.

[43.] Ibid.

[44.] Op Cit. Tomlinson, 2000.

[45.] Op Cit. Doel, 1995.

[46.] Duncan, A. J. *Is the Use of the Military in Complex Humanitarian Aid Operations a Political Quick Fix or can it be the Cornerstone that Leads to Long Term Solutions?*, MA Military Studies No. 4 Dissertation, Cranfield University, RMCS, Shrivenham, 1998.

[47.] Ibid.

[48.] Nobbs, C. **'G5/Civil Affairs: A Short term Fix or a Long term Necessity?'** in *British Army Review*, No. 115, April 1997, pp. 53 – 56.

[49.] Op Cit. Duncan, 1998.

[50.] Skeats, Major S. R. *Operating in a Complex Environment: How can the British Military Improve Interagency Cooperation in Peace Support Operations?*, MDA No. 12 Dissertation, Cranfield University, RMCS Shrivenham, 1998.

[51.] Ibid.

[52.] Op Cit. McInnes, 1998.

[53.] Ibid.

[54.] Cross, Brig T. **'Logistic Support for UK Expeditionary Operations'** in *Royal United Services Institute Journal*, February 2000, pp. 71 – 75.

[55.] Ibid.

[56.] Ibid.

[57.] Moore, D. and Antill, P. **'Where Do We Go From Here? Past, Present and Future Logistics of the British Army'** in *British Army Review*, Autumn 2000, pp. 64 – 72.

[58.] Op Cit. Cross, 2000; Smart, P. **'Support to the Front Line'** in *Royal United Services Institute Journal*, February 2000, pp. 67 – 70; Moore, D. and Antill, P. **'British Army Logistics and Contractors on the Battlefield'** in *Royal United Services Institute Journal*, October 2000, pp. 46 – 52; Reeve, Major D. **'Contractors in British Logistics Support'** in *Army Logistician*, May – June 2001, pp. 10 – 13.

[59.] Op Cit. Doel, 1995.

Case 3.7 // Expeditionary Operations

Key Issues Affecting the Provision of Logistic Support to the UK Armed Forces in Expeditionary Operations

Christianne Tipping

Head of the Defence Leadership and Management Programme, RUSI.

Introduction

IBM commissioned RUSI to produce a paper exploring the issues which govern the provision of logistic support to the UK's armed forces in expeditionary operations in the short to medium term. The paper seeks to answer the following research question:

What, in broad terms, are the issues which govern effective logistics support to the UK's armed forces in expeditionary operations and to what extent will present policy and practice address these in the period to 2013?

The methodology adopted was a review of open source documentation, complemented by a series of discussions with subject matter experts in both the Ministry of Defence (MoD) and the defence industry. A large quantity of information was gathered and it has not been possible to cover all of the issues that emerged. Instead, the paper focuses on the themes which arose repeatedly in discussions. While much of the paper is grounded in fact derived from official documentation, the methodology used means that the content has also been informed by the personal opinions of subject matter experts. As such, the paper cannot be considered to be absolutely objective although the author has sought to be even-handed with respect to the representation of, and the weight given to, the perspectives of all contributors.

Context and Nature of Expeditionary Logistics

Achieving success in the Cold War scenarios, for which the UK armed forces were configured for almost fifty years, required a logistics support system that could deliver mass rather than velocity; stockpiles of matériel were prepositioned close to the likely area of operations and there was little need for a fast, responsive supply chain. After the end of the Cold War, the Government sought to realise a peace dividend and put pressure on the MoD to become more efficient and reduce spending. Options for Change in 1990 saw significant restructuring and downsizing across the armed forces and the 1991 PROSPECT study gave rise to changes in the size and structure of the MoD Headquarters.

The 'New Management Strategy' introduced in the early 1990s sought to improve financial accountability by aligning management responsibilities with delegated budgetary authority, and switching the focus from the cost of consumption (inputs) to the cost of activity (outputs). 'Front Line First: The Defence Cost Study', a programme of cuts introduced in 1994, brought further downsizing, base closures and restructuring; it also advocated contractorisation as a means of driving out cost inefficiencies with economies of scale being achievable where support services could be provided on a tri-service basis. The Strategic Defence Review of 1998 laid out the defence policy of the new Labour Government stipulating what the armed forces would be expected to do and outlined new force structures, including the formation of the Defence Logistics Organisation; though originally not supposed to be cost-driven, savings were nevertheless demanded. The 2004 'Delivering Security in a Changing World Future Capabilities Paper' brought further reductions in manpower and additional restructuring across the services.

Having significantly decreased the UK's military mass from 1990 onwards, successive governments proceeded to adopt a more interventionist approach to foreign policy, and the armed forces were required to gear for expeditionary operations. It was assumed that being smaller and having less mass to move, the military would simply cope with the requirement for a faster, more responsive logistics system. Operation Telic in 2003 proved that this was not the case. The mounting and sustainment of expeditionary operations create significant logistical challenges as they require men and matériel to be transported strategic distances, to destinations where there may be little in the way of infrastructure, with varying demands for an often unknown duration.

The so-called 'Four Ds' vary with the location, scale, intensity and maturity of each operation and the logistics support system has to be robust and flexible enough to meet these varying requirements. The longer the decision to commit to an operation is delayed, the greater the adverse impact on the logistics build up. Sometimes there may be good reason for delay – for example, the need to protect operational security – but often the delay stems from political inertia.

Though work was underway to address logistics inefficiencies before 2003, Operation Telic highlighted that the armed forces had failed to make the necessary changes to logistics processes to be able to meet the demands of a large-scale operation with short deployment timelines. The National Audit Office report on Operation Telic drew attention to specific areas of shortfall in the UK's logistics performance relating to: stock shortages; inadequate asset tracking; inadequate logistics communications; failure to deliver priority items; and inadequate control over the coupling bridge (CB) between the UK and the deployed operating bases (DOBs). To address these issues and generate coherence across existing projects, the Defence Logistics Transformation Programme (DLTP) was launched in 2004 with the aim of increasing the effectiveness, efficiency and flexibility of logistics support across the whole of the MoD. However, with the customary departmental demand for savings, it seems inevitable that the DLTP was initially more focused on efficiency rather than effectiveness. The transformation – with a renewed focus on effectiveness brought by the current Chief of Defence Matériel (CDM) – continues today in the guise of the Defence Logistics Programme (DLP) launched in 2006 and within which the Defence Logistics Strategy is enshrined.

The DLP is built around five themes: Comprehensive Capability Planning; Flexible Command and Control (C2); Minimised Demand on Logistics; Optimised Support Network; and Unifying Logistics Ethos. Each theme has associated strategic objectives and clearly defined future states. There is a comprehensive programme of activity in train which reaches out to 2020 with key milestones to be achieved along the way. A great deal of effort has been put into improving logistics support and significant improvements have been made since Operation Telic. CDM, as the Logistics Process Owner, is responsible for providing coherence across the logistics system and it is his intent that all levels of operational command have confidence that the right support will be delivered when needed. This means that logistics must reliably deliver matériel, services and information to enable commanders to make better decisions and have greater freedom of action.

The on-going transformation of logistics is taking place against a backdrop of two significant operations and substantial budgetary pressures. While a tight budget provides a strong incentive to drive out inefficiencies, it also makes it more likely that pressing operational needs will be met at the expense of delays or cuts to future programmes that contribute to the longer term aim of achieving greater logistics effectiveness. Many members of the logistics community expressed concern about the impact that this year's Planning Round (PR08) decisions may have. Furthermore, there has been a significant change in the defence landscape since the publication of the Defence Industrial Strategy (DIS) in 2005. The emphasis has shifted from contracting through competition to partnering (where appropriate) with a renewed focus on managing capabilities and equipment on a through-life basis. The formation of Defence Equipment and Support (DE&S) should facilitate the realisation of an end-to-end (E2E) approach from concept to disposal as data acquired from supporting in-service equipment is fed back to inform new programmes. The realignment of budget programming responsibilities will also contribute to enabling a through-life approach as acquisition and support costs are brought together in a single plan.

Industry too faces a challenging business environment with reduced numbers of platforms being ordered, longer life-cycles of platforms (which create industrial capability sustainability issues as well as necessitating through-life equipment upgrades), greater technological complexity, increased risk being transferred from the MoD customer, and the need to reconfigure for the provision of through-life capability services. The latter is needed by a customer seeking better availability, reliability and maintainability, and through-life cost reductions.

Having outlined the context in which the delivery of logistics support takes place, the following paragraphs examine some of the issues that members of the logistics community both within the MoD and industry consider to be critical.

The Evolving Role of Contractors

Commercial contractors have played a role in the delivery of logistics support for centuries so the concept is not new. However, the nature of contracting is changing as the MoD seeks to move away from traditional supply or spares inclusive contracts to Contracting for Availability (CFA), also referred to as Contractor Logistic Support (CLS). Under the former arrangements, spares constituted a revenue stream for industry which therefore had no incentive to strive for reliability; as such it was costly for the MoD. Under the CFA model, spares represent a cost to the contractor whose focus is therefore on the ultimate rather than intermediate outcomes. This reduces risk to the MoD and provides a positive incentive for greater sustained profit to industry. However, CFA will only work if the prime contractor is able to cascade the risk down the supply chain to where it can best be managed. A difficulty that many prime and tier one contractors encounter is that there is often reluctance within small tier two and three companies to accept risk.

Contracting often constitutes a financially attractive option as contractors usually have lower overheads than the MoD and are able to scale their workforce flexibly to meet the needs of a specific contract. Furthermore, as more complex equipment enters service, the maintenance of that equipment may require specialist skills which are only available from a contractor. However, availability-based contracts must be acceptable to front line commanders who ultimately have to carry the operational risk; the drive for efficiency must not trump effectiveness. This is even more so where Contracting for Capability (CFC) is being considered – in other words, where a contractor provides an entire capability such as air refuelling. Operational commanders need assured delivery and so such contracts have to be watertight. In its report entitled 'Transforming Logistic Support for Fast Jets', the National Audit Office (NAO) recommended that the MoD needed to be certain it had sufficient commercial, cost modelling and project management skills internally to be able to negotiate and write complex contracts.

In addition to providing a fair level of risk-based profit to industry, the contractual frameworks for CFA must recognise the interdependencies of all stakeholders. For example, where aircraft CFA contracts have been set up, what was traditionally called first line servicing continues to be undertaken by military technicians on a squadron. As such, even though the contractor is responsible for availability, he does not control all aspects of the process. Moreover, he is dependent on good quality, accurate and timely information being passed back from the squadrons in order to assess consumption and wastage rates, plan spares and drive continuous improvement but there is invariably no shared data environment, particularly with suppliers to the primes. At present, Key Performance Indications (KPIs) tend to be written for the contractor to deliver against but, given the dependencies throughout the stakeholder network, there may be merit in having KPIs against which to measure the performance of all players – this would also reflect the transition from a transactional type relationship to one based on partnership and would drive the necessary behavioural changes on the MoD side.

While the use of contractors can usefully release military capability for employment elsewhere, it is essential that the armed forces' ability to prosecute operations is not compromised. Contractor Support to Operations (CSO) provides 30-40 per cent of UK defence overseas sustainment effort, a figure that has risen dramatically in the past few years. Some commanders who have recently returned from operational theatres posit that over-reliance on contractors on operations is a cause for concern, creating problems of control, co-ordination, security and management. There is also often uncertainty regarding the likelihood of some contractors remaining in theatre if the environment becomes hostile. While risk may not be significant at the level of an individual contract, the cumulative risk associated with having multiple contracts can be substantial. For risk to be minimised, the military must be clear about what it wants contractors to do in theatre; the maturity of the theatre and the tempo of operations are important considerations in determining whether CSO is appropriate.

Contractors deploying on operations have to understand the operational environment – how to get civilians into theatre and how to manage and protect them once they are there. Tactical tempo is often lost if the civil-military interface goes wrong and there are few courses of action open to a commander dealing with a failing contractor when there is no military capability to backfill. Problems arise more commonly with contractors employed under Host Nation Support arrangements, but such contracts are often important for nation-building and also help to minimise the deployed footprint so have to be managed pragmatically.

Realisation of the potential benefits of CLS will depend on the successful management and exploitation of information across the whole enterprise. Where insufficient and/or inaccurate supply and maintenance data comes back from the front line it hampers any effort by the contractor to take early action to address potential performance issues or even just to create a spares plan. There is recognition of the need to automate data capture to remove the burden of data inputting from busy first line technicians, but transmission will still be via MoD systems which are often very slow. Furthermore, the transmission of logistics data back to the UK is often a comparatively low priority for an operational commander. Increased collaboration between the MoD and industry will require wider availability of commercial infrastructure and the use of common standards. This requirement will be addressed in part by the Logistics Coherence Information Architecture (LCIA) but will still leave issues of data sharing across organisational boundaries to be resolved.

The move towards CFA/CFC has been reinforced by the DIS which seeks to transform the traditional defence procurement business model by shifting emphasis from competition to long-term partnering arrangements. This necessitates a collaborative approach between the MoD and industry. The DIS demands a considerable change in mindset and culture on both sides. Partnering between customer and supplier can deliver millions of pounds in savings over time but decisions to partner must be driven by the business need to both partners rather than purely as a cost-saving measure for the customer. Achieving best value means that MoD must have a clear understanding of what it requires from industry; this has been an area of weakness in the past.

Optimising the Joint Supply Chain

The Joint Supply Chain (JSC) is defined in JDP 4-00 Logistics for Joint Operations as 'the network of nodes (resources, activities and distribution options) that focus on the rapid flow of matériel, services and information between the Strategic Base and deployed Force Elements (FE) in order to generate, sustain and redeploy operational capability'. The JSC reaches from the factory to the receiving unit in theatre. It is a constituent part of the Defence Support Chain (DSC) which is defined as 'the entire matériel chain from procurement and provision of an item to the point of consumption for usage and including alternative, or indirect, methods of supply for all commodities'. The Director General Joint

Supply Chain (DG JSC) is responsible for delivering a reliable JSC which can be sustained at an appropriate velocity and deal with variability of demand in line with the intensity and scale of an operation. This is an unenviable task, given the number of organisations whose activities must be integrated if logistic support is to be delivered successfully.

Organisations contributing to the JSC at the operational level are J4 Permanent Joint Headquarters (PJHQ), J4 Joint Force Headquarters (JFHQ), Joint Force Logistic Component Headquarters (JFLogC HQ), Defence Supply Chain Operations and Movements (DSCOM) (incorporating the Defence Logistics Operations Centre (DLOC) and the Directorate of Movements Operations) as well as personnel from the Front Line Commands and Contractors in Support to Operations (CSO). At the nonoperational level are DG JSC, DE&S and industry. Bringing coherence across this complex network is no small undertaking and activity is guided by the JSC Blueprint and the JSC Board Plan, both of which are approximately half-way through implementation. Current work on the JSC is focused less on the physical movement of people and matériel and more on optimising flows of information to make logistic support more efficient.

In the deployed environment, the JFLogC takes responsibility at the force level for distribution, redistribution and recovery in line with the demands of the operational support and maintenance cycles. This is increasingly important as current and future operations will involve industry taking matériel as far forward as is safe. The JFLogC has played an important role in bringing together the joint elements and some argue the value of a separate logistics component has already been proven.

A number of the issues associated with coupling bridge (CB) failures in Operation Telic have been addressed, especially with respect to organic lift; the UK is now better able to set up, run and maintain substantial CBs through improved use of air- and sealift and better prioritisation of loads. Organisational structures have also been improved and a performance management cell has been established in DSCOM to assess supply chain performance against a set of metrics to support continuous improvement. The roll-on roll-off ferries acquired under a private-finance initiative (PFI) arrangement enable better planning as they offer assured delivery.

At present some 60 per cent of total airfreight is carried into theatre by charter companies which represent a substantial risk; the acquisition of two additional C-17s to take the fleet to six will alleviate some of the pressures, but will not eliminate the need for air charter. The introduction of the Purple Gate through which all matériel from must pass before being dispatched to theatre will improve the flow of items from the Main Operating Base (MOB) to the point of use. A contractor mounting cell embedded in the DLOC will effectively act as a Purple Gate for civilian personnel; attendance became compulsory for all contractors deploying into theatre from 31 January 2008. This will not only better prepare individuals for deployment, but should enable the deployed operational commander to have accurate information about how many contractors are in theatre, what they are doing there and what the contract terms are. This is important because the presence of deployed contractors places additional demands on the logistics infrastructure, for example with respect to the use of the air bridge.

Within the JSC, the interfaces and processes work well in the UK. However, in the operational space, there are some problems maintaining the integrity of the JSC within the single-service elements where a degree of stovepiping remains. The physical aspects of the JSC are being used, but each service is essentially running its own supply chain out to the Forward Operating Bases (FOBs). This is partly a consequence of personnel operating legacy single service information systems (IS) which hampers the introduction of standardised processes. Two programmes – Management of the Joint Deployed Inventory (MJDI) and Management of Matériel in Transit (MMiT) – are crucial to making the logistics operational space truly joint but their introduction and rollout will depend on PR08 decisions; if they survive the planning round, MJDI rollout would begin in 2010 and MMiT in 2009. Once tri-service systems are in place, processes and training can become joint which will contribute significantly to the behavioural and cultural changes that will be required for E2E embedding of the JSC.

Improving Logistic Shared Situational Awareness and Visibility of Matériel Flow through the JSC

For the JSC to be efficient, and for the war fighter to have confidence in the system, asset visibility is essential, particularly for high-value, highly active, repairable items which have significant impact on operational output. At present, visibility of assets, stocks and matériel in transit presents a challenge. Programmes to improve matériel management systems had high priority in the 2007 iteration of the Logistic Decision Support and Management Capability Management Plan but difficult funding decisions have to be made in PR08. Consignment Tracking (CT) improved significantly with CONVIS Stage 1 – which married VITAL/RIDELS with the TAV(-) active RFID tracking system. At present, the CT system does not quite reach as far as the FOB; improvements will be realised with the release of CONVIS Stage 3 (planned delivery in 2008) which will bring all services onto a single system. Matériel movement management will be further enhanced by MMiT and the Air Movements Operations (AMO) programme (currently scheduled to be introduced late 2009) will improve the visibility of freight.

MJDI will deliver the most significant logistical improvement in the operational space by increasing confidence through better inventory management and JAMES2 (initial operating capability currently planned for late 2010) will enable more efficient and effective management of equipment in the land environment. Improving inventory and asset management systems will also help decrease the logistics footprint by removing some of the causes of overstocking in the forward area which should, in turn, allow operational commanders greater agility.

Contractors providing logistics support currently lose sight of equipment and spares once they have left the UK and problems also arise when contractors take non-codified spares with them when they deploy forward, so there is still some way to go before the system could be described as E2E. There is a stated desire to achieve E2E visibility across the whole of the defence enterprise but even the US has not yet been able to achieve this. However, E2E visibility requires more than a technological solution – it also needs efficient and effective manual systems, processes and training for controlling flow of matériel through the supply chain. The complexity of an E2E system would require the incorporation of a filtering function so that the right information at the right level could be delivered to meet each user's specific requirements. For the deployed user, it is arguably more important to have joint situational awareness than it is to have E2E visibility – increased confidence that demands will be satisfied on time will ultimately reduce the need to know exactly where items are while in transit.

Planning for Operations

The strategic planning for operations undertaken within MoD and the operational level planning undertaken by PJHQ should both be shaped by information from DE&S. Moreover, it is essential that logistics planning be undertaken as an integral element of operational planning and not as an afterthought. Good planning requires accurate forecasting of requirements which must be informed by data from previous operations and exercises; if the right data has not been captured, forecasting becomes inaccurate which can ultimately impact on the operation. Significant headway has been made with operational level supply chain planning but there is more to be done to improve feedback to the long-term planning process.

Force structures are set against the military tasks laid down in defence planning assumptions (DPAs) but it is openly acknowledged that DPAs have been exceeded for many years. In both Iraq and Afghanistan, force structures and lay down have had to be altered to fit the specifics of the campaign; operational imperatives will always override policy guidance. Judgement is then required to determine at what stage those changes should be made permanent. Anticipated tasks shape equipment choices which often simply will not be up to the task in a new and unfamiliar deployed environment. Urgent Operational Requirements (UORs) will inevitably emerge and these can present many challenges in the logistic support domain. There is clear policy guidance to the effect that UORs need to be compliant with the JSC element of the Support Solutions Envelope. However, timeliness is often a key driver and so risk will be taken on the support solution; it is difficult to configure a support solution when there is no data on likely usage, consumption rates and little by way of reliability trials. In terms of supportability, UORs are not brought in with through-life capability management in mind and this ultimately gives rise to difficult decisions about whether to bring items onto permanent inventory and support them beyond the operation for which they were procured. To do so not only diverts resources from other planned programmes but may impact upon equipment plan coherence and introduce fleet within fleet problems.

Most multinational activity takes place in theatre, but coalition nations are not very good at planning multinational logistics cooperation. Effort needs to be directed towards building multinational logistics capability as every nation cannot do everything and so must have trust in coalition partners. In the same way that the appropriateness of the use of contractors in theatre varies with the intensity and maturity of the operation, so the appropriateness of adopting a multinational approach varies against these same criteria. At the start of an operation, it is prudent to adopt a national approach to logistics support – indeed it is probably the only option. As an operation becomes more enduring, it is sensible to explore ways of decreasing the logistics footprint and achieving economies of scale by cooperating with other nations. Various options can be considered – nations can adopt lead nation or role specialist status or arrangements can be put in place for collective contracting.

Improving Information Systems and Information Management

The recurring theme which links the aforementioned issues, and which emerged in every interview, was the need to provide better logistics information. The historical absence of a coherent logistics information management policy and a haphazard approach to investment in systems and applications allowed logistics IS to be developed on an ad hoc basis to meet specific (often tactical level) information needs. There was no strategic information need driving the design of applications, no definition of the effect that must be achieved through the IS and no mandated requirement for designers to collaborate or to ensure compatibility with other applications. The lack of common standards, poor infrastructure and unregulated approach to data assurance combined to deliver an inefficient, fragmented logistics information system in which functionality is often duplicated and there is no single repository for data as it is locked into various IS systems which are unable to communicate with each other. In practice this means users may have to interrogate multiple systems to get the information they need to complete a task. This is clearly inefficient in terms of time and it may also be costly to maintain multiple legacy systems.

Service Oriented Architecture (SOA) has the potential to resolve many of the technical issues in the information area; this is especially so when much of the data held on legacy systems is brought together in an enterprise data warehouse thereby enabling it to be 'written once, read many times'. SOA is more than a technical tool but many organisations fail to realise its full benefits as they never undertake a fundamental re-examination of the business layer (objectives, structures, processes, and human factors) and are therefore unable to achieve organisational optimisation. A great deal of time, resource and energy has been devoted into rationalising and integrating legacy systems and this bottom-up approach may enable the Logistics Applications IPT (LAIPT) to reduce costs associated with supporting legacy systems.

However, the presence of the enterprise data warehouse means that application rationalisation no longer has to be the focus of main effort – time and resource may be better directed elsewhere. A top-down approach is needed to complement bottom-up rationalisation and to ensure that future IS capability does not simply enable the whole support chain to merely do what it already does only a little better than it currently does it.

There is an opportunity to redesign business processes to deliver greater effect, but this will require much more integration between the business and information communities within the MoD. Improved mutual understanding within these communities will be vital to achieving the desired outcomes of the LAIPT's proposed Future Logistics Information Systems (Delivery Partner) (FLISDP) initiative which seeks to appoint a commercial partner to maintain existing LogIS applications and develop future solutions. At present, there is little incentive for equipment IPTs to adhere to cross-platform information approaches – their principal task is to deliver better equipment availability at lower support cost and little benefit may accrue to them from investing in improving information. Governance mechanisms need to be put in place to ensure that equipment IPTs adhere to a common information approach.

The Joint Coherence Project 06 sought to establish the basis for a common approach. It defined a control framework of rules, tools and standards within Defence Logistics

Information; key deliverables were a high level E2E functional model, a logistics information architecture (LCIA) and logistics information standards. The LCIA was developed in response to changing operational concepts (NEC, Directed Logistics) and logistics transformation issues (such as CFA, lean support and through-life management). It delivers a top-down presentation of functions, information categories and information flows, and sets out the generic logistics information that has to be exchanged in order to execute logistics and through-life management effectively. It also maps functions to support domains for particular contract types and maps information categories to international logistics standards.

In essence, LCIA helps business users to identify their information needs; this constitutes a significant step forward. The full LCIA model is extremely detailed and possibly a little daunting for potential users, so the LCIA team has taken on the task of assisting IPTs with building an LCIA model to satisfy their specific business needs. The output of the model is the Logistics Information Plan (LogIP) and the process generates real understanding of the information of the business. It ensures the project aligns with logistics policy and is coherent with other programmes, it facilitates a better understanding of how technical systems can support the business, and it can be used to inform trade-off decisions. The LogIP itself serves as an input to the contract, informs the requirements documents, and supports the through-life management plan.

Pulling together the LCIA demonstrated that competitors within the defence industry were able to work with each other and the MoD to develop a joint framework. However, in comparison to other sectors, the defence industry is some way behind with SOA. If nothing else, LCIA provides a common language which should assist the trust-building process. Trust is critical to overcoming the issues associated with data sharing as organisational boundaries become evermore blurred across the whole of the support chain. The question of who should share what with whom rarely has a straightforward answer. Making all information available might seem unhelpful as much of what is collected and stored can appear to be of little value at first sight. Nevertheless, the opportunity to exploit real-time usage information should be brought into a direct capability management framework for use by industrial partners, as in the case of aero-engines. Only then will the value be evident to the customer, because suppliers will then be incentivised to deliver effective capability management services.

In the same vein, there are some complex issues around information assurance that need to be addressed as an E2E information model develops, not least the need to agree a common understanding of what is meant by 'information assurance'. Furthermore, there is a need to clarify who pays for assurance, who owns the information (and the liability for it), and who carries the risk. Assurance applies a measure of compliance against standards but standards change fairly frequently so there will always be a legacy tail. Determining ownership is a critical issue as the owner must assume responsibility for quality and security; an added difficulty in the defence context is classification especially given the cultural tendency in the UK to over-classify 'just in case'.

Over-classification increases cost and introduces inefficiency but will not be readily eliminated until there is a way of estimating or measuring the true impact of loss of data. Often, all that is actually required is for data to stay within a trusted path which can be written into specifications for approach to handling of data across boundaries.

A Five-Year Look Ahead

With respect to the development of logistics support, five years is a fairly short period. The nature of logistics is unlikely to change in that timeframe, but the boundary between MoD and industry will have moved and a greater proportion of logistics support, or the mechanisms that enable it, will be delivered by industry. In the context of logistics transformation, the likely position in 2013 is highly dependent on the outcomes from PR08. In such a tight planning round it would be surprising if all logistics information programmes were to pass through unscathed. Moreover, there is little likelihood that defence funding will increase any time in the near future, so even if programmes survive PR08 unchanged, there are future planning rounds to get through. If programmes are pulled or delayed, there is a risk that the MoD will be forced to adopt a piecemeal, 'make do' approach to improving logistics information.

However, if the critical programmes are approved or retain their funding, significant improvements to in-theatre inventory management and asset visibility will have been realised. The deployment of MJDI, MMiT and CONVIS will have reduced data requirements significantly and a culture of asset tracking will be embedded. As logistics IS converge, improvements in processes and the adoption of a common language will follow which will, in turn, support further integration and coherence across the JSC. According to its current plan, the MoD should also be halfway through implementation of SOA and, in getting there, should have a better appreciation of the dynamics driving the business.

Progress will continue to have been made against the DLP targets and the JSC Board Action Plan. There will be a greater understanding across defence of the need for agility in the logistics domain and logisticians will be more at ease providing support to expeditionary operations as the vast majority of the uniformed cadre will have been deployed.

Conclusion

Expeditionary operations place variable and often significant demands on the logistics support system, which has had to undergo substantial redesign and restructuring since the end of the Cold War. Logistics transformation continues apace but against a backdrop of two significant operations and severe budgetary pressures. The delivery of logistics support is becoming more complex as the increasing use of CLS blurs the traditional military-industrial boundary. While there is often a persuasive case for adopting CLS, the issues relating to operational risk must be evaluated and managed appropriately. Moreover, if the full benefit of CLS is to be realised, improvements must be made to the flow of information between the front line and industry, with a clear need to automate data capture to remove the burden of data inputting from busy first line technicians. Performance metrics for CLS must take due account of stakeholder interdependencies and accountabilities.

Significant headway has been made in relation to developing processes and organisational structures to enable optimisation of the JSC. Many of the failures highlighted by Operation Telic have been rectified and further JSC enhancement will emerge as DG JSC continues to drive through the implementation of the JSC Blueprint and JSC Board Action Plan. The single-service stovepipes that remain in the deployed environment will be broken down by the introduction of MJDI and MMiT, ensuring that the E2E integrity of the JSC is maintained.

Improved visibility of assets and matériel in transit will enhance shared situational awareness and increase the war fighters' confidence in the supply chain. In addition, improvements to inventory and asset management systems will help to reduce overstocking thereby reducing the logistics footprint. The achievement of full E2E JSC visibility is still some way off but it is clear that a technological solution alone will not deliver this. User confidence that the system can be relied upon to provide an accurate picture of the situation is arguably more important that having E2E visibility.

Delivering effective logistics support to operations requires logistics planning to be undertaken in concert with operational planning. Planning is generally well done, but weaknesses remain in relation to capturing data and feeding it back into the long-term planning process. Planning for multinational co-operation in coalition operations is also not as good as it could be, although models for the actual delivery of logistics support on a multinational basis are fairly well established.

The need to provide better logistics information to support decision-making and shared situational awareness has been a recurring theme. SOA will help to overcome many of the technical problems associated with logistics IS, but will not provide a complete business solution nor enable optimisation of the organisation unless a thorough review of the business layer is undertaken. While it may be sensible to continue with rationalisation and integration of applications as a means of reducing costs associated with supporting legacy systems, the presence of the enterprise data warehouse means that application rationalisation no longer has to be the focus of main effort – time and resource may be better directed elsewhere. LCIA will help users to identify the information needs and map information flows and, importantly, provides a common language. Issues associated with data sharing across organisational boundaries will need to be addressed as will those relating to information assurance.

The 2013 position with respect to logistics support will be affected significantly by the outcome of PR08. If programmes are cut, progress against the DLP and JSC Board Action Plan will be extremely limited and few tangible improvements would be evident in the deployed environment. If, however, key programmes such as MJDI, MMiT and CONVIS survive PR08, substantial operational benefit will be derived.

Recommendations

- IPT leaders and the policy staff who provide guidance to them must ensure that a decision to contract for availability or capability is driven by effectiveness (in through-life capability management services) rather than the need to reduce programme costs.
- CLS contractual frameworks and performance metrics should reflect the interdependencies and accountabilities of all stakeholders, not just the industrial partner.
- A decision to enter into CSO arrangements must only be taken with a full understanding of the risks including the cumulative risk of managing multiple CSO contracts in theatre.
- A 'top down approach' to complement 'bottom up' rationalisation should be developed to ensure that the full potential of SOA can be realised and governance structures should be put in place to ensure that equipment IPTs adhere to a common information approach.
- Key IS programmes related to improving asset visibility, asset management and inventory management must be supported to enable E2E coherence across the JSC and to increase the operational commanders' confidence in the logistics support system.

- The management and exploitation of information across organisational boundaries, and indeed across the whole of the defence enterprise, must be improved if the full benefits of CLS are to be realised.
- Rationalisation of legacy systems and applications should continue where there is a business case for doing so; if time and resource are scarce, consider whether effort should be directed elsewhere.
- Failure to invest sufficiently in PR08 must be prevented as it would result in further sub-optimisation and proliferation of information capability for logistics, possibly resulting in a recurrence of the issues encountered in Operation Telic.

Bibliography

Development, Concepts and Doctrine Centre, Logistics for Joint Operations, Joint Doctrine Publication 4-00 (London: Ministry of Defence, 2006), available at <http://www.mod.uk/DefenceInternet/MicroSite/DCDC/OurPublications/JDWP/>.

The Defence Logistics Support Chain Manual, Joint Service Publication 886 (London: Ministry of Defence, 2008), available at <http://www.mod.uk/DefenceInternet/MicroSite/DES/OurPublications/Jsp886TheDefenceLogisticsSupportChainManual.htm>.

Defence Logistics Programme Update (2007).

Defence Logistics Programme (2006).

Joint Supply Chain Board Action Plan (January 2008).

Logistics Decision Support and Management Capability Management Plan (2007).

DE&S Business Strategy (April 2007).

Guide to Using the Logistic Coherence Information Architecture (London: Ministry of Defence, 2007), available at <http://www.mod.uk/DefenceInternet/MicroSite/DES/OurPublications/LogisticsCoherenceInformationArchitecturelciaUserGuide.htm>.

Defence Industrial Strategy, Defence White Paper (London: The Stationery Office, 2005).

Ministry of Defence, 'Acquisition Operating Framev

Author's notes from RUSI-DEM Focused Logistic

Author's notes from RUSI C4ISTAR Conference, October 2007.

Author's notes from Shephard Defence IT Conference, February 2008.

Author's notes from interviews with Ministry of Defence and industry representatives.

National Audit Office, Operation TELIC - United Kingdom Military Operations in Iraq (2003).

National Audit Office, Transforming Logistics Support for Fast Jets (2007).

The author wishes to extend her heartfelt thanks to all who so generously shared their knowledge through presentations or interviews. Opinions and comments have intentionally been presented without attribution.

This Occasional Paper was originally sponsored by IBM.

Notes

Learning and Development

Dr David M Moore and Peter D Antill

Centre for Defence Acquisition, Cranfield University, Defence Academy of the UK.

Case 3.2: Army Foundation College

Relevance

This case provides an opportunity for group discussion upon a vitally important area. The UK has been a leading proponent of PFI/PPP. The imperative behind PFI/PPP has been the need for more efficient and effective use of Government funding in respect of defence. Arguably, PPP can be seen as a concept of partnering involving long-term relationships between Government and the Private Sector which can be seen as being more flexible than PFI, although they are built upon essentially the same philosophy.[1]

The basis of decision-making has been value-for-money with a view from the UK Government that the Private Sector can bring disciplines to bear to achieve greater efficiency when private interests are engaged, partly through innovative design and delivery solutions and partly because there is no cost to the Government until a satisfactory project outcome is achieved. It has been argued that finance from private sources can be a better way of managing financial risk and that this correlates more efficiently with the risk of a specific project when private capital markets are engaged. As RUSI's Whitehall Paper No. 63 points out however:

"... affordability in the short term is also clearly an incentive to Government to engage private finance allowing projects involving significant capital expenditure to be taken forward when Government capital resources are limited. A further incentive, not fully acknowledged by the UK Government, is the desire to keep capital expenditure and the attendant capital costs introduced by Resource Accounting and Budgeting off departmental balance sheets."[2]

However, there is a major issue in that Private Sector finance models at present are focused upon ensuring that a supplier can deliver the required quality of service or capability throughout a long-term contract when the customer may not be able to identify precise requirements and predict and contract for changes requiring modification over contract duration of between ten and twenty-five years. Any adjustments to the requirements will be at cost and can undermine the definitive benefits of clear risk allocation and cost control.

Other challenges are the up front costs of setting up PFIs (making them only suitable for long term larger projects), management of technological risk by the supplier particularly where advanced technology is involved, and peculiarities of the defence environment in which military risk and guarantee of supply may be dominant issues and in which system complexity prevails.[3]

Thus, whilst this has been a major strategic approach for the UK Government, there are issues that need to be taken into consideration. The relevance of this case, is that PFI/PPP has been, and continues to be, at the forefront of thinking, especially at a time of austerity and an even greater emphasis on cost reduction, nevertheless the challenges and issues arising are also extremely relevant. Anyone considering the use of such an approach must be fully conversant with these matters.

Context

This case, is indicative of many instances where an approach to outsourcing takes into account long-term funding by a Private Sector organisation.

In the past, UK defence projects that have been supported by private finance could be classified as follows:

* **Operational:**   Military capability and logistic support
                             Training
                             Communications and information systems

* **Non-Operational:**   Education and training
                             Communications and information systems
                             Infrastructure and logistic support

This is a classic non-operational category where there are considerable similarities to projects that may be undertaken within the commercial environment. At one stage, the MoD was considering (as indeed was the UK Government) the use of such financing approaches to a wide range of activities that included, for the MoD, operational commitments. It is unlikely however, that these can grow much further in respect of, for example, land operations in Afghanistan. It is much more likely, that these will be applied in non-operational environments, although, there is a grey area that links the benign environment to one that is hostile and an example of this, is the Future Strategic Tanker Aircraft. In general, the classic PFI model is not best suited to information technology projects because the rate of technological change in this sector affects requirements as well as service provision and does not permit rigid long term contracts.

Even then, these is no certainty in the optimal approach in that in the years 2008 – 2011 these have been so many so many socio-economic and political changes that decision-making on contracts such as the one in this case, are subject to huge changes in requirement. As an example, in realm of defence education and training, the Defence Academy of the United Kingdom, as a result of expenditure cuts, global recession and the *Strategic Defence and Security Review of 2010*, considerable changes have had to be made to course syllabi, whilst the proposed Defence Training Academy which was to have been a joint services training establishment located at RAF St Athan in South Wales has been cancelled (at least for the time being).[4]

Issues

Although this case might appear, at first glance, to be relatively superficial, the more one engages in discussion with others, the more issues, and their possible resolution, will be identified. For example, an initial feature is whether to use an existing location or a green field site. This alone can prompt discussion, around matters such as costs concerned with a new build, as opposed to the renovation and upgrading of current facilities, transport and the existing skills base etc. Leading on from this are matters such as the growth in student intake and sustaining it at the highest level. On the one hand is the matter of policy, on the other, is the practical approach that has to be taken to ensure effective utilisation. It is possible for example, that one of the ways to achieve an optimal solution is the use of a consortium.

However, to do so will require careful selection and then the maintenance of relationships between organisations and personnel who have differing criteria, stakeholders, ethos, behaviours and cultures. For example, the military, civil service, service providers, commercial organisations, academia – all have differing values and espoused beliefs. At the very heart of this case and indeed any other PFI/PPP will be the contract. Within the contract, great attention must be paid to the following (which are only indicative examples):

- Allocation of places – awards (type), awards – processes and mechanisms, class sizes, student to staff ratio, consistency of candidate consideration etc.
- If residential, living accommodation is required and there will be issues of security, health and safety, washing facilities, leisure facilities etc.
- Provision of infrastructure, e.g. IT, Internet access, lecture theatres, breakout rooms, training areas etc.
- Catering and cleaning – often overlooked, yet a basic underlying factor from which student satisfaction and academic achievement can be built.
- Essential to ensure the smooth operation of such an organisation will be focused, customer-based administration support.
- Academic support – who will provide the academic staff, will they be trainers or educators?
  Will they come from a private organisation or a recognised university?
  How will they interface with the military personnel?
- Security – as a military establishment, one might expect military personnel to undertake this role, however, this is not normally the case and could be provided by the MoD Guard Service with support from the MoD Police.

At a higher level, consideration must be given to the various risks involved. These range from the financial, to the operational implementation. It is essential that sound project management approaches be utilised to ensure the effective planning, not only of the design and construction (or alteration) of the premises, but also to ensure clarity of requirement, definition of outcomes, and the clear period of transition through to ultimate commission. In respect of risk, other matters arise, for example:

- What is the break even point in respect of usage?
- Will there be any value in what remains at the end of the contract (physical premises and IPR)?
- Will there be legal requirements in respect of health and safety and can public policy be met whilst operating in a Private Sector environment?

It should be noted, that there are a number of ways in which private finance could be utilised. Innovation is the key and some options are:

- Operating leases where lease terms are arranged on assets that could include a range of accompanying services within contracts ('wet' leases).

- Capability contracts which are an evolution of the 'wet' operating lease in which the customer specifies delivery of a capability rather than an asset.
- Pooling assets between nations as a further way of spreading the high costs of defence equipment across a number of different customers.

This case is based upon a benign environment; one in which risk can generally be seen to be commercial in nature. There is much discussion about its use closer to frontline operations and this case can be used to raise the issues that can be seen as relevant. Where commercial companies are operating in the frontline, albeit in a support capacity, there is the potential for commercial and military goals to conflict. There are three issues relating to the Private Sector's involvement in providing operational capability, which are:

- The responsibility for military risk.
- The guarantee of delivery.
- Interoperability with other (national and international) military capability.

Military risk can be managed up to a point through government indemnity although definitions of combat and the frontline have become blurred by the risk of asymmetric attack. Guarantee of delivery can be managed in the contractual process and by means of Sponsored Reserves. The behavioural aspects of interoperability present a problem that can be best addressed through an incremental approach and exploitation of test cases such as FSTA. If new innovative private finance schemes are to work in the future, the clear understanding of where risks lie is essential.

Only then will it be possible to establish necessary, substantial specialist markets with an understanding of the specific risks of the defence sector for debt and equity. All parties in an acquisition that is funded innovatively need to understand the full range of risks and the perspectives of the other stakeholders including the differing perspectives amongst the various customer stakeholders. There is potential in a combined public and private sector approach to profit and risk sharing which could help to develop and protect the secondary markets for defence assets that would be necessary, in many cases, for effective operating leases and capability contracts.[5]

Case 3.3: Support Chain Management in the RAF

Relevance

It is important to note the time at which this case is set. The UK, in common with other countries, had been preparing for a major conventional, symmetrical conflict. To ensure operational effectiveness, it had followed a policy of large stock holdings that would ensure rapid resupply and engineering support to all the services. Since then, the likely nature of military operations, especially for western countries like the UK has changed out of all recognition. Allied to this, is a continual pressure for what was once known as a 'peace dividend' i.e. pressure to spend less money on defence and more on other areas of the Public Sector. This has been exacerbated by global crises and the continual contraction of the UK's manufacturing base. Hence, although this case represents a scenario from several years ago, it is still relevant in today's socio-political-economic climate. This case should also be read in conjunction with Cases 3.5 and 3.7.

Context

At the time of this case, the Royal Air Force undertook its logistic support separately to the other two services. In common with other parts of the MoD, it was subject to a considerable number of initiatives introduced by Central Government across the Public Sector to enhance value for money in public spending. While the concept of 'value for money' has been hotly debated, in practice, in such circumstances they will involve changes in the way in which such activities have been previously undertaken.

In order for such changes to be effectively implemented, the way in which they are managed is crucial. In this context, a further challenge is that logistics was not seen as one of those 'high status' areas. Furthermore, in the whole area of organisational culture (an understanding of which is key to managing change) events had been such that enthusiasm was waning for any further initiative. This can be summed up in the following two quotations:

"We trained hard ... but it seemed that every time we were beginning to form up into teams we would be reorganised, [and] I was to learn later in life that we tend to meet any new situation by reorganising: and what a wonderful method it can be for creating the illusion of progress while producing confusion, inefficiency and demoralization."[6]

"History knows many more armies ruined by want and disorder than by the efforts of their enemies."[7]

This of course seems cynical but one should not underestimate the strength of feeling of those individuals in the organizations affected (again this has resonance with the situation currently faced by the MoD and indeed the Public Sector as a whole). A final point in relation to the context is that the whole concept of operations and the potential threats the UK was likely to face was changing, from a highly symmetrical one in a well-understood and defined theatre, to one characterised by uncertainty in terms of both the nature and the location of possible conflict and asymmetry in terms of the operational methods of the potential opponent.

Issues

In considering these, the reader should first consider what strategy should be employed in line UK policy and in coordination with the strategy employed by the other services. Once clarity and agreement have been achieved, there is a need to consider organizational structures and the manner of implementation. Many conceptual aspects must be taken into account in this setting – these will include:

- Business Process Re-engineering.
- Structure of organising committees/teams.
- The use and limitations of Information Technology.
- The strategic and tactical approaches to inventory management.
- The use of in-house or outsourced resources for support.
- Strategic and tactical approaches to procurement.
- Re-organisation and relocation of functions and personnel.
- Relationships (at all levels).
- The use of Requirements Managers (the implicit and explicit acceptance of the need for stockholding).
- A lack of relevant training and development.
- Clear integration between engineering and supply activities.

This case provides an opportunity to consider key issues in managing change in a relevant defence context. It does not require in-depth detail to appreciate what went wrong; it as resonance with situations that have arisen since the time of the case and that, surprisingly, still seem to arise. In reality, this situation was never 'put back on track' rather, it was overtaken by other initiatives and changes in the concept of operations and the way which support would subsequently be undertaken. For example, following the 1998 Strategic Defence Review (SDR), logistic support was remodelled on a joint service basis under the auspices of the Defence Logistics organisation (DLO), which was itself eventually integrated with the Defence Procurement Agency (DPA – formerly the Procurement Executive) to form Defence Equipment and Support (DE&S) in April 2007. An indication of how things have moved on and continue to change is that the activities at RAF Witten are about to cease.

Case 3.5: Strategic Mobility

Relevance

This is not so much a case rather it is a set of perspectives that were written in the early 2000s. The reader should not regard these as absolutely, nor are they absolutely wrong; they are taken from a personal perspective and as such, provide a useful summary of the situation that was developing for the UK in the early part of the 21st Century. What is important is that these perspectives serve as a baseline for consideration of the way forward for defence organisation such as those in the MoD in the UK.

Initially the paper takes into account practical issues for example, the UK Armed Forces becoming increasingly tri-service. For other countries, being 'joint' not only provides similar opportunities but even greater challenges as there may be more than three services (for example, in the United States, the US Marine Corps and US Coast Guard are considered separate services). Increasingly, the use of commercial organisations to provide logistic support to military operations has been growing since the end of the Cold War and in this sense, logistics is seen both as a supply chain activity and a reliability/ maintenance issue. For example, for the United States, the number of contractors compared to the number of soldiers is around the 1:1 ratio[8] in Afghanistan, a figure that has come down compared to both the Korean and Vietnam Wars.[9]

A final point here, concerns definitions. Should one care to go back to the 1960s and look at definitions of activities that would improve managerial performance, there would be many functional activity definitions. Few, if any, would emphasise the need for inter-functional joint thinking. Through the 1970s, 1980s and 1990s, there has been a gradual but consistent move towards recognising that commercial and military success comes with an integrated, multi-disciplinary approach to their operations.

For the UK, this integrated approach or 'total process perspective' has meant that defining activities within boundaries becomes increasingly difficult. Definitions have not always kept up with this perspective. Some people will talk about logistics when they mean procurement, others would call both activities supply chain management; yet more might say that support chain management recognises logistics engineering. The UK utilises the word 'acquisition' to encompass all these activities in an integrated manner. However, other countries, utilise the word acquisition to mean procurement whilst logistics is very much the support function. An example is the United States. This is identified in order for the reader to appreciate the developing nature of the subject, yet to contrast the differing approaches taken by various countries.

The definition of acquisition as currently used by the MoD states that:[10]

"Acquisition is how we, the Ministry of Defence (MOD) work together with industry to provide the necessary military capability to meet the needs of our Armed Forces now and in the future.

It covers the setting of requirements; the selection, development and manufacture of a solution to meet those requirements; the introduction into service and support of equipment or other elements of capability through life, and its appropriate disposal.

Acquisition is supported by business processes such as:

- *Setting and managing requirements*
- *Negotiating and managing contracts*
- *Programme, project and technology management*
- *Investment approvals*
- *Safety management."*

Context

Whilst defining logistics is a part of the context for this paper, it is not the main aspect. At the time of writing, it was not anticipated that the UK would be involved in long term, large scale, long-distance operations (i.e. far from the Home Base), such as Iraq and Afghanistan hence planning for such activities, whilst limited, could be accomplished using or at least adapting many of the themes identified in this paper. For countries such as the UK, where commitment to operational theatres such as Afghanistan is paramount, the concepts that emerge and develop from this paper can enable effective logistic support to be undertaken.

This becomes more pertinent when one takes into account, that for many years, countries such as the UK have had a need to balance financial imperatives against ongoing operational imperatives. Simplistically, financial efficiency has had to be improved, yet operational effectiveness could not be seen to be impaired.

Issues

The paper identifies a number of elements of integration. Whilst these are not exhaustive, it provides the basis for consideration of the emerging thinking in respect of this topic. In particular, it develops a conceptual view of how risk and effectiveness can be prioritised in a practical, real world application.

The reader should use the elements to develop thinking around the depth to which these topics can be applied in current operations involving long-range logistics and joint integration.

Case 3.6: Humanitarian Logistics

Relevance

This paper was originally written circa 2002. It was written because the authors perceived a correlation between the way in which the military conduct logistic support to operations and the logistic support to disaster relief operations. Whilst indicative of the situation at the time of writing, both the military and disaster relief organisations have moved on in the way they conduct such support. Nevertheless, the overall situation regarding such logistic support still pertains.

The actors identified here, still exist although as will be seen from other case studies, military logistics is increasingly based on the use of contractors, and a smaller, more intelligent customer base making decisions about who should provide particular services. Similarly, humanitarian aid organisations have developed a recognition that the best practices implemented by commercial organisations and increasingly adopted by the military, could favourably impact upon the immediate relief provided in a given scenario.

Context

The players undertaking humanitarian aid are identified and whilst there can be no doubting their moral and ethical commitment to assisting fellow human beings, there has arguably been something of a naivety in the manner in which logistic improvements have been implemented. This is apparent in a lack of coordination between the particular organisations involved. However, contemporary consideration by humanitarian aid organisations recognises this and there is now a greater movement towards cooperation and coordination, especially in terms of 'command and control'.

Despite this, this paper enables the reader to gain an understanding of the setting at a time when pressure for improvements in logistic performance had not fully fallen upon humanitarian aid organisations.

Issues

Fundamental to conceptualising disaster relief logistics is the similarity between the way that defence organisations deploy and the logistics necessary to deliver and sustain them in theatre. To the uninitiated it may seem an anachronism that there are similarities between operations that may involve warfighting and those whose aim is ensuring human survival. Apart from the similarities, the military may have a major part to play in the initial response to a disaster, as it is the military who will often have the manpower and equipment to support the initial response.

The paper identifies many of the similarities and differences that have to be considered – it notes these I diagrammatic form as well as discussing them in more detail. There are many complexities (both internal and external) that must be taken into account when considering enhanced logistics performance in a disaster relief scenario where the military have a central role to play. Not least of these, is that of cultural dissonance where the modus operandi of those involved will be completely different. As it draws towards a conclusion, the paper notes that emerging military logistics practise will develop beyond that detailed in here and the various cases in this book, indicate how that has, and continues, to take place. It is a particularly useful point of reference and identifies that military operations are not always about warfighting and that the principles involved not only apply in humanitarian relief but can be utilised by such organisations.

;peditionary Operations

Relevance

This case links closely to Cases 3.3 and 3.5. It was written relatively recently (in 2008) and coalesces the key issues as we move towards contemporary application of logistics in the defence environment.

It recognises that expeditionary warfare is here for the longer-term as far as the UK and many of its allies are concerned. Whilst recognising the background to where we are now, it also provides a look forward to issues that are likely to arise in the next three to five years. It brings the reader up to date in terms of organisational structural changes that impact upon the Defence Logistics Programme and the inherent Defence Logistics Strategy. In particular, it recognises the ongoing transformation of logistics with a particular emphasis on the existing and continually evolving role of contractors. The need for optimisation on a joint basis of the MoD's supply chain is discussed and the importance of matters such as visibility of materiel highlighted. The paper brings to the fore the strategic planning for operations as well as the operational level planning undertaken by the Permanent Joint Headquarters (PJHQ). Information technology is identified as an enabling feature for integration and enhanced logistics performance.

Finally, a five-year look ahead (2008 – 2013) provides an indication of potential progress towards improved performance. In turn, the conclusions and recommendations provide an indication of the issues and challenges that still lie ahead. Nevertheless, taking into consideration all of the cases that have been presented thus far, the reader can see how much change and development has taken place to enable the UK to have the logistic support systems that now exist.

Notes

[1] National Audit Office. *Allocation and Management of Risk in Ministry of Defence PFI Projects*, HC343, London: TSO, 30 October 2008. Available at <http://www.nao.org.uk/publications/0708/mod_pfi_projects.aspx>

[2] RUSI. *The Innovative Use of Private Finance in Defence Acquisition*, Whitehall Papers No. 63, London: Routledge, 2004. Executive Summary available at <http://www.rusi.org/publications/whitehall/ref:P41DE7C295078C/>

[3] Ibid.

[4] *See* <http://www.bbc.co.uk/news/uk-wales-south-east-wales-11571849>.

5. Op Cit. Whitehall Paper No. 63 – Executive Summary.

6. Petronius, Arbiter, Greek Navy, 210 BC, quoted in Moore, D. and Antill, P. (2000) **'Integrated Project Teams: The MoD's New Hot Potato?'** in *The RUSI Journal*, Volume 145, No. 1, pp. 45 – 51.

7. Richelieu, in Van Creveld, Martin. *Supplying War*: Cambridge, Cambridge University Press,1995, p. 17 quoted in Moore, D. and Antill, P. (2000) **'Integrated Project Teams: The MoD's New Hot Potato?'** in *The RUSI Journal*, Volume 145, No. 1, pp. 45 – 51.

8. Gjelten, T. (2009). *'Ratio of Contractors to Troops at War? 1 to 1'*, located at <http://www.npr.org/templates/story/story.php?storyId=121590071>

9. Grier, P. (2008). *'Record number of US contractors in Iraq'*, located at <http://www.csmonitor.com/USA/Military/2008/0818/p02s01-usmi.html>

10. Ministry of Defence. *'What is Acquisition?'* on the Acquisition Operation Framework website, located at <http://www.aof.mod.uk/aofcontent/strategic/guide/sg_whatisacq.htm>

Index